Fractals, Chaos, Power Laws

Ich sage euch:
man muss noch Chaos in sich haben,
um einen tanzenden Stern gebären zu können.
Ich sage euch:
ihr habt noch Chaos in euch.

Yea verily, I say unto you:
A man must have Chaos yet within him
To birth a dancing star.
I say unto you:
You have yet Chaos in you.

—FRIEDRICH NIETZSCHE,
Thus Spake Zarathustra

Fractals, Chaos, Power Laws
Minutes from an Infinite Paradise

Manfred Schroeder

W. H. Freeman and Company
New York

Library of Congress Cataloging-in-Publication Data

Schroeder, M. R. (Manfred Robert), 1926–
    Fractals, chaos, power laws: minutes from an infinite paradise/
Manfred Schroeder.
        p.   cm.
    Includes bibliographical references and index.
    ISBN 0-7167-2136-8   ISBN 0-7167-2357-3 (PBK)
    1. Symmetry (Physics)   2. Self-similarity   3. Recursion   4. Scaling   I. Title.
QC174.17.S9S38   1990
530.1—dc20   90-36763

Dedication page: Portrait of Georg Cantor. (Akademie der Wissenschaften zu Göttingen)

Printed in the United States of America

Seventh printing, 1999

To Georg Cantor

From his paradise no one shall ever evict us

—DAVID HILBERT, defending Cantor's set theory

Also by Manfred Schroeder

Number Theory in Science and Communication

Contents

Color Plates

Plate 1   Three heavenly bodies meet. (A) As a double star (yellow and red, approaching from the left) and a single star (blue, closing in from the right) get close to each other, their orbits become wildly chaotic. After some time, they separate again, the double star orbiting off to the upper left and the single star receding to the lower right. (B) In this three-body rendezvous there is a switch of partners. The yellow member of the incoming double star exchanges its red partner for the single blue star, dancing away with its new mate to the lower left. The discarded red partner wanders off alone to the upper right. (Courtesy of Wolf Dieter Brandt.)

Plate 2   This fanciful self-similar leaf was generated by iterated affine transformations. (Courtesy of Holger Behme.)

Plate 3   Newton's iteration has three basins of attraction ("countries," shown in red, green, and blue). They meet at a fractal border with the following bizarre property: wherever two countries meet, the third is also present. Paradoxically there are no border *lines*, only three-sided border *posts*. Would international borders so designed promote peace? (Courtesy of Holger Behme.)

Plate 4   "Rainbow to Infinity," combines a large number of logarithmic spirals in different colors. (Courtesy of Holger Behme.)

Plate 5   "Clouds over Eastern Europe," a photograph taken by the author from his home near the crumbling "wall."

Plate 6   Real-world fractals. (A) Peeling paint at the Berkeley Swim Club, a multiply connected fractal. (B) Fractal growths in microorganisms: red algae on a rock in Point Reyes National Seashore, California.

Plate 7   Plants and flowers produced by turtle algorithms. (A) Rose campion, also called dusty miller (*Lychnis coronaria*), raised by Przemyslaw Prusinkiewicz and James Hanan. (B) Vervain (Verbena). (C) Lilac twig, grown by Prusinkiewicz and Hanan. (D) "The Garden of L," planted by Prusinkiewicz, Hanan, David Fracchia, and Deborah Fowler. (E) Plant with basipetal flowering sequences. (F) Flower field, fertilized with stochastic L-systems. (Plates 7A–F copyright © 1988 by P. Prusinkiewicz, University of Regina.)

Plate 8   (A) The Mandelbrot set (black) of complex numbers c for which the iteration $z_{n+1} = z_n^2 + c$ with $z_0 = 0$ stays bounded. Colored areas signify c-values for which the iteration is unbounded. For c-values from the main "cardioid" area of the **M** set, the iteration has a period length equal to 1. The circular disk to the left of the cardioid has a period length of 2, and so on. Each disk or "wart" comprises c-values for a given finite period length. The **M** set is a connected set, but it is not known whether it is locally connected. (B) A blow-up in the complex plane reveals a wart and filaments connecting it to smaller replicas of the **M** set. (C) Another enlargement in the complex plane shows an eleven-armed "whirlpool" in which smaller **M** sets swim. The arms themselves embrace smaller whorls whose parts, no doubt, contain still smaller whorls. (D) Self-similar descent: zeroing in on one of the "baby" **M** sets frolicking in the whirlpool reveals its patent parentage, the "mother" **M** set shown in A. (Courtesy of Holger Behme.)

Plate 9   Self-organized genetic drift between 16 different "species," $n = 1$ to 16 (shown in different rainbow colors). (A) Initially, the different species are randomly intermingled on a square lattice. At every click of the evolutionary clock, each lattice point occupied by species n will change any of its four nearest neighbors that belong to the species $n - 1$ to its own species number ($n = 0$ corresponds to $n = 16$). (B) At a later stage, different genes dominate larger and larger coherent areas, but many fine-grained neighborhoods persist. (C) Still later in the evolutionary process, a new kind of genetic pattern emerges: spirals with periodically repeating species. (D) Although the genetic interactions are strictly local, large spirals are the surviving dominant pattern. As in quasicrystals and numerous other natural phenomena, local rules engender long-range order and global designs. (Courtesy of Holger Behme.)

PREFACE

> Symmetry, as wide or as narrow as you
> may define its meaning, is one idea by
> which man through the ages has tried to
> comprehend and create order, beauty,
> and perfection.
>
> —HERMANN WEYL

The unifying concept underlying fractals, chaos, and power laws is self-similarity.
Self-similarity, or invariance against changes in scale or size, is an attribute of
many laws of nature and innumerable phenomena in the world around us.
Self-similarity is, in fact, one of the decisive symmetries that shape our universe
and our efforts to comprehend it.

Symmetry itself is one of the most fundamental and fruitful concepts of
human thought [Wey 81]. By symmetry we mean an *invariance* against change:
something stays the same, in spite of some potentially consequential alteration.
Mirror symmetry, that is, invariance against "flipping sides," is perhaps the most
widely noticed symmetry. Nature built many of her organisms in nearly symmetric ways, and most fundamental laws of physics, such as Newton's law of
gravitation, have an *exact* mirror symmetry: there is no difference between left
and right in the attraction of heavenly (and most earthbound) bodies. However,
the nonconservation of parity in radioactive decay—that is, the *violation* of
point symmetry in the "weak" interactions—has finally taught even the physicists to take the distinction between right and left seriously.

Another important symmetry is invariance with respect to geometric translation. Our trust in invariance under transpositions in space and time is, in fact,
so unlimited that we believe that the laws of nature are the same all over the
cosmos—and that they have been, and will remain so, for all time.

An equally momentous symmetry is invariance with respect to rotation.
A circle is invariant under rotation around its center by any angle. A square

can be rotated only through angles that are multiples of $360°/4 = 90°$; it is said to have a fourfold symmetry axis. A regular hexagon has a sixfold symmetry. While the rotational symmetry of a flower or a starfish may be imperfect, the exact isotropy found in the fundamental laws of nature is one of the most powerful principles in elucidating the structure of individual atoms, complicated molecules, and entire crystals. Transposition, rotation, and mirror symmetries, acting together, shape crystals from diamonds to snowflakes. And the same three symmetries govern much of what we find pleasing in ornamental designs.

An even more astounding symmetry is the exact identity of like elementary particles. There simply is no difference between an electron here and an electron there—on a distant star, for example. In fact, the perfect identity of photons, the particles of light, has disqualified them from being counted as so many identifiable individuals, resulting in a new kind of particle statistics, discovered by S. N. Bose and rendered palatable by Einstein—a way of counting not heretofore encountered in a world filled with tangible objects.

It was one of the greatest mathematicians of our century, Emmy Noether, who first pointed out the connection between the symmetries of the fundamental laws of physics with respect to displacements in space and time and rotations, on the one hand, and the conservation of linear momentum, energy, and angular momentum on the other. (Noether taught at Göttingen, where David Hilbert, overcoming obstinate prejudice, had finally secured a faculty position for her. In the dismantling of German science in 1933, she was forced to leave Göttingen. She died at Bryn Mawr in 1935.)

Other symmetries have had equally profound consequences in our understanding of the universe we inhabit. Invariance against uniform motion has given us special relativity, a fusion of space and time into space-time and, as its best-known consequence, the equation $E = mc^2$. The equivalence of acceleration and gravity postulated by Einstein is the basis of his general theory of relativity, which further revolutionized our appreciation of space, time, and matter.

Yet, among all these symmetries flowering in the Garden of Invariance, there sprouts one that, until recently, has not been sufficiently cherished: the ubiquitous invariance against changes in size, called *self-similarity* or, if more than one scale factor is involved, *self-affinity*. The enormously fruitful concepts of self-similarity and self-affinity pervade nature from the distribution of atoms in matter to that of the galaxies in the universe. And in mathematics, too, self-similarity is deeply entrenched. Some 300 years ago the German philosopher and polymath Gottfried Wilhelm Leibniz used the scaling invariance of the infinitely long straight line for its definition. Cantor sets and Weierstrass functions are other early examples—albeit less smooth—of self-similar structures in mathematics, later joined by Julia sets and other marvels of set theory.

It is perhaps symptomatic that with *set theory* still another abstract branch of mathematics has penetrated the real world. There simply seems to be no limit to Eugene Wigner's "unreasonable effectiveness" of mathematics. Indeed, who would have thought that such utterly mathematical constructions as *Cantor sets,* invented solely to reassure the skeptics that sets could both have zero measure and still be uncountable, would make a real difference in any practical realm, let alone become a pivotal concept? Yet this is precisely what happened for many natural phenomena from gelation, polymerization, and coagulation in colloidal physics and chemistry to nonlinear systems in innumerable branches of science. Percolation, dendritic growth, fracture surfaces, electrical discharges (lightnings and Lichtenberg figures), and the composition of quasicrystals are best described by set-theoretic constructs.

Or take the weird functions Karl Weierstrass invented a hundred years ago purely to prove that a function could be both everywhere continuous and yet nowhere differentiable. The fact that such an analytic pathology describes something in the real world—nay, is *elemental* to understanding the strange attractors of nonlinear dynamic systems (such as the double swing and the three-body problem)—gives one pause.

The word *symmetric* is of ancient Greek parentage and means well-proportioned, well-ordered—certainly nothing even remotely chaotic. Yet, paradoxically, self-similarity, the topic of this tome, alone among all the symmetries gives birth to its very antithesis: *chaos*, a state of utter confusion and disorder. As we shall endeavor to show, the genesis of chaos is, in fact, closely related to self-similarity and its inherent lack of "smoothness."

Perhaps not surprisingly, self-similarity entails numerous paradoxes in measurements of time, length, and even musical pitch. Think of Zeno's tardy turtle, pursued—but never overtaken—by swift Achilles. Why do certain lengths increase without bound when we measure them with ever smaller yardsticks? How would Euclid have explained plane geometric figures whose areas scale not as the squares of their apparent perimeters but as some lesser power, such as 1.77 and other fractional exponents? What should we think of musical sounds that, when scaled *up* in frequency, sound—incredibly—*lower* in pitch? How are such monstrosities possible? And how can we describe them in a consistent, meaningful manner?

Here a particularly felicitous thought by Felix Hausdorff comes to the rescue. His and Abram Besicovitch's new ways of looking at dimension dethrones it from its integer position and propels it into the realm of real numbers, giving us one of the sharpest tools—the *Hausdorff dimension* and its ramifications—with which to attack the strange sets that self-similarity breeds.

And while recalling some of the glorious names of the past, we should never forget our great contemporary, the inimitable Benoit B. Mandelbrot, who, single-handed, rescued set theory's most brittle functions and "dustiest" sets from near-oblivion and planted them right in the middle of our daily experience and consciousness. Yes, for all these years, we *have* been living with fractal arteries, not far from fractal river systems draining fractal mountainscapes under fractal clouds, toward fractal coastlines. But, kin to Molière's would-be gentleman, we lacked the proper prose—*fractal*, noun and adjective—that Benoit B. begot.

But our story also has a silent and immobile hero: the digital computer. There can be little doubt that computers have acted as the most forceful forceps in extracting fractals from the dark recesses of abstract mathematics and delivering their geometric intricacies into bright daylight. In fact, the impact of fractal images, often of unimagined beauty and appeal, has given computer graphics a surprising new dimension.

Synopsis

We open our treatise with one of the most charming uses that similarity was ever put to: the young Einstein's proof of Pythagoras's theorem. By adding just a single straight line, in the right place, to a right triangle and applying plenty of similarity, the popular theorem is proved without further prodding.

We then invade the unlimited domain of *self*-similarity as manifest in fractals, multifractals, and the scaling laws of physics, psychophysics, and boundless other fields.

In phase spaces, we encounter deterministic chaos and strange attractors. Percolation and other phase transitions lead us to critical exponents and a hierarchy of different dimensions. Following Poincaré, we immerse ourselves in the self-similarities of iterated mappings, from baker transformations and Bernoulli shifts to logistic parabolas and circle maps. Neither tori, *cantori*, nor Arnold tongues will faze us as we (sur)mount devil's staircases to unwind among the rational winding numbers festooning Farey trees.

And when we talk about nonlinear dynamics we must remember some of the great contributors of recent vintage: Siegel, Moser, Lorenz, Wilson, Feigenbaum, and—last but not least—the great Russian "school" exemplified by such names as Lyapunov, Arnol'd, Sinai, Chirikov, Alexeev, Anosov, Pesin, and the recently deceased master mathematician Kolmogorov.

Cayley trees, also known as Bethe lattices, will provide us with a fitting point of departure for many a practical fractal, such as our bronchial and vascular systems. Cellular automata concern us as models of both biological growth and chemical reactions.

We are also strangely attracted to symbolic dynamics, kneading (and needing) the Morse-Thue sequence and, especially, the Fibonacci rabbit sequence and their discrete self-similarities that, indiscreetly, tell us so much about period doubling, mode locking, frustrated Ising spins, and fivefold symmetric quasicrystals. Many of these subjects were shrouded in mystery and beset by paradoxes before the sharp scalpels, fashioned by scaling and renormalization theories, revealed the underlying tissue and made them tractable. In fact, it is no accident that viable fundamental field theories in physics are renormalizable, as they must be if they are to shun sham scales.

And, of course, we will not hesitate to run down random fractals, from Brownian motion to diffusion-limited aggregation and stock market hiccups (some hiccups of late!). The poor gambler's ruin and the St. Petersburg paradox will provide further food for fractal reflections.

These, then, are some of the exciting, and sobering, themes sounded in the present volume. The aim of this exposition is to enhance the reader's understanding of self-similarity, perhaps the most pregnant of all of nature's symmetries, and to illustrate the wide-ranging applications of scaling invariance in physics, chemistry, biology, music, and—particularly—the visual arts, as manifested in the recent renaissance of computer graphics through fractal images and their iterative beauty.

Acknowledgments

This book owes its existence to many sources. Apart from a brief encounter, in my dissertation, with chaos among the normal modes of concert halls, a "nonintegrable" system if there ever was one, my main stimulus came from the early demonstrations by Heinz-Otto Peitgen and Peter Richter of fractal Julia sets. Their beautiful images, and the intriguing mathematics which underlies them, as epitomized in their book *The Beauty of Fractals*, have made a lasting impression on me.

My first meeting with Mandelbrot's work was his analysis of word frequencies in natural and artificial languages, which touched upon my own interests in computer speech synthesis and recognition. Mandelbrot's monumental monograph *The Fractal Geometry of Nature* influenced me immeasurably, as it did so many other people.

I have also greatly benefited from Robert Devaney's books and lectures on chaos, Jens Feder's *Fractals*, Michael Barnsley's *Fractals Everywhere*, and Dietrich Stauffer's charming *Introduction to Percolation Theory*.

I learned a lot about new developments during the 1988 Gordon Conference on Fractals, organized by Richard Voss and Paul Meakin, which brought together many of the world's outstanding experts in the field.

At the University of Göttingen, it was mainly through the work of Werner Lauterborn that I saw chaos in action in nonlinear dynamic systems from cavitation bubbles to Toda chains. My own students at the Drittes Physikalisches Institut provided both stimulus and rectification during a series of lecture courses on self-similarity, fractals, and chaos. Heinrich Henze and Karl Lautscham made skillful additions to the demonstrations that accompanied these lectures. Holger Behme, Wolf Dieter Brandt, and Tino Gramss contributed many of the computer graphics in this book.

The late Walter Kaufmann-Bühler of Springer Verlag, New York, and Ronald Graham, a longtime colleague at Bell Laboratories, provided early encouragement.

Hildegard Franks and Esperanza Plata in Murray Hill, New Jersey, supported by Maija Sutter and Irena Schönke in Göttingen, converted nearly illegible handwriting to readable typescript. Liane Liebe and Gisela Kirschmann-Schröder took loving care of the illustrations. Anny Schroeder, as during a previous itinerary, furnished much needed logistic support.

Jerry Lyons persuaded me to deliver myself and the manuscript into the hands of W. H. Freeman and Company, and I have not regretted heeding his appeal one bit. I greatly enjoyed working with him; with Georgia Lee Hadler, senior editor; with designer Nancy Singer; and with illustration coordinator Bill Page. The professional competence of my copy editor, Richard Mickey, proved beyond compare.

Last, but not least, I thank Hans-Werner Strube, Allan Hurd, Victor Klee, and Paul Meakin for their expert reviews of the manuscript and their manifold corrections and cogent comments.

Murray Hill and Göttingen   *Manfred Schroeder*
May 1990

1

Introduction

> *I want to know how God created this*
> *world. I am not interested in this or that*
> *phenomenon, in the spectrum of this or*
> *that element. I want to know His*
> *thoughts; the rest are details.*
> —ALBERT EINSTEIN

Nature abounds with periodic phenomena: from the motion of a swing to the oscillations of atoms, from the chirping of a grasshopper to the orbits of the heavenly bodies. And our terrestrial bodies, too, participate in this universal minuet—from the heart beat and circadian rhythms to monthly and even longer cycles.

Of course, nothing in nature is *exactly* periodic. All motion has a beginning and an end, so that, in the mathematical sense, strict periodicity does not exist in the real world. Nevertheless, periodicity has proved to be a supremely useful concept in elucidating underlying laws and mechanisms in many fields.

One reason for the universality of simple harmonic motion is the linearity—or near-linearity—of many physical systems and the invariance with displacement in space and time of the laws governing their behavior.

But there are numerous other phenomena in which linearity breaks down and, instead of periodicity, we get aperiodic or even chaotic motion: the smooth waves on a well-behaved lake turn to violent turbulence in the mountain brook, and the daily sunrise, the paradigm of predictability, is overshadowed by cloud formations, a haven for *chaos*—albeit *deterministic* chaos.

But no matter how chaotic life gets, with all regularity gone to bits, another fundamental bulwark often remains unshaken, rising above the turbulent chaos: *self-similarity*, an invariance with respect to *scaling*; in other words, invariance not with *additive* translations, but invariance with *multiplicative* changes of scale. In short, a self-similar object appears unchanged after increasing or shrinking its size. Indeed, in turbulent flows, large eddies beget smaller ones, and these spawn smaller ones still—and so on *ad infinitum* (almost). In general, one of the conspicuous consequences of self-similarity is the appearance of exceedingly fine-grained structures, now generally called *fractals* after Benoit B. Mandelbrot, the father of fractals [Man 83].[1]

Many laws of nature are independent, or nearly so, of a scaling factor. The fact that scaling usually has a limit (Planck's constant, when things get too small, or the speed of light, when objects fly too fast) does no harm to the usefulness of "thinking self-similar," just as the lack (outside mathematics) of strict periodicity is no great impediment in the real world. In a sense, self-similarity is akin to periodicity on a *logarithmic* scale.

Self-similarity, strict or otherwise, reigns in many fields in many guises, and in this book we shall explore some of the many manifestations of self-similarity in the world around us. Among the topics treated are the following:

• Scaling laws and their exponents in physics, psychophysics, and physiology

• Random walks in the stock market and under the microscope; floods, forest fires, the distribution of galaxies, and other "accidents" with statistical self-similarity

• Scaling invariance, self-similarity, and some of their mathematical models, such as Cantor sets and Julia sets

• Fractals and their characterization by Hausdorff, and other noninteger dimensions; fractal paradoxes and their resolution; Weierstrass functions and Hilbert curves; Koch flakes, Sierpinski gaskets, and other non-Euclidean constructions in two and more dimensions; fat fractals and multifractals

• Iterated mappings and a selection of the ensuing self-similarities

• The logistic parabola and other unimodal maps with universal scaling laws; period-doubling bifurcation to chaos; the Feigenbaum constant, symbolic dynamics and the Morse-Thue sequence; Sharkovskii's universal ordering of orbits

• Complexification of the quadratic map and the Mandelbrot set

• Devil's staircases, Farey trees, Arnold tongues, and modelocking; the "rabbit sequence" and the quasi-periodic route to temporal and spatial chaos; Ising spins and quasicrystals

• Laplace's triangle and cellular automata

---

1. References in brackets are listed alphabetically at the end of the book. The numbers refer to the year of publication.

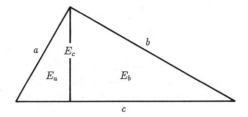

Figure 1 Pythagoras's theorem: sketch for proof by the 11-year-old Einstein based on similarity.

In this chapter some of these topics are introduced informally, together with the leading dramatis personae.

Einstein, Pythagoras, and Simple Similarity

I will a little think.
—ALBERT EINSTEIN, in America

When Jacob Einstein taught (Euclidean) geometry to his 11-year-old nephew Albert, the young Einstein—even then striving for utmost parsimony—felt that some of Euclid's proofs were unnecessarily complicated.[2] For example, in a typical proof of Pythagoras's theorem $a^2 + b^2 = c^2$, was it really mandatory to have all those extra lines, angles, and squares in addition to the basic right triangle with hypotenuse, c and sides a and b?

After "a little thinking," the sharp youngster came up with a proof that required only *one* additional line, the altitude above the hypotenuse (see Figure 1). This height divides the large triangle into two smaller triangles that are *similar* to each other and similar to the large triangle. (Triangles are similar if their angles are the same, which is easily seen to be the case in Figure 1.)

Now, in Euclidean geometry, the area ratio of two similar (closed) figures is equal to the *square* of the ratio of corresponding *linear* dimensions. Thus, the areas E_a, E_b, and E_c (E as in German *Ebene*) of the three triangles in Figure 1 are related to their hypotenuses a, b, and c by the following equations:

$$E_a = ma^2 \tag{1}$$

---

2.   I have the story from Schneior Lifson of the Weizmann Institute in Tel Aviv, who has it from Einstein's assistant Ernst Strauss, to whom it was told by old Albert himself.

$$E_b = mb^2 \qquad\qquad (2)$$

$$E_c = mc^2 \qquad\qquad (3)$$

where m is a dimensionless nonzero multiplier that is the *same in all three equations*.

Now a second look at Figure 1 will reveal that the area of the large triangle is, of course, the sum of the areas of the two smaller triangles,

$$E_a + E_b = E_c$$

or, with equations 1 to 3,

$$ma^2 + mb^2 = mc^2$$

Dividing this identity by the common measure m promptly produces Pythagoras's renowned result

$$a^2 + b^2 = c^2$$

proved here by an 11-year old person[3] by combining two fertile scientific principles that were going to stand the grown-up Einstein in good stead: simplicity and symmetry, of which self-similarity is a special case. Yet the true beauty of Einstein's proof is not that it is so simple, but that it exposes the true essence of Pythagoras's theorem: *similarity* and *scaling*.

The resemblance of equation 3 to Einstein's later discovery, his famous $E = mc^2$, is of course entirely fortuitous. The equivalence of mass m and energy E, which is at the basis of nuclear power in all its guises, is a consequence of Lorentz invariance. This invariance, which underlies special relativity, was predicted by Einstein in 1905 after, it seems, several false starts and a "little more thinking" (see Figure 2).

A Self-Similar Array of Self-Preserving Queens

One of numerous chess problems is the placement of as many queens as possible on a chessboard of a given size so that no queen "attacks" (shares a row, column, or $\pm 45°$ diagonal with) any other queen. For a $k \times k$ square board, there can be at most k nonattacking queens. But are k peacefully coexisting queens always possible? What if k is very large? Doesn't the complexity of placing the queens grow exponentially with the size of the board? As we shall see, the placement is actually very simple, even for arbitrarily large boards, if we focus our attention

---

3. Really a "nonperson" at that stage, considering the neglect he suffered in his Munich high school [Pye 85].

Figure 2 Einstein on the verge of discovering his famous formula $E = mc^2$—a cartoonist's view [Har 77]. (© 1991 by Sidney Harris)

on boards for which k is a pure power of an integer and judiciously exploit the principle of self-similarity in the construction of the solution. (Again, we describe an object or a structure as self-similar if it looks the same when we magnify the object or a properly chosen part of it.)

Figure 3 shows a pattern of queens, the 5×5 board sustaining five non-attacking pieces. (This particular placement could have been obtained by a *greedy algorithm*: Starting on the lower left and proceeding column by column, always place the next queen in the lowest position still "eligible.")

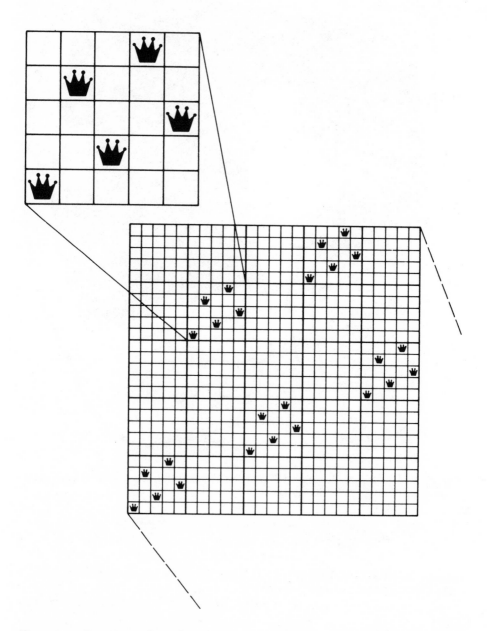

Figure 3   Five nonattacking chess queens on a 5 × 5 board (top) and solution for the 25 × 25 board derived from the 5 × 5 board by similarity.

From a solution on the 5×5 board, we can immediately construct a possible placement for the 25×25 board, which can be considered to be composed of $5 \times 5 = 25$ boards of size 5×5. We simply leave most of those twenty-five 5×5 boards empty, except those five that correspond to the positions of the queens in the original board. Figure 3 illustrates the procedure without the need for more words.

To fill a 125×125 board with peaceful queens, simply think of it as twenty-five boards of size 25×25, five of which are filled in the by now familiar pattern with the 25×25 solutions while the remaining twenty boards are left void. Continuing in this manner, we have, after n steps, a $5^n \times 5^n$ board with 5^n pieces.

This process can be extended *ad infinitum* to yield an immaculately *self-similar* distribution of self-preserving queens. Indeed, selecting one of the five occupied subboards of side length one-fifth of the entire board and magnifying it by a factor of 5 will precisely reproduce the entire board. The factor 5 is called the *scaling* or *similarity factor* of the board.

What numbers other than 5 can be used as scaling factors in such self-similar schemes? Can we exploit self-similarity for the construction of boards whose side is not a pure power of an integer (as 5^n is)? The interested reader can find further clues in the illuminating article by Clark and Shisha [CS 88].

A Self-Similar Snowflake

Repeating a given operation over and over again—on ever smaller scales—culminates, almost inescapably, in a self-similar structure. Here the repetitive operation can be algebraic, symbolic, or geometric, as in the case of the five dormant queens whom we have just allowed to come alive and multiply without limit, proceeding on the path to prefect self-similarity.

The classical example of such a repetitive construction is the *Koch curve*, proposed in 1904 by the Swedish mathematician Helge von Koch. The basic principle and the final result are equally charming: Take a segment of straight line (Figure 4A, the *initiator*) and raise an equilateral triangle over its middle third as shown in Figure 4B. The result is called the *generator*. Note that the length of the generator is four-thirds the length of the initiator.

Repeating once more the process of erecting equilateral triangles over the middle thirds of straight line segments results in Figure 4C. The length of the fractured line is now $(\frac{4}{3})^2$. Iterating the process infinitely many times results in a "curve" of infinite length, which—although everywhere continuous—is *nowhere differentiable*. It is approximated, as far as pen and ink permit, in Figure 4D.

Similarly lamentable "functions," continuous but without tangents, were first defined a century ago by the German mathematician Karl Weierstrass, just to show his skeptical colleagues (a horrified Hermite among them) that such functions did indeed exist. But other authorities, not least the great Austrian

(A)

(B)

(C)

(D)

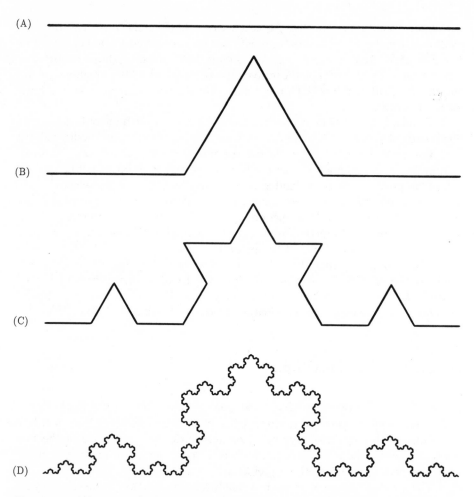

Figure 4   Initiator (A) and generator (B) for the Koch curve, the next stage in the construction of the Koch curve (C), and high-order approximation to the Koch curve (D).

physicist Ludwig Boltzmann, saw the light: Boltzmann wrote to Felix Klein (in 1898) that nondifferentiable functions could have been invented by physicists because there are problems in statistical mechanics "that absolutely necessitate the use of nondifferentiable functions." And his French colleague Jean Perrin went even further when, in 1906, he presaged present sentiment about such mathematical monsters, saying that "curves that have no tangents are the rule, and regular curves, such as the circle, are interesting but quite special." How politely put! Now, following Mandelbrot, we simply call such nondifferentiable curves *fractals*.

A New Dimension for Fractals

The universe is not only queerer than we
suppose but queerer than we can suppose.
—J. B. S. HALDANE

Applying the Koch generator (see Figure 4) to an equilateral triangle and painting the interior black results in a solid star of David (Figure 5A). Infinite iteration converges on the Koch snowflake (intermediate stages of the construction are shown in Figure 5B). How long is its perimeter? After n iterations it has increased $(\frac{4}{3})^n$-fold over the perimeter of the initial triangle. Thus, as n approaches infinity, the perimeter becomes infinitely long. To characterize the perimeter's size, we can therefore no longer use its length. We have to invent a new measure that can distinguish between fractals manufactured from different generators. But while inventing new measures, we want to stay as close as possible to what we have always done when measuring lengths.

For a smooth curve, an approximate length $L(r)$ is given by the product of the number N of straight-line segments of length r needed to step along the curve from one end to the other and the length r: $L(r) = N \cdot r$. As the step size r goes to zero, $L(r)$ approaches a finite limit, the length L of the curve.

Not so for fractals! The product $N \cdot r$ diverges to infinity because, as r goes to zero, we enter finer and finer wiggles of the fractal. However, asymptotically, this divergence behaves according to a well-defined homogeneous power law of r. In other words, there is some *critical exponent* $D_H > 1$ such that the product $N \cdot r^{D_H}$ stays finite. For exponents smaller than D_H, the product diverges to infinity, while for larger exponents the product will tend to zero. This critical exponent, D_H, is called the *Hausdorff dimension* after the German mathematician

(A)   (B)

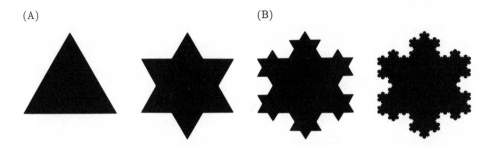

Figure 5   Initiator and generator for the Koch flake (A) and intermediate stages in the construction of the Koch flake (B).

Felix Hausdorff (1868–1942). Equivalently, we have

$$D_H := \lim_{r \to 0} \frac{\log N}{\log (1/r)}$$

For the nth generation in the construction of the Koch curve or snowflake, choosing $r = r_0/3^n$, the number of pieces N is proportional to 4^n. Thus,

$$D_H = \frac{\log 4}{\log 3} = 1.26 \ldots$$

The fact that D_H lies between 1 and 2 is somehow satisfying, because an infinitely long curve is, in some metric sense, more than just a one-dimensional object—without being a two-dimensional area, since the curve does not cover a region in the plane. In fact, we shall soon see that Hausdorff's definition of dimension, which, as we now know, can take on fractional values, makes much sense in many ways. Of course, for a *smooth* curve, $D_H = 1$; and for a smooth surface the number N of covering disks is proportional to $1/r^2$ and therefore $D_H = 2$. Here r is the diameter of the N little disks needed to cover the area. Similarly, for a compact three-dimensional volume, D_H comes out equal to 3.

Surprisingly, however, for D_H to equal 2, we do not need an area; a topologically *one*-dimensional entity, a line, suffices. A well-known example is the asymptotically self-similar Hilbert curve (see Figure 6A), which comes arbitrarily close to each point in the unit square. Its construction is illustrated in Figure 6B. The final result is, of course, self-similar. Blow up any appropriately chosen subsquare by a linear factor 2^n and it will resemble the entire figure.

Since the nth generation of the Hilbert curve consists of $2^{2n} - 1$ segments of length $1/2^n$, its Hausdorff dimension equals 2, as behooves an area-filling curve. Figure 7 shows an artistic variation on the Hilbert curve theme. Can you recognize that the underlying image is a human face?

Adjacent points on the Hilbert curve are adjacent in the unit square, but not vice versa! This property distinguishes the Hilbert curve from broadcast TV scans, which are discontinuous at the line ends,[4] and from Cantor's totally discontinuous mapping of the unit square onto the unit interval, whereby the point in the square $x = 0.x_1, x_2, x_3, \ldots$; $y = 0.y_1, y_2, y_3, \ldots$ is mapped to the point on the line $0.x_1, y_1, x_2, y_2, x_3, y_3, \ldots$.

When Cantor first saw that, in this manner, an area could be reversibly mapped to a line, he wrote "I see it, but I don't believe it." But evolution, in constructing our brain, discovered millenia ago that in order to fill a volume

---

4. Interestingly, some sophisticated image-scanning techniques do follow Hilbert's prescription for a space-filling curve. The reason is that points adjacent in time along a "Hilbert scan" are also adjacent in space in the scanned image, making for simpler image processing.

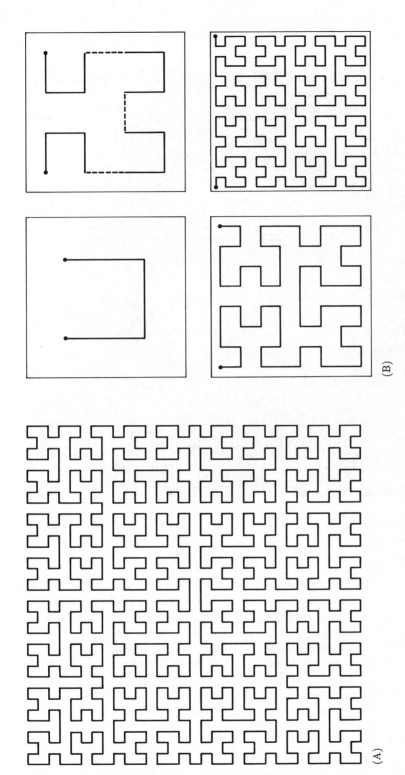

Figure 6 (A) Toward the self-similar Hilbert curve; (B) steps in construction of the Hilbert curve.

Figure 7 Contorted Hilbert curve: an artist's version, to be viewed from a distance. (Courtesy of W. J. Möller.)

while preserving two-dimensional adjacency, it had to construct the gray matter of our cortex in a folded manner resembling a three-dimensional Hilbert curve.

Hilbert curves in higher-dimensional spaces have also found interesting applications in information theory: the so-called *Gray codes* [Gil 58], so named after their inventor. In a binary Gray code for the integers, only a single bit of the code changes between one integer and the next. Thus, the four integers 0 to 3 are encoded by two binary bits as follows: $0 = 00$, $1 = 01$, $2 = 11$ and $3 = 10$ (and not as in the standard binary code, where $2 = 10$ and $3 = 11$, creating a two-bit jump between the codes for 1 and 2). Figure 8 shows successive stages for the construction of a Hilbert curve in three-space, visualizing generalized Gray codes [Gil 84].

While Cantor's mapping of an area to a line is discontinuous in both directions, the Hilbert curve is discontinuous in only one direction. There are other

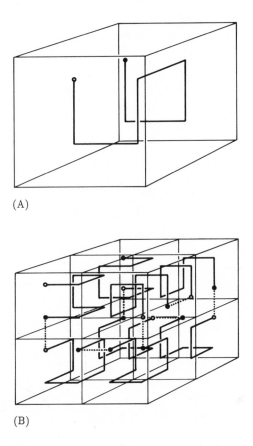

(A)

(B)

Figure 8   Constructing a three-dimensional version of the Hilbert curve (A); Hilbert curve for illustrating Gray code (B).

mappings from an area to a line, due to Bernhard Bolzano (1781−1848) and Giuseppe Peano (1858−1932) but none are continuous in *both* directions.

A Self-Similar Tiling and a "Non-Euclidean" Paradox

Look at the seven fractal "tiles" shown in Figure 9A. They are obtained from seven hexagons (see Figure 9B), by breaking up each side into a three-piece zig-zag as shown on one of the sides. If the inner angles of the three pieces are 120°, then the lengths of the three segments will be $1/\sqrt{7}$ times the length of the unbroken side.

Iterating the breaking up process *ad infinitum* results in a fractal tiling pattern, of which Figure 9A is an approximation. As a result of this construction, the

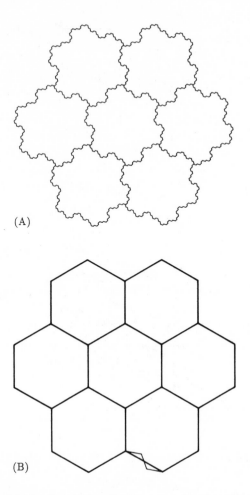

Figure 9 (A) Fractal tiles that fit together to cover the plane. The seven tiles shown, taken together, are similar to a single tile, giving rise to a "non-Euclidean" paradox. (B) Tiling hexagons: the initiator of the fractal tiles shown in part A. One generator, consisting of three straight-line segments, is also shown.

perimeter of the entire figure, consisting of seven fractal "hexagons," is similar to each of the seven hexagons.

Thus, we have found a *self-similar* tiling of the plane based on "hexagons," where each tile is surrounded by six like tiles. (Note that while regular hexagons do tile the plane, the tiling is not self-similar. A hexagon surrounded by six like hexagons is *not* a larger hexagon.)

Simple inspection of Figure 9A shows that the perimeter of the large fractal hexagon contains the perimeter of the small fractal hexagons precisely three times. Thus, following Euclid's scaling rule for the areas of similar geometric figures, the total area should be $3^2 = 9$ times the area of one of the small fractal hexagons. But it isn't! The area ratio is only 7.

What went wrong? Where did we go astray? Has Euclid finally been caught napping? Well, the ancient Greeks (with the possible exception of one of the Zenos) can continue to rest in peace. Fractal geometric objects like the one illustrated in Figure 9A were never on exhibit in Euclid's school (nor were they used to tile its floors). Euclid probably never considered nondifferentiable functions or bounded curves of infinite length. But later generations of mathematicians did, and since Hausdorff we know that the dimensions of such curves are not necessarily equal to 1 but perhaps exceed 1. For example, the Hausdorff dimension D_H of the perimeter of our fractal hexagons is $\log 3/\log \sqrt{7} = 1.12915 \ldots$. Thus, in adapting Euclid's scaling idea, we should raise 3 (the perimeter "ratio") not to the power 2 to obtain the area ratio, but—since the perimeter already has dimension $1.12915 \ldots$—to the power $2/1.12915 \ldots = 1.77124 \ldots$. This gives an area ratio of 6.999999999 on my pocket calculator—close enough to the true area ratio of 7 to 1 that is immediately apparent in Figure 9A. Thus, we can reformulate Euclid's scaling theorem about similar areas and obtain a more generally valid result, applicable to fractals and nonfractals alike:

For similar figures, the ratios of corresponding measures are equal when reduced to the same dimension on the basis of their Hausdorff dimensions.

It is because of properties like this that Hausdorff dimension is such a useful concept. It is *one* proper extension of the concept of dimension to fractal objects, which model, however approximately, a great many phenomena in the real world surrounding us—and *in* us. Just think of the human vascular system, or your lungs with their hierarchical branchings, leading to astonishingly large surface areas that are well described by fractal geometries and Hausdorff dimensions.

Thus, the idea of the Hausdorff dimension has resolved a potentially disastrous paradox by widening our concept of dimension to include fractional and even transcendental values. We shall resume this theme in the body of this book and get to know other fractal paradoxes such as that of a musical chord that sounds *lower* in pitch when reproduced at a higher tape speed! (See pages 96–98 in Chapter 3.)

At the Gates of Cantor's Paradise

I place myself in a certain opposition to
widespread views on the nature of the
mathematical infinite.
—GEORG CANTOR

The Hausdorff dimension D_H is useful not only for characterizing fractal curves of infinite length but also point sets, or "curves" of *zero* length. Not surprisingly, for such point sets D_H is typically less than 1. A famous example is Georg Cantor's original self-similar "middle-third–erasing" set with which he demonstrated, to the astonishment and disbelief of the contemporary mathematical community, that there are sets having measure ("length") zero with *uncountably* many members.

Cantor constructed his highly counterintuitive set as follows. He started with the closed unit interval [0, 1], that is, a straight-line segment of length 1 including the two endpoints. He then "wiped away" the open middle third $(\frac{1}{3}, \frac{2}{3})$ and repeated the process on the remaining two segments of length $\frac{1}{3}$ (see Figure 10).

Repeating the middle-third wiping-out process over and over again leaves not a single connected line segment; the total length or measure of the remaining set is zero. Yet, as we shall later see, the leftover "dust" still contains infinitely many, in fact *uncountably* many, "points." In fact, one can already appreciate this from the arithmetic description of the Cantor set: its members are precisely all those fractions in the interval [0, 1] that eschew the digit 1, such as 0.2 or 0.2022.

How do we characterize the content of a set whose length measure is zero? Again Hausdorff offers help. After n wiping stages, we are left with $N = 2^n$ straight-line pieces, each of length $r = (\frac{1}{3})^n$. Thus, the Hausdorff dimension D_H equals $\log 2/\log 3 = 0.63\ldots$, a value between 0 and 1, as expected because the Cantor dust is more (a lot more!) than just a point (dimension 0) and much less than a length of line or curve (dimension 1). As in the case of the fractal Koch curve, the value of D_H is not an integer; in fact, it is a transcendental number.

Figure 10   Construction of the "middle-third–erasing" Cantor set. It has zero measure, yet is uncountable. Its fractal dimension equals $\log 2/\log 3 = 0.63\ldots$.

In the course of this book we shall encounter "innumerable" other instances of dusty, Cantor-like sets in a wide variety of settings—such as the celebrated Sierpinski gasket, which we will introduce next.

The Sierpinski Gasket

Fog on Fog.
—HERMANN WEYL, commenting on
Cantor's transfinite numbers

Are there Cantor-like dusts spread out in two dimensions? Yes, there are. Start with the equilateral triangle shown in Figure 11A and remove the open central upside-down equilateral triangle with half the side length of the starting triangle. This leaves three half-size triangles. Repeating the process on the remaining (right-side-up) triangles leaves, after n iterations, $N = 3^n$ triangles of side length $r = r_0(2^{-n})$ (see Figure 11B). The Hausdorff dimension D_H for the set resulting from an infinite iteration of this procedure, called the *Sierpinski gasket* after the prolific Polish mathematician Waclaw Sierpiński (1882–1969), equals log 3 /log 2 = 1.58 . . . , an irrational number smaller than 2, in spite of the fact that the gasket is embedded in two dimensions.

It is interesting to note that the Sierpinski gasket combines self-similarity with another important, but classical, symmetry: rotation. Indeed, the gasket is congruent to itself when rotated around its center by an angle of 120° (or any integer multiple of 120°). Such symmetries, combining infinite scaling and finite rotation, can be observed in many fractals—and the prescient works of Maurits Escher (see Figure 12).

Incidentally, the "fractal" dimension of a fractal set is not necessarily a non-integer. For example, the Hausdorff dimension of the self-similar board of non-attacking queens (see pages 4–7) equals 1.

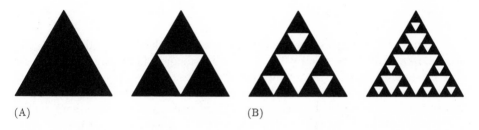

(A)   (B)

Figure 11   (A) Generator for the Sierpinski gasket. (B) Toward the Sierpinski gasket: a two-dimensional uncountable set with zero measure and fractal dimension log 3 /log 2 = 1.58 . . . .

Figure 12   An image by Escher that combines rotational symmetry and infinite scaling [Esc 71].

Figure 13 illustrates a three-dimensional generalization of the Sierpinski gasket. Its construction starts with a regular tetrahedron (a pyramid bounded by four equilateral triangles) from which a half-size upside-down regular tetrahedron has been cut out. This process is repeated on the four remaining tetrahedra and all subsequent tetrahedra to yield the spidery tower shown in Figure 13.

The Hausdorff dimension of this self-similar construction follows immediately from the first step: with $N = 4$ remaining pieces of size $r = \frac{1}{2}$, we have $D_H = \log 4/\log 2 = 2$, a fractal dimension that happens to be an *integer*, but a full unit less than the embedding Euclidean dimension $d = 3$.

Figure 13 A three-dimensional version of the Sierpinski gasket. Its fractal dimension, log 4/log 2 = 2, has an integer value (2), albeit smaller than the dimension of the supporting space (3).

Sierpinski gaskets in two or more dimensions model many natural phenomena and man-made structures. Think of the Eiffel Tower in Paris, designed by Gustave Eiffel. If, instead of its spidery construction, it had been designed as a solid pyramid, it would have consumed a lot of iron, without much added strength. Rather, Eiffel used trusses, that is, structural frames whose members exploit the rigidity of the triangle. (A triangle, in contrast to a rectangle, cannot be deformed without deforming at least one of its sides.) However, the individual members of the largest trusses are themselves trusses, which in turn are made from members that are trusses again. This self-similar construction guarantees high resilience at low weight. The structures of Gothic cathedrals, too, betray great faith in this principle of achieving maximum strength with minimum mass.

And Buckminster Fuller (1895–1983) and his skeletal domes popularized the fact that strength lies not in mass but in *branch points*. In fact, counter to intuition, the Sierpinski gasket and like constructions consist of nothing *but* branch points. (A branch point on a curve has more than two points arbitrarily close to it.) Certain boundary sets (of strange attractors, for example) share this property with the Sierpinski gasket (see pages 38–40, where this exclusive branching is exploited to "settle" an international boundary problem).

The Sierpinski gasket is good for another counterintuitive surprise. For Euclidean bodies in d dimensions, the volume V is proportional to R^d, where R is some linear measure of size. Surface area S varies as R^{d-1}. Thus, $S \sim V^{(d-1)/d}$. For example, for $d = 3$, $S \sim V^{2/3}$. In fact, for the sphere, $S = 4\pi R^2 = (36\pi)^{1/3} \cdot V^{2/3}$.

However, for fractal objects this simple Euclidean relation often breaks down. As we have seen, the Hausdorff dimension of the Sierpinski gasket equals log 3 /log 2 $\approx$ 1.58. What is the Hausdorff dimension of its edges? It is easy to see that every time we reduce the yardstick by a factor 2, the number of edge segments goes up by a factor 3. Thus, the Hausdorff dimension of the edges, the "surface" of the Sierpinski gasket, is also log 3/log 2: "volume" and "surface" have the same dimension. We can also see this by expressing the mass $M(R)$ of the gasket, that is, the number of points inside a circle of radius R, as a function of the radius: on average we find $V(R) \sim R^{1.58}$. But for the total edge lengths $S(R)$ inside the circle we find the same dependence: $R^{1.58}$. As a consequence, for the Sierpinski gasket, area V and edge length S are proportional to each other: $V \sim S$, a paradoxical result indeed.

We shall have the pleasure of meeting the Sierpinski gasket again, both in its original form and, in Chapter 17, in a discrete version, the Laplace triangle modulo 2. In the Meantime, let us relish some of its refreshing implications, such as the board game invented by the mythical Sir Pinski.

Sir Pinski's Game and Deterministic Chaos

Consider the following "parlor game" played by two or more persons:

• Each player picks an initial point inside an equilateral triangle.

• Then the player doubles its distance from the nearest corner along a straight line from that corner, thereby arriving at a point p_1.

The player who can repeat the distance doubling most often without falling outside the triangle wins the game. As we shall see, there are *uncountably* many, but still *very few*, initial points that guarantee winning or a tie, "very few" in the sense that a random choice has a zero probability of infinite survival under the rules of the game.

Figure 14 shows the equilateral triangle with an initial choice, marked by 0, and its three successors or "images" marked 1, 2, and 3. Note that the point 3 already lies outside the triangle; the initial choice is therefore not a good pick. How can we avoid such bad points? We will answer the question first geometrically and then arithmetically.

Note that point 2, which lies inside the small, white (upside-down) triangle, is mapped outside the large triangle. In fact, a little reflection (in more than one sense of the word) will show that all points inside the small white triangle will be mapped to the outside. Thus, the white triangle is *out* as a good starting area—and so, of course, are its preimage and the preimages of the preimage and so on *ad infinitum*. In other words, any point that, sooner or later, is mapped into the white triangle is a loser.

But what *are* the preimages of the white triangle? A little more reflection will show that they consist of three half-size upside-down triangles, one inside

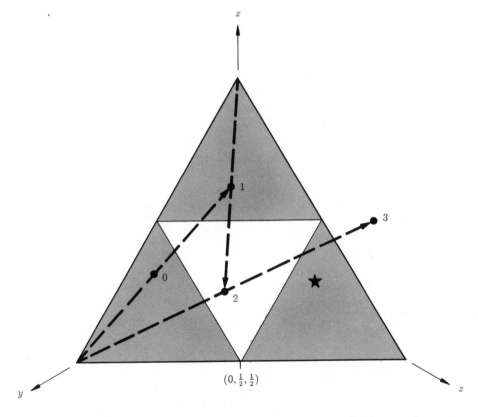

Figure 14　Sir Pinski's chaos game: How many times can you double your distance to the nearest vertex without leaving the large equilateral triangle?

each of the remaining three dark triangles. And the preimages of these three preimages are *nine* upside-down triangles, again scaled down in side length by a factor 2 and cut out of the centers of the nine remaining quarter-size triangles.

Thus, in delimiting good initial choices, we find ourselves constructing a *self-similar* figure, the well-known Sierpinski gasket (see Figure 15), a Cantor set embedded in two dimensions, with zero area and Hausdorff dimension equal to $\log 3/\log 2 \approx 1.58$. Picking an initial point at random, however, will almost certainly land us in white territory, a prelude to the disaster of being eventually mapped outside the big triangle.

In order to avoid potential disputes resulting from poor drafting, the just described Sir Pinski game should be played arithmetically; that is, the initial points and all their images should be stated by their coordinates in a suitable coordinate system. Although two coordinates suffice to locate a point in the plane, a more convenient system, matched to the symmetry of the triangle, uses three coordinates, x, y, and z, as shown in Figure 14. The corners of the triangle have the

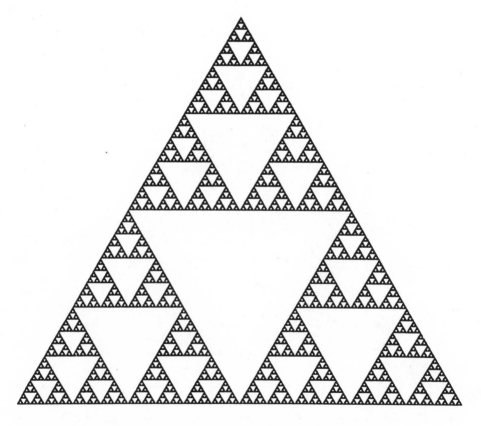

Figure 15   Sierpinski gasket, the winning set of Sir Pinski's game.

value 1 for the corresponding coordinates, and the opposite sides have the value 0. Thus, the midpoint on the horizontal side, for example, has the coordinates $x = 0$ and $y = z = \frac{1}{2}$.

Of course, a set of three coordinates in the plane is redundant, and the coordinate values cannot be chosen independently, because of the constraint $x + y + z = 1$. The points *inside* the triangle are further subject to the constraints $x > 0$, $y > 0$, and $z > 0$.

Now, how does our mapping, defined by doubling the distance to the nearest corner, look arithmetically? Suppose for our initial choice (x_0, y_0, z_0) the nearest corner is the lower left (y) corner. Then the image of (x_0, y_0, z_0) is $(2x_0, 2y_0 - 1, 2z_0)$. The factors of 2 occurring in this mapping suggest using binary notation for the coordinates. Multiplication by 2 is then a simple left shift of the digits. Thus, the point 0 in Figure 14, which has the approximate coordinates $(\frac{5}{16}, \frac{39}{64}, \frac{5}{64}) = (0.0101, 0.100111, 0.000101)$, suffers the following fate:

$$x_0 = 0.0101 \qquad y_0 = 0.100111 \qquad z_0 = 0.000101$$

$$x_1 = 0.101 \qquad y_1 = 0.00111 \qquad z_1 = 0.00101$$

$$x_2 = 0.01 \qquad y_2 = 0.0111 \qquad z_2 = 0.0101$$

$$x_3 = 0.1 \qquad y_3 = -0.001 \qquad z_3 = 0.101$$

Here y_3 is negative—that is, (x_3, y_3, z_3) lies outside the triangle—and the player who picked the point (x_0, y_0, z_0) is eliminated from the game.

Once outside the triangle, the images "escape" to infinity. Arithmetically, what are the *good* initial choices that stay inside the triangle and thus are never eliminated? If we can find a complete answer to this question, we will also have discovered an *arithmetic* description of the Sierpinski gasket to boot!

Then what led to the unwanted negative value of y_3 in the mapping? The answer is that y_2, the largest preceding coordinate, was smaller than $\frac{1}{2}$; in other words, the first fractional binary digit of y_2 was a 0 and not a 1. Hence, a good initial point must not have 0s in all three coordinates for any of its fractional binary places. This rule is violated by (x_0, y_0, z_0), which has only 0s (not a single 1) in the third binary places. Taken together with the constraint $x + y + z = 1$, this means that good points (x, y, z), that is, members of the Sierpinski gasket, have precisely one 1 and two 0s in every binary place of x, y, and z.

There is a charming similarity here with the ternary representation of the original (middle-third–erasing) Cantor set, which contains only 0s and 2s and no 1s. In fact, the connection between this arithmetic representation of the Sierpinski gasket and Cantor's construction is quite close: the first missing 1 of a Cantor number (right behind the "ternary point") corresponds to the deletion of the interval $(\frac{1}{3}, \frac{2}{3})$ from the unit interval. The absent 1 in the second ternary place corresponds to the subsequent elimination of the two intervals $(\frac{1}{9}, \frac{2}{9})$ and $(\frac{7}{9}, \frac{8}{9})$, and so on.

What does the absence of three 0s in the binary representation of the Sierpinski gasket mean geometrically? Three 0s in the first binary place to the right of the binary point would mean that neither x, y, nor z exceeds $\frac{1}{2}$. Geometrically speaking, this corresponds to the central half-size upside-down triangle, left white in Figure 14, which is thus excluded from the Sierpinski gasket—as indeed it is in the first step of the geometric construction of the gasket. We could also argue that a 1 in the first binary place of x, y, or z means that either x, y, or z is greater than (or equal to) $\frac{1}{2}$. Geometrically, these three cases correspond to the three half-size right-side-up triangles (shaded in Figure 14).

What would three 0s in the *second* binary place correspond to geometrically? A little triangular reasoning will reveal that they correspond to quarter-size upside-down triangles cut from the centers of the three half-size shaded triangles left over after the first cutting operation. In general, three 0s in the nth binary place imply the elimination of 3^{n-1} upside-down triangles of side length 2^{-n} from the 3^{n-1} right-side-up triangles left standing after $k^n - 1$ cutting operations. Thus, the binary representation of the Sierpinski gasket corresponds, place by place, to its geometric construction. The two descriptions are equivalent.

A proper Sierpinski point is $(\frac{1}{3}, \frac{2}{3}, 0) = (0.\overline{01}, 0.\overline{10}, 0)$, for example, which lies on the left side of the triangle, one-third up from the lower left corner. Our distance-doubling mapping will make it alternate, with period length 2, with the point $(\frac{2}{3}, \frac{1}{3}, 0)$ as is clear both geometrically and from the period length of 2 of the binary fractions for $\frac{1}{3}$ and $\frac{2}{3}$.

Are there periodic points with period length 3? If so, our mapping should be equivalent to a 120° rotation. To find such points, we simply have to consider binary fractions with period length 3. And indeed, $(0.\overline{010}, 0.\overline{001}, 0.\overline{100}) = (\frac{2}{7}, \frac{1}{7}, \frac{4}{7})$, which is marked by a star in Figure 14, is such a point. The other two points of its orbit are $(\frac{4}{7}, \frac{2}{7}, \frac{1}{7})$ and $(\frac{1}{7}, \frac{4}{7}, \frac{2}{7})$, reached by twice rotating 120° counterclockwise. The only other period-3 orbit is obtained by interchanging two coordinates—for example, by starting with $(\frac{1}{7}, \frac{2}{7}, \frac{4}{7})$, whose two successors are found by 120° rotations clockwise: $(\frac{2}{7}, \frac{4}{7}, \frac{1}{7})$ and $(\frac{4}{7}, \frac{1}{7}, \frac{2}{7})$.

Periodic points exist for all period lengths. Thus, for example, the point (1, 0, 0), the upper corner of the triangle, has period length 1; that is, it (and the other two corners) are *fixed points*. We shall later encounter this scenario and similar mappings again, and we shall derive a formula for the number of different orbits of a given period length. (This derivation will involve the *Moebius function* from number theory, a function whose multifarious functions in higher arithmetic one should know about; it "twists things around," much like the much better known Moebius strip.)

Under the rules of the game, the image of a point in the Sierpinski gasket (called a *Sierpinski point*) is a Sierpinski point. The Sierpinski points therefore form what is called the *invariant set* of the map: once a Sierpinski point, always a Sierpinski point. If you start with a Sierpinski point with irrational coordinates, its orbit may look completely chaotic, but the succession of image points is fully determined by the coordinate values of the initial point. This is why such behavior, which abounds in nature, is called *deterministic chaos*: the rules governing the "game" are unambiguously deterministic, but the results are ultimately unpredictable because, ironically, the *real* world does not admit the vast majority of *real* numbers, namely, all those with infinite precision.

Three Bodies Cause Chaos

One will be struck by the complexity of this figure which I do not even attempt to draw. Nothing more properly gives us an idea of complication of the problem of three bodies and, in general, of all the problems in dynamics where there is no uniform integral
—HENRI POINCARÉ

A more attractive property of the Sierpinski points is that they are all repulsive points, or *repellors*; that is, a point arbitrarily close to a Sierpinski point will not stay near its images under our mapping, let alone be attracted to it; rather, its distances will *diverge* from the corresponding images of the Sierpinski point. In fact, the divergence will be exponential. The reader with a personal (or impersonal) computer is encouraged to try this and see for him- or herself. The reason for this exponential divergence is easy to see because our mapping corresponds to left shifts of the binary digits that encode the coordinates of the points. Thus, sooner or later, the first "error" bit will arrive at the binary point, which means that the initial difference, no matter how small, will have been magnified to half the height of the triangle. After that, all succeeding bits are random errors; the motion of an initially almost periodic point will become *chaotic*. In fact, this simple example contains the very essence of chaos and accords fully with its definition: small initial errors grow exponentially until they "dominate" any regular motion.

Although we may still not be aware of it, chaotic motion is much more widespread in nature than regular motion [Wis 87]. In fact, the jury is still out on whether planetary motion, the repository of regularity, is not chaotic in the long run. Certainly, Pluto and several other heavenly bodies already "stand" convicted of causing (or suffering) chaos. The smoke rising from a motionless cigarette in still air, first forming a regular ("laminar") flow, becomes a turbulent swirl only a few inches above the ashtray (see Figure 16). And what happens when two stars (a "double star"), encircling each other elliptically (good behavior!), meet a third star? Their regular motion turns wildly chaotic, (see Color Plate 1A). But, just as with human triangular relations, in the end two stars may pair off again to resume a regular orbit, as is the case in Color Plate 1A. However, one member of the initial couple may have switched partners in the course of the chaotic confusion during the three-body encounter (see Color Plate 1B).

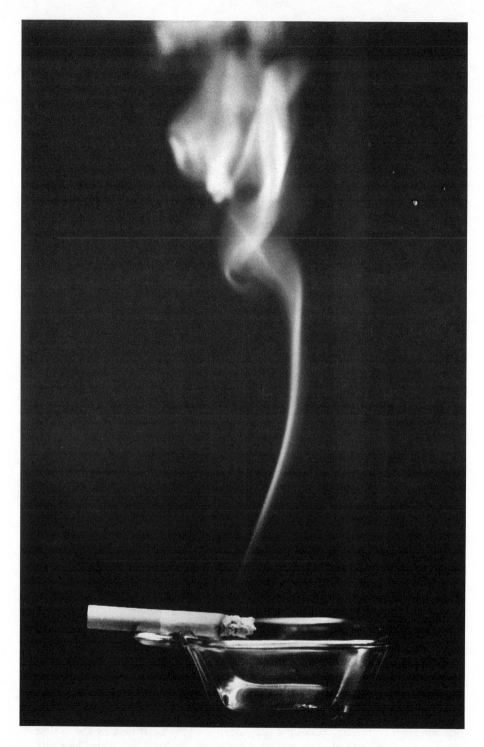

Figure 16   The laminar and the turbulent in cigarette smoke.

Of course, Newton's laws of gravitation, which govern the motions of our three heavenly bodies, are completely deterministic. But the far-future fate of the three partners can depend very sensitively on their initial positions and velocities. Here we have another case of deterministic chaos. In fact, the ultimate stability of the sun's planetary system, including Earth, humankind's common spaceship, has still not been rigorously proved—even without meddling from Nemesis, the hypothetical dark and distant sister of our sun.

We shall encounter more of this chaos, so intimately related to self-similarity, in the body of this book. And for the insatiable reader who has become addicted to disorder, there is James Gleick's recent bestseller *Chaos* [Gle 87] to devour.

Strange Attractors, Their Basins, and a Chaos Game

Determinism, like the Queen of England,
reigns—but does not govern.
—MICHAEL BERRY

Let us look at the *inverse* mapping of Sir Pinski's game (see pages 20–25) and see whether it holds any surprises (or can teach us a lesson or two). Inverses are generally good to look at for a variety of reasons. For one, repellors turn into attractors (and vice versa). And new concepts arise, such as *basins of attraction* and *strange attractors*.

In the inverse of Sir Pinski's map, we again pick a point inside (or outside) an equilateral triangle, but now we *halve* the distance to the most *distant* corner. We can be pretty sure that halving will never lead to a divergent explosion. But what points will we converge on?

First we remember that in Sir Pinski's game a Sierpinski point will remain a Sierpinski point, and since the map has a unique inverse (except for points with equal distances to two or all three corners), the same will be true for the inverse map.

But what happens to all the other points inside the triangle? Arithmetically, the inverse map looks as follows. If x is the smallest coordinate (i.e., if the x corner is the most distant one), then the inverse map of (x, y, z) is $((1 + x)/2, y/2, z/2)$. In binary notation the division by 2 means a right shift and adding $\frac{1}{2}$ means inserting a 1 in the place to the right of the binary point.

Let us start with a non-Sierpinski point, for example, $(\frac{1}{8}, \frac{1}{4}, \frac{5}{8}) = (0.001, 0.010, 0.101)$, and follow its course. By our rule it will map into $(0.1001, 0.001, 0.0101)$, which will go into $(0.01001, 0.1001, 0.00101)$, and so forth. Note that with each mapping we insert exactly one 1 and two 0s in the first place behind the binary point. With each further mapping this triplet will move one binary place to the right. Thus, asymptotically, no matter where we start, we approach a Sierpinski number, which has exactly one 1 in *all* binary places. (And if we start with a Sierpinski number, we will, of course, stay with the set.)

In fact, we will converge on one of the two period-3 cycles: $(0.\overline{001}, 0.\overline{010}, 0.\overline{100})$ and its two successors, or $(0.\overline{001}, 0.\overline{100}, 0.\overline{010})$ and its orbit—*which* of the two is determined by the coordinate values of the initial point (whether its ordered values constitute an even or odd permutation of x, y, z).

This insight, too, can be exploited in a game in which one has to predict, as closely as possible, the twelfth iterate, say, given only a rough initial location. If for the initial point $x_0 > y_0 > z_0$ holds, then after $3n$ mappings the image will approach the period-3 point with $x > y > z$, namely, $(0.\overline{100}, 0.\overline{010}, 0.\overline{001})$, within less than 2^{-3n} (for an initial point inside the triangle). This point is thus the attractor for the $60°$ sector defined by $x > y > z$ (whose apex is the center of Figure 14). This sector is its *basin of attraction* for the threefold iterated inverse Sir Pinski map. The five other period-3 points have the remaining five $60°$ sectors as their basins of attraction. The boundaries of these basins are smooth (in fact, straight) lines, in contrast to many other basins that we shall get to know, which have fractal rims.

The inverse Sir Pinski game is kin to a "game" called *chaos game* invented by Michael Barnsley, as described in his recent book *Fractals Everywhere* [Bar 88]. In Barnsley's chaos game, players "roll" a three-sided die, marked x, y, and z, and halve the distance of a preselected point inside a given triangle to the corresponding corner, also marked x, y, or z. (We leave the construction of the die to the reader as an exercise.) Alternatively, a random number generator with three possible outcomes will do.

What is the basin of attraction of the chaos game? The Sierpinski gasket (affinely transformed if the given triangle is not equilateral)! The proof follows directly from our analysis of Sir Pinski's game. However, the orbit of any initial point, as its iterates approach the attractor, will be completely chaotic. Such an attractor with infinitely many points that form a Cantor-like set is called a *strange attractor*—strange, because familiar attractors consist of either single points (fixed points), finitely many points (periodic orbits), or continuous manifolds that give rise to periodic or aperiodic orbits.

Strange attractors are encountered in many (nonlinear) physical, chemical, and biological systems that are "not integrable" and therefore show ultimately unpredictable, *chaotic* behavior. In fact, the usual "textbook" cases, nicely integrable, are now recognized as singular exceptions; the *real* world outside the textbooks, including romantic attraction, remains largely unforeseeable, moving along strange attractors, sometimes *very* strange attractors indeed.

However, not all is lost; the world is not complete chaos. Strange attractors often do have structure: like the Sierpinski gasket, they are self-similar or approximately so. And they have fractal dimensions that hold important clues for our attempts to understand chaotic systems such as the weather.

Strange attractors have recently found another, most surprising application. Barnsley has shown in his abovementioned book that many ordinary images, be they black-and-white or in color, can be approximated by a superposition of the strange attractors of a limited number of affine transformations, each transformation occurring with a given probability. An affine transformation in the plane is specified by a rotation, a scaling, and a displacement for each of the two coordinates. Since affine transformations in the plane are thus completely specified by six real numbers, an entire picture can be specified by some multiple of seven numbers, say $7 \cdot 13 = 91$ numbers.[5]

To understand the approximation of images by strange attractors better, we note first that the Sierpinski gasket consists of three triangular regions, each of which is a contractive affine transformation of the entire gasket. (A contractive transformation decreases the distances between all pairs of points.) Geometrically, these affine transformations correspond to moving a point along a straight line to half the distance to one of the three corners of the Sierpinski gasket. Numerically, the transformations are given by inserting a 1 behind the binary point for x, y, or z and one 0 each for the other two coordinates while shifting all binary digits of the point (x, y, z) one place to the right.

To generate the entire gasket, we start with an arbitrary initial point (x_0, y_0, z_0) somewhere near the gasket and select one of the three transformations—rotation, scaling, or displacement—at random to give us a point (x_1, y_1, z_1). On successive mappings, these three transformations are selected independently with given probabilities p_1, p_2, and p_3.

Because of the rules for inserting 1s and 0s, it is clear that the iterates will soon grow closer and closer to Sierpinski points, that is, members of the gasket. Because of the randomness of selecting the different mappings, the iterates will not become stuck in a "periodic rut" but will "hop" around and "illuminate" the entire gasket.

For equal probability, or $p_1 = p_2 = p_3 = \frac{1}{3}$, each of the three parts of the gasket will be visited with equal likelihood. By choosing other values for the p_k, we can produce different degrees of illumination or shadings for different parts of the attractor.

This process for generating images can be further generalized as follows. Instead of selecting the three corners of an equilateral triangle, we can pick any three points in the plane. In fact, we can specify *any* number of completely general affine transformations, each with its own probability of being chosen. But even with these generalizations, it is still surprising that, using fewer than 100 parameters, realistic looking scenes from nature can be generated by this "strangely attractive" method.

The promises for highly effective image data compression by *iterated function systems*, as the method is called by Barnsley, are mind-boggling—once image

---

5.   In number theory, 91 is jocularly known as the smallest composite number that *looks* like a prime, the reason being that there is no simple rule (other than division) to recognize its two factors. But note that 91 times 11 equals 1001, so that for numbers above 1000 divisibility by 7, 11, or 13 can be tested by subtracting the appropriate multiple of 1001. Thus, 9399 is divisible by 13 because 390 is.

decomposition in terms of attractors is computationally expedited. Color Plate 2 illustrates an image generated in this fashion.

Percolating Random Fractals

The Sierpinski gasket is an example of a two-dimensional *deterministic* fractal. Picking a point inside the triangle from which the gasket has been carved, we know immediately whether it is a member of the fractal set or "falls through the cracks." Many man-made deterministic fractals, like the Sierpinski gasket, are visually attractive and algebraically intriguing. However, most of *nature's* fractal gaskets are best modeled by *random* fractals, generated by stochastic processes. Among the many cases that have been diagnosed from the point of view of random fractals is the spread of epidemics and forest fires. Other examples of such fractals are random resistor networks, polymer bonds, and, apparently, the ice floes drifting through the Bering Sea.

To make things as discrete and simple as possible, consider a large square lattice whose lattice points are "occupied" independently with probability $p < 1$ (see Figure 17). The "occupants" could be trees, people, atoms, or whatever; it does not matter. The fraction of the lattice points that are unoccupied or "empty" equals $1 - p$. An important question is the following: Do the occupied sites form a *continuous* path from the lower edge of the lattice to the upper edge? A continuous path is defined as a path that goes from an occupied site to a neighboring occupied site. (The neighbors of a site are the sites immediately to the north, east, west, or south of it.) If such a path exists, the lattice is said to *percolate* (as in a coffee percolator, from Latin, "to flow through"). If the occupied sites were occupied by air and the "empty" sites by ground coffee, then the water could indeed percolate through the coffee.

The smallest density p of occupied sites for which the infinite lattice percolates is called the critical density or *percolation threshold* p_c. In spite of its simple definition, the exact percolation threshold of sites on the square lattice is still unknown. Massive Monte Carlo simulations put it at approximately 0.59275, with more digits constantly being appended as increasingly more powerful computers are brought to bear.

In addition to *site* percolation, there is *bond* percolation, in which *all* the sites are occupied but the "bonds" from a site to its immediate neighbors occur with probability $p < 1$. Missing bonds have a probability of $1 - p$. Percolation here means a connected path of *bonds* through the lattice. The bond percolation threshold for the infinite square lattice is known exactly: $p_c = 0.5$. But it took two decades of simulation and theory to prove this simple-looking result. Thus, in a large random network of electrical resistors based on a square lattice, electric current could flow between two opposite sides if at least half the bonds were

Figure 17   Square lattice with randomly occupied sites below percolation threshold.

conducting. The resistors through which the current actually flows are called the *backbone* of the cluster; the other resistors are called *dangling bonds*.

At the site percolation threshold ($p \approx 0.5927$ for the square lattice), the occupied sites of the infinite lattice form clusters of connected sites of all sizes. In fact, their distribution follows a simple power law: the number $n(s)$ of clusters having s occupied sites is proportional to $s^{-\tau}$ with $\tau = 187/91 = 2.\overline{054945}$ for the square lattice [Sta 85]. The power law $n(s) \sim s^{-\tau}$ means that the ratio of the numbers of clusters of two different sizes is independent of cluster size s; it depends only on the size ratio.

Figure 18 shows clusters of many sizes from pairs of sites ($s = 2$) to a "spanning cluster" that connects the upper and lower edges. A large lattice, 10 times larger than the one shown in Figure 18, would show exactly the same cluster distribution, except that it could accommodate clusters 10 times larger. Thus, percolating clusters are self-similar or *free of scale*, from the distance of neighboring sites to the size of the entire lattice. However, below the percolation threshold, the upper cutoff length for self-similarity is not the size of the lattice but rather the *correlation* length ξ, defined as the length over which the probability

Figure 18   Square lattice, probability of occupied sites equal to percolation threshold. Clusters occur on many size scales. These clusters form a statistically self-similar pattern.

of two sites belonging to the same cluster has decayed to $1/e \approx 0.368$. For distances smaller than ξ, the occupied sites form a fractal; above ξ Euclidean geometry prevails, with the number of occupied sites $M(R) \sim R^d$, where d is the Euclidean embedding dimension. At the percolation threshold, ξ diverges to infinity and the probability that two sites, even at an arbitrarily large distance, belong to the same cluster is bounded away from zero.

Percolating clusters, being self-similar fractals, ought to have fractal dimensions. The first measure that comes to mind is the so-called *mass exponent* D_m, which measures the number of occupied sites (the "mass") $M(R)$ within a circle of radius R: $M(R) \sim R^{D_m}$. For Euclidean objects, of course, D_m equals the Euclidean dimension; for example, the area $M(R) = \pi R^2$, or, in other words, $D_m = 2$, for a filled disk. But for fractals, D_m is generally smaller than the Euclidean dimension (also called the *embedding* dimension) that contains ("embeds") the fractal. For the triadic Cantor set, for example, the fraction of the included set $M(R)$ grows on average as $R^{0.63}$; that is, D_m equals the Hausdorff dimension $D_H = \log 2 /\log 3 \approx 0.63$. Similarly, for the two-dimensional Sierpinski gasket, $M(R) \sim R^{1.58}$; that is, D_m again equals the Hausdorff dimension $D_H = \log 3/\log 2 \approx 1.58$.

What is the mass exponent of the percolating cluster on a two-dimensional lattice? Theory gives a value of $D_m = 91/48 = 1.89583$, in good agreement with the best values found by simulation [Sta 85].

The integer 91 that appears in the numerical values for both τ and D_m suggests that they are related. In fact, $\tau - 1 = 2/D_m$. We shall return at some length to the interesting relationships between characteristic exponents in the chapters on percolation (Chapter 15) and phase transitions (Chapter 16). In Chapter 10, we shall see that in many cases the mass exponent equals the *correlation dimension* D_2, one—albeit an important one—in an infinite hierarchy of fractal dimensions.

Power Laws: From *Alvarez* to *Zipf*

Homogeneous power laws, like Newton's universal law of gravitational attraction $F \sim r^{-2}$, abound in nature—dead and alive alike. Since homogeneous power laws, upon rescaling, remain homogeneous power laws with the same exponent (-2 in Newton's case), such laws are, by definition, self-similar. In other words, Newton's law is *true on all scales*, from the wavelength of light to light-years; it has no built-in scale of its own. Newton's gravitational universe, if we so wished, could be compressed or inflated at will.[6]

The same inverse square law that governs gravitation also describes the falloff of radar power with distance. This simple fact was exploited by German submarines during World War II. By measuring the increase in radar intensity, they could gauge the rate of approach of an enemy plane and dive undersea for safety before the plane could attack.

This tactic worked very well for Grand Admiral Karl Dönitz until the American physicist Luis Alvarez (1911–1988) had a foxy vision, code-named Vixen. Alvarez suggested reducing the radar power so that it would be proportional to the *third* power of the range of the submarine. Thus, while the plane was approaching, the power incident on the unsuspecting U-boat was actually *decreasing*, giving the false impression that the radar plane was flying *away*. A grand idea indeed! (For the attacking plane, however, the received radar power reflected from the boat would still increase as it closed in [Alv 87].)[7]

---

6. Recently, though, some doubt has been cast on the unlimited validity of Newton's law. A still mysterious "fifth force" appears to knock on Newton's underpinnings, adding terms that introduce a natural length scale of a few hundred meters [AZLPAGNCFFMSSBCGHHHKSW 89]. At very small scales, Newton's law runs into the *Planck length* (10^{-35}m), which reminds us that eventually gravitation needs to be properly quantized and endowed with uncertainty.

7. This scheme of Alvarez is somewhat reminiscent of Genghis Khan ("Universal Ruler") and the wily Mongol tactic perfected by the horsemen of the Golden Horde. While seemingly galloping away from their pursuers, they would actually allow them to close in and then suddenly stand up in their stirrups, turn around in their saddles, and launch their arrows at the dumbfounded enemy.

Another wide-ranging example of a homogeneous law is the one that connects the areas A of similar plane figures with their diameters, their perimeters, or any other of their linear dimensions l: areas are proportional to linear dimensions squared, or $A \sim l^2$. Of course, this is not true for areas on *curved* surfaces; the radius of curvature introduces a length scale that destroys "truth on all scales." In fact, as everyone knows, distances and areas on the surface of a sphere are limited to a maximum size, given by the radius of the sphere.

In contrast to gravitation, interatomic forces are typically modeled as *in*-homogeneous power laws with at least two different exponents. Such laws (and expontential laws, too) are not scale-free; they necessarily introduce a characteristic length, related to the size of the atoms.

Power laws also govern the power spectra of all kinds of noises, most intriguing among them the ubiquitous (but sometimes difficult to explain) $1/f$ noise. Thus, the noise in many semiconductor devices is not "white" (i.e., independent of frequency) and not "brown" (with a $1/f^2$ frequency dependence, like Brownian motion), but has an in-between exponent, which is why it is sometimes called *pink noise*. Pink noise is also a preferred test signal in auditory research, because it has constant power per *octave* (not per hertz) and is thus well matched to the inner ear's frequency scale.

And, as we shall see in the course of our excursion into the world of fractals, power-law exponents do not have to be integers; they can be, and often are, *fractions*.

Not surprisingly, we find homogeneous power laws not only in the inanimate world; they inhabit living nature, and particularly human perception, too. Thus, over much of the auditory amplitude range, subjective loudness L is proportional to the physical sound intensity I raised to the three-tenths power: $L \sim I^{0.3}$. This means that merely to double the loudness of a rock group of five musicians, say, we have to increase their number *tenfold*, to 50 players of equal power output. (This minor calculation explains the resounding enamoration of popular music makers with electronic amplifiers.)

By the same token, if we want to halve the loudness of a continuous "rumble" emanating from a busy highway, we have to reduce the acoustic noise output by a factor of ten! This may sound difficult, but it is not, at least not from a purely physical point of view: tire noise—the main culprit at steady highway speeds—decreases drastically with decreasing vehicle speed. In fact, the noise intensity is approximately proportional to the *fourth* power of speed.

On the other hand, a tenfold increase in the average intensity of traffic noise caused by a tenfold increase in traffic *density* can raise the rate of complaints by irate residents perhaps a *hundred* fold: one loud truck every 5 minutes may be tolerable, but one every 30 seconds could be a nightmare and would certainly make outdoor conversation nearly impossible. And what is true for trucks is just as true for low-flying aircraft.

Power laws are also ubiquitous in economics. In fact, nearly 100 years ago, the Italian economist Vilfredo Pareto (1848−1923), working in Switzerland, found that the number of people whose personal incomes exceed a large value follows a simple power law [Par 1896, Man 63a]. Other instances of power laws in economics and the fallacies of trading schemes based on them are discussed by Mandelbrot [Man 63b, 63c].

One of the more surprising instances of a power law in the humanities is *Zipf's law* connecting *word rank* and *word frequency* for many natural languages. (The word with rank *r* is the *r*th word when the words of a language are listed with decreasing frequency.) This law, enunciated by George Kingsley Zipf (1902–1950), states that, to a very good approximation, relative word frequency *f* in a given text is inversely proportional to word rank *r*:

$$f(r) \approx \frac{1}{r \ln (1.78R)}$$

where R is the number of different words [Zip 49]. Laws like $f(r) \sim 1/r$ are called *hyperbolic laws*. If we assume $R = 12{,}000$, for example, we find that the relative frequencies of the highest-ranking words (*the, of, and, to,* and so on, in order of rank) are approximately 0.1, 0.05, 0.033, 0.025, and so on.

Figure 19 shows the close match between Zipf's homogeneous power law and actual data. Claude Shannon, the creator of information theory, has used Zipf's law to calculate the entropy of a source of English text that sputters words independently with Zipf's probabilities [Sha 51]. This entropy is given approximately by

$$H = \frac{1}{2} \log_2 (2R \ln 2R) \qquad \text{bits per word}$$

For $R = 12{,}000$, we get $H \approx 9$ bits per word, while $R = 300{,}000$ yields an entropy of about 11.5 bits per word. Of course, this is only an upper bound, because words (though perhaps independent of actions) are not independent of each other—except in random "poetry." This interdependence of words ("redundancy") in a meaningful text, of course, reduces the entropy.

Considering that the average length of English words is about 4.5 letters, or 5.5 "characters" including one space between words, we see that the entropy of English text is roughly bounded by 2 bits per character.

Zipf's hyperbolic law, which is applicable not only to the language as such but also to individual writers, has some rather curious consequences. To wit, for a good writer with an active vocabulary of $R = 100{,}000$ words, the 10 highest-ranking words occupy 24 percent of a text, while for basic (newspaper?) English with one-tenth the vocabulary ($R = 10{,}000$), this percentage barely increases (to about 30 percent). Of course, any writer would find it difficult to avoid words like *the, of, and,* and *to.*

Zipf has endeavored to derive his law from *Human Behavior and the Principle of Least Effort* (the title of his 1949 treatise). But Mandelbrot, in an early effort, has shown that a monkey hitting typewriter keys *at random* will also produce a

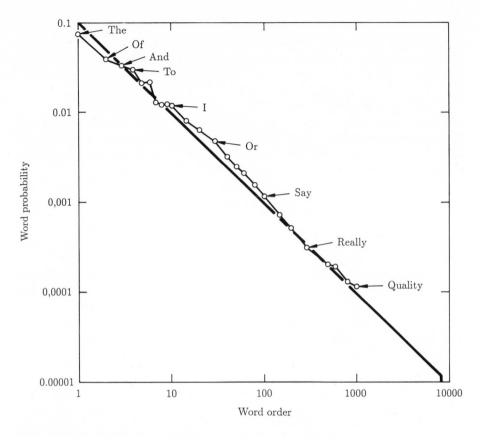

Figure 19   Word frequency as a function of word rank follows Zipf's law.

"language" obeying Zipf's hyperbolic law [Man 61]. So much for lexicographic *Least Effort*!

A detailed analysis shows that if the monkey's typewriter has N equiprobable letter keys and a space bar (with probability p_0), then his words (defined as letter sequences between spaces) have relative frequencies.

$$f(r) \sim r^{-1+\log(1-p_0)/\log N}$$

With $N = 26$ and $p_0 = \frac{1}{5}$, say, the exponent of r equals -1.068, only slightly less than -1. In general, the monkey words can be modeled as a Cantor set with a fractal dimension D that equals the reciprocal of the exponent of $1/r$. In our example,

$$D = \frac{1}{1 - \log(1 - p_0)/\log N} \approx 0.936$$

For a nine-letter alphabet and $p_0 = \frac{1}{10}$, the exponent equals -1.048, corresponding to a Cantor "dust" with $D \approx 0.954$. An arithmetic model for the (infinitely many) words of this nine-letter "language" is all the decimal fractions between 0 and 1 in which the digit 0 never occurs (not counting 0s at the end of terminating fractions).

Here are a few "three-letter" words of this language: .141, .241, .643, .442, .692, .121. Of course, .103, .707, and .0$\overline{3}$ are nonwords because they contain 0s.

Such languages do not have an average rank, but the *median* word rank of our "exemplary" language is an astonishing 1,895,761; that is, it takes the 1,895,761 most frequent words of the language to reach a total probability of one-half. (By contrast, the median word rank of English texts lies between 100 for typical media output and 500 for highly literate writers.) Thus, the monkey, while strictly clinging to Zipf's law, produces a rather wordy (and otherworldly) language.

Another, equally surprising "speech pathology" of the monkey language is the impossibility of constructing a dictionary for it, because its words form an *uncountable* Cantor set. (We would perhaps not be put off by an infinitely thick dictionary, as long as its entries could be sequentially numbered—but we could never countenance an uncountable compendium.)

If the monkey language has a fractal dimension, does it have any self-similarities? It certainly has. Multiply all words of the "decimal language" by 10 and drop the integer part (or, in general, just strike out the leftmost "letter" of each word) and you have another monkey word (most likely forming a nonsensical word sequence). In fact, the words of such languages grow on self-similar trees. Take any branch, no matter how high it is and seemingly small: it is *identical* to the entire tree.

And here we see the difference from natural languages most clearly: commonly spoken and written languages do *not* grow on self-similar trees—or, if we insist on hanging them from such trees (perish the thought), most branches would be dead.

Indeed, in natural languages many letter combinations are nonwords. Nevertheless, numerous English words are *homographs* (identical spellings) of words in other languages. And I do not mean such trivial cases as the uni(n)formed GENERAL, which means the same "thing" in many idioms. No, the interesting instances are "incognates" (unrelated words) such as the English word STRICKEN, which means *to knit* in German, or FALTER (a German *butterfly*) and LINKS (the German *left*). And what about such triplets as ART, which is a German word for KIND, which may mean MINOR in German, which in turn is a technical term in the theory of determinants (in either language). Finally, a fivefold string: ROT-RED-TALK-STEATITE-SOAPSTONES. Who can conceive sextuplets?

There are "letterally" hundreds of Anglo-German words like that, and I once composed a (short) German story using only English words. When I showed this story to a German-speaking Hungarian in the United States, his bored comment was "nothing but random poetry"—even after repeated proddings to look at the text with an open mind, as in visual texture (figure-background) discrimination, one of his research interests. When, half a year later, I showed the same text once more to the same Hungarian friend, this time in Germany, he read it and commented "Interessant! Interessant!!" Talk about the impact of context in human perception! (I leave it as an exercise to the linguistically inclined reader to compose a novel that makes *sense* in both German and English, or any other pair of languages in which at least the letter frequencies are not too different.)

How about the French woman who was amazed at the quantities of "soiled underwear" offered for sale in the United States when she first came upon the common come-on *Lingerie Sale*?

Sometimes a double-duty word engenders a double entendre, or rather a twofold misunderstanding. Shortly after I moved to Göttingen, the building superintendent of the physics institute, who collected my foreign parcel post from customs, went around the campus confiding that "Professor Schroeder is importing *poison* from the U.S.A. It even says so right on the packages: Gift!" Gift indeed, the German word for poison, and cognate to the English gift, because *gift* is something one gives (occasionally, anyhow), as in the surviving *Mitgift*, the bride's dowry.

When I told this tale to the (research) chemist Francis O. Schmitt of the Massachusetts Institute of Technology, he parried with the perfect misunderstanding in reverse. One of his students had once reported from a postdoctoral stay in Germany how generous indeed the indigenous chemical industry was: every other bottle in his lab was labeled GIFT! So, in certain parts of the world, better not to swallow the "presents."

Of course, not all homographs are quite so harmless. Consider *Not*, the German *emergency*. An Australian friend of mine (a linguist, no less) once found himself trapped inside a building in Austria (was the place on fire?), but every door that he approached repulsed him with a forbidding "verboten" sign saying NOTAUSGANG!—not exit? My increasingly frantic friend, desperately seeking *Ausgangs*, knew enough Latin and German (besides his native English) to properly decode *aus-gang* as *ex-it*. But in the heat of the emergency, he never succeeded in severing the Gordian knot: *Not* is not *not*.

Newton's Iteration and How to Abolish
Two-Nation Boundaries

As every pupil learns, the equation $z^2 = 1$ has not one but two solutions: $z = +1$ and $z = -1$. But suppose we did not know this; we could then start with some initial guess z_0 and use Newton's *tangent method* of finding a "closer" approximation z_1. In our case, Newton's method gives $z_1 = (z_0^2 + 1)/2z_0$. For positive z_0, the approximation z_1 will lie closer to the solution $+1$. For example, for $z_0 = 0.5$ we get $z_1 = 1.25$. In fact, all z_0 whose real part is larger than zero, upon repeated application of the formula, "migrate" toward $+1$. Similarly, z_0 with negative real parts will converge on -1. Thus, the line in the complex number plane for which the real part of z_0 vanishes (i.e., the imaginary axis) is the boundary between the two *basins of attraction* of the two solutions $+1$ and -1, respectively. Easy as pie.

What about $z^3 = 1$? It has, of course, three solutions: $z = 1$, $z = \omega$, and $z = \omega^2$, where ω is the standard abbreviation for exp $(i2\pi/3)$. Starting again with an initial guess, Newton's method now gives $z_1 = (2z_0^3 + 1)/3z_0^2$ for the next approximation. Iterating this formula, we expect to converge on one of the three solutions (1, ω, or ω^2), depending on the sector in which the initial value z_0 is located. In other words, we expect the three basins of attraction to partition the complex plane into three $120°$ pie-shaped pieces. But nothing could be further from the truth, as the English mathematician Arthur Cayley (1821–1895) first noted with utter surprise in 1879. (We shall encounter Cayley again when we consider self-similar trees.)

The real behavior of the harmless looking iteration $z_{n+1} = R(z_n) = (2z_n^3 + 1)/3z_n^2$ is complex almost beyond belief. For one, there are no pie-shaped pieces for the basins of attraction of 1, ω, and ω^2. In fact, there is, in the entire complex plane, not a single connected piece of boundary between two basins. Suppose we have a point z_0 that, upon iteration, converges on $+1$, and suppose further that we have another point nearby that converges on ω; then there is always a third point, even nearer to z_0, that iterates toward the third solution, ω^2. It is as if international jealousy (or prudence) abhorred two-nation boundaries and a third country *always* interposed itself between two others.

This kind of incredible behavior, and of such a simple equation at that, has stunned not only mathematical laity but many a hard-boiled professional too, until, from 1918 on, Gaston Julia (1893–1978) and Pierre Fatou (1878–1929) showed that, for iterations of rational functions in general, the boundary points of one basin of attraction are the boundary points of *all* basins. These boundary points form a set that is now called a *Julia set* in Gaston's honor (the complementary set of complex numbers is appropriately called a *Fatou set*). Thus, iterations that have more than two basins of attraction cannot have basin boundaries that are simple connected line segments. Such boundaries must, per force, be fractals consisting of totally disconnected point sets—an infinitely fine sprinkling of uncountable numerical "dust".

Color Plate 3 shows, in red, green, and blue, the three basins of attraction of 1, ω, and ω^2, respectively, in the complex Gaussian plane. In the center of the figure ($z = 0$), we see a kind of cloverleaf where the three basins (each represented twice) meet in a single point. The central cloverleaf has three preimages, again cloverleafs, albeit somewhat distorted. These three cloverleafs have nine even smaller cloverleafs as preimages, and so on *ad infinitum*, in a beautiful display of self-similarity. It is in this manner that all boundary points become boundary points of all three attractors, precisely as the point $z = 0$. In fact, the Julia set *is* the set of preimages of $z = 0$. But the true dustiness of the set can never be shown with man-made machinery of finite resolution. In fact, a Julia set that is not the entire complex plane has *no* interior points. So it's all or (almost) nothing for Julia sets.

Instead of attractors, we can define a Julia set also in terms of repellors. Indeed, the Julia set J_R of a rational function R comprises all of its (uncountably many) repellors. This makes intuitive sense because J_R is the *boundary* of R's basins of attraction, but does not belong to the attractive basins themselves. However, the fact that the forward orbit of any repellor should "visit" *all* other repellors is a bit surprising.

Interestingly, not only the forward orbit of a repellor, but its backward orbit too (generated by the inverse map), is dense in J_R. Since repellors become attractors for the inverse map, Julia sets can be computationally constructed in a stable, albeit nonuniform, manner from a single repellor subjected to the inverse map; small errors in the computation will not explode, as they would for the forward map. All this is beautifully explained in *The Science of Fractal Images* by Peitgen and Saupe [PS 88].

We shall come upon Julia sets again on pages 243–248 in Chapter 11, where we will deepen our knowledge of these fascinating and often fractal sets.

Could Minkowski Hear the Shape of a Drum?

When, in late 1910, the great Dutch physicist Hendrik A. Lorentz delivered the Wolfskehl lectures[8] at Göttingen, he threw in a conjecture that Hilbert (his host) immediately predicted to be unprovable in his lifetime. Lorentz's conjecture, which is important in thermodynamics (for calculating the specific heat of solids), blackbody radiation, and concert hall acoustics, says that the number of resonances $N_3(f)$, up to some large frequency f, depends only on the *volume* V of the resonator and not on its shape.

Someone in the audience by the name of Hermann Weyl (who later succeeded Hilbert in Göttingen) didn't share the great man's pessimism. In fact, within a short while, Weyl succeeded in proving that, asymptotically, for large f and for resonators *with sufficiently smooth* but otherwise *arbitrary boundaries*,

$$N_3(f) = \frac{4\pi}{3} \, V\left(\frac{f}{c}\right)^3$$

---

8.   Paid for from the proceeds of the (still unclaimed) Wolfskehl Prize, administered by the Göttingen Academy of Sciences and to be awarded for the settlement (one way or another) of Fermat's last theorem. The original amount of the prize was 100,000 gold marks, but inflation, engendered by two world wars, reduced this to 7600 deutsche marks.

where c is the velocity of sound (or light, in the case of blackbody radiation). The corresponding formula for two-dimensional resonators (think of drums or surface waves on a lake) is

$$N_2(f) = \pi A \left(\frac{f}{c} \right)^2$$

where A is the surface area of the resonator. The result is asymptotically correct, to order f^2, again independent of the shape of the boundary (perimeter).

These stunning formulas were later improved by correction terms involving lower powers of f [HBM 39]. For example, for a given boundary condition, the correction term for $N_2(f)$ is

$$\Delta N(f) = \frac{1}{2} P \frac{f}{c}$$

where P is the length of the resonator's perimeter.

What happens if we drop Weyl's smooth-boundary restriction? What if the perimeter is a fractal, with fractal dimension $D > 1$? M. V. Berry surmised [Ber 79] that

$$|\Delta N(f)| = \left(L \frac{f}{c} \right)^D$$

where L is a length constant and D is perhaps the Hausdorff dimension of the perimeter. This is a reasonable assumption because the exponent of f in any of the terms of these formulas, including the correction terms, equals the Euclidean dimension (3, 2, or 1) of the content measure (volume, area, or length) of the resonator. Thus, for a fractal perimeter that has infinite length and fractal dimension D, the corresponding power of f might very well be f^D.

Berry's conjecture that D was in fact the *Hausdorff* dimension turned out to be wrong in some cases. Rather, as Lapidus and Fleckinger-Pellé have shown, the proper fractal dimension is that of *Minkowski* [LF 88]—another kind of nontrivial dimension, introduced by Hermann Minkowski (1864–1909)[9] for different purposes (and extended to fractals by Bouligand); it does not always coincide with the Hausdorff dimension.

The definition of the Minkowski dimension D_M for a curve (be it fractal or smooth) is roughly as follows. Let the center of a small circle with radius r follow the curve to measure the *Minkowski content*, that is, the area $F(r)$ of the resulting

---

9. Minkowski, like Hilbert, was born in Königsberg, where their lifelong friendship began. Minkowski fused geometry and number theory and gave special relativity its proper four-dimensional space suit ("space-time") in preparation for its voyage, under Captain Einstein, into general relativity and modern cosmology.

"Minkowski sausage" traced out by the circle (see Figure 20). Divide the area $F(r)$ by $2r$ and let r go to zero. For a smooth curve, the result will be the length of the curve. But for a fractal "curve," the result may "explode," that is, exceed any finite limit. In fact, the quotient $F(r)/2r$ will be proportional to r^{1-D_M}, which—for $D_M > 1$—will diverge to infinity for $r \to 0$. The value of D_M that measures this explosion is defined as the *Minkowski-Bouligand dimension*. Equivalently, we can define D_M by

$$D_M := \lim_{r \to 0} \frac{\log F(r)}{\log (1/r)} + 2$$

provided the limit exists. (For some fractals, it is in better taste to distinguish between the two sides of the sausage.) For a smooth curve, $F(r) \sim r$ and $D_M = -1 + 2 = 1$, as expected.

Figure 20   A "Minkowski sausage" defines the "content" of a curve.

The preceding formula for D_M is reminiscent of that for the Hausdorff dimension, but note that in place of the number of "covering pieces" $N(r)$ we have an "area," $F(r)$, the content of the Minkowski sausage. And there is a $+2$ added to the ratio of the logarithms. (This $+2$ could be made to disappear, however, by replacing $F(r)$ by $F(r)/r^2$ inside the logarithm.)

It is conjectured that for all strictly self-similar fractals the Minkowski dimension is equal to the Hausdorff dimension D. If they are different, D_M exceeds D.

Why is it that the Minkowski, rather than the Hausdorff, dimension controls the number of resonant modes associated with the boundary? Intuitively, the reason is simple. Normal modes need a certain *area* (or volume) associated with the boundary (not the number of covering pieces as for the Hausdorff dimension).

What happens if the resonator domain itself is a fractal, not just its boundary, that is, if the solid isn't solid but has holes on all scales? How does such a fractal "sponge" vibrate? We might conjecture that the foregoing equation for $N_2(f)$ would then have to be modified to

$$N_2(f) = \left(a\frac{f}{c}\right)^{\bar{\bar{d}}}$$

where $\bar{\bar{d}}$ is an appropriate fractal dimension, called the *spectral dimension*, and a is some characteristic length. But we have to be careful here, because a fractal with $D < 2$ embedded in two dimensions is often but a "dust," and how can a *dust* support normal modes? However, as we shall see in the chapter on *percolation* (Chapter 15), there exist, at and above the *percolation threshold*, infinite *connected* clusters of "atoms" that have a finite mass and can support normal modes of vibration. For wavelengths exceeding the typical cluster size, called the *correlation length*, the density of modes is that of a homogeneous body; that is, it is proportional to f^{d-1}, where d is the (integer) Euclidean dimension of the space in which the percolating network is embedded. But for wavelengths *below* the correlation length, the normal modes "see" the self-similar fractal structure of the clusters and the mode density exponent drops from $d-1$ to $\bar{\bar{d}} - 1$, where the spectral dimension $\bar{\bar{d}}$ typically has a fractal value that differs from that of other fractal dimensions (the Hausdorff and Minkowski dimensions, for example).

By analogy to the particles of light—the ubiquitous *photons*—normal modes of vibration, familiar from musical instruments, are commonly called *phonons* when quantized. Phonons are crucial to our understanding of many physical phenomena, including the specific heat of solids and superconductivity, at both low and high temperatures—perhaps even *room* temperature (in Alaska, with windows wide open, no doubt). Phonons live in crystal lattices and feel at home in amorphous substances too. Phonons in fractal media, when they exist, are now often called—what else?—*fractons*. Fractons are believed to play an increasingly important role in our understanding of a vibrant nature.

A related subject is the diffraction of waves from fractal structures ("diffractals"). Since far-field or "Fraunhofer" diffraction is essentially a Fourier transform, the self-similarities (deterministic or statistical) of the scattering fractal must be fully reflected in the diffraction pattern of the incoming radiation, be it electromagnetic, audible, or ultrasound, electrons, neutrons, or neutrinos. (Is neutrino diffraction by the fractal structure of the universe observable?) Clearly, wave diffraction is a sensitive tool not only for classical bodies, but for fractal matter too. Fractal diffraction is also pressed into (military) service to simulate radar clutter with (confusing) detail on many length and size scales.

What happens to the density of normal modes for vibrating fractals whose fractal dimension *exceeds* their Euclidean dimension? Imagine a violin string whose local matter density varies in a Cantor-like way: the middle third of the string, say, has twice the density of the remaining two thirds, which in turn have their central mass densities increased by a factor of 2, and so forth *ad infinitum*. For a "classical" string of length L with uniform mass density, the number of normal modes is given by

$$N_1(f) = 2L\,\frac{f}{c}$$

In analogy with Weyl's formulas, we expect the number of modes of the "jazzy" fractal string to vary as

$$N_1(f) = \left(b\,\frac{f}{c}\right)^{D_M}$$

with $D_M > 1$. Here b is again a characteristic length.

This brings up an interesting and, as it turns out, important question: Can we calculate the (variable) thickness of the string from its resonance frequencies? Such *inverse problems* occur in many guises in many fields. (For example, can we find the location of a tumor inside the brain from the x-ray shadow it casts in different directions? The answer, within limits, is yes—by *computer tomography*.)

For the violin string, unfortunately, the answer is no. However, if we know the resonance frequencies for *two* independent boundary conditions, then we have all the information necessary to calculate the mass distribution of the (lossless) vibrating string.

The solution of the string problem became important at one point in the author's research on basic mechanisms of human speech production (a prerequisite for better-sounding talking computers, without the unfeeling "electronic accent" that can still be heard today when machines "talk"). One would, of course, learn a lot about human speech production if one could deduce the shape of the vocal tract (tongue position, for example) from the recorded speech sounds (which reflect the vocal-tract resonances). Such a capability would also help the deaf and hard-of-hearing as an adjunct to lipreading, because these people could then "see" the (computed) positions of the tongue on a video monitor.

Regrettably, this is difficult for the reasons just stated, namely, that *two* sets of resonances are needed. However, it turns out that determination of the input impedance of the vocal tract, measured at the lips, together with certain other assumptions, permits one to calculate the vocal-tract area function and therefore the motions of the tongue as the subject articulates different speech sounds [Schr 67].

Thus, the question whether we can hear the shape of the vocal tract has to be answered with caution. Yes, we *can* hear it (that is, after all, how we perceive speech)—but the solution is not unique: there are always several different tongue positions that sound alike. This *articulatory ambiguity* is in fact exploited by the ventriloquist, who manages to keep the lips immobile while using other articulators to "take up the slack" from the lips.

Enough of Lorentz, Hilbert, Weyl, Minkowski—and ventriloquists!

Discrete Self-Similarity: Creases and Center Folds

Repetition is a seldom-failing source of self-similarity, beginning with such simple things as paper folding: Take a piece of paper and fold it once. This creates a V-shaped (left-turn) crease (see Figure 21A, generation 1). Fold it over again (parallel to the first fold) and you get three creases, V V Λ: the original center fold V, surrounded by a V on the left and a Λ-shaped fold on the right (Figure 21A, generation 2). Another folding in the same direction yields the crease sequence V V Λ V V Λ Λ (Figure 21A, generation 3). Further folding creates crease sequences of increasing lengths. Each new generation is obtained from the previous one by interpolating alternating V's and Λ's around its letters, beginning with a single V. Thus, the fourth generation reads V V Λ V V Λ Λ V V V Λ Λ V Λ Λ. In an alternative construction, generation $n + 1$ is obtained from generation n by copying it, appending a center fold V, and then appending generation n read backward with V and Λ interchanged [DMP 82]. This operation is equivalent to "pivoting" generation n around the center fold V (which is what the folding in fact does).

But where is the self-similarity in this crazy succession of creases? Let the untold truth unfold! Pick every other "letter" in V V Λ V V Λ Λ, say, beginning with the second letter (V), an operation appropriately called *unfolding*, and you get the "mother" sequence V V Λ, which, by our alternative construction, must also be the initial part of the daughter sequence V V Λ V V Λ Λ. Thus, the *infinite* folding sequence, obtained in this manner, is precisely self-similar: taking every second (even-numbered) crease recreates the entire sequence. Discrete self-similarity could hardly be simpler.

Can you construct a direct, nonrecursive formula for the nth letter in the universal crease word? Suppose n is written as a binary number and the first digit to the left of the first 1 is . . .

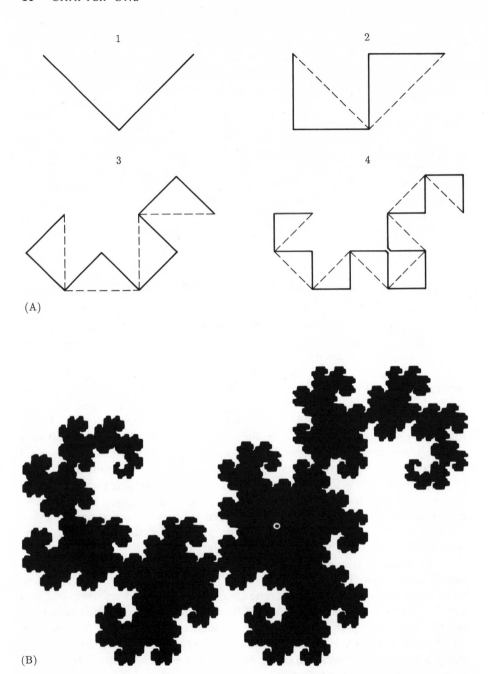

Figure 21 (A) Basic dragon curve, generated by right-angle creases. (B) Self-similarity revealed in later generations of the dragon curve. (C) Center creases (marked by dots) fall on a self-similar logarithmic spiral.

(C)

Figure 21 *(continued)*

The self-similarity inherent in the crease sequence can be brought to visual life by making all creases right angles (see Figure 21A). The fractal so generated is known as the *dragon curve*, because later generations (see Figure 21B) resemble a dragon. Two such dragons produce the *twin* dragon [DK 88]. The dragon curve is self-similar (see Figure 21C). The successive center folds, marked by little dots, fall on a logarithmic spiral, one of the basic (and smooth!) self-similar objects, with many interesting applications (see pages 89–92 in Chapter 3).

A *twin* dragon (see Figure 22A) comes alive in a noteworthy number system using a *complex* base. With the advent of digital computers, the binary system, using only the two digits 0 and 1, became the most widely used notation for numbers.[10] Nowadays computers deal a lot with complex numbers, that is, numbers having two "components": a real part and an imaginary part. Complex numbers thus require *two* sets of binary numbers. It would be nice, of course, if

---

10.   True, Claude Shannon once built a computer called THROBAC based on the *Roman* numerals (I, II, III, IV, and so on), but this exercise in masochism somehow did not catch on (in contrast to Shannon's information theory, which continues to shine brightly).

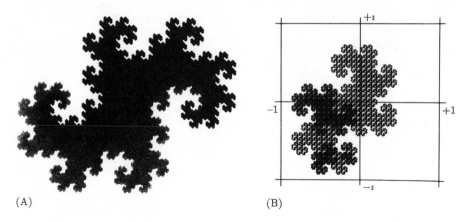

(A)   (B)

Figure 22 (A) Twin dragon. (B) The proper fractions in the binary number system for complex numbers using $1 - i$ as a base: a mirrored twin dragon. Twin dragons tile the plane [DK 88].

complex numbers, too, could be written as *single* binary numbers, but that seems impossible (eschewing such foul play as interleaving digits).

Yet there exists a complex number system using only the two digits 0 and 1, but its base is not 2. Obviously, the base must be a complex number, and its magnitude must not exceed $\sqrt{2}$. Otherwise, two binary digits would have to cover a magnitude range larger than 2. If we further call for evenhandedness between the real and imaginary parts, then a best-base bet would be $(1 - i) = \sqrt{(2)} \exp(-i\pi/4)$ (or one of the other three primitive eighth roots of 16).

Of course, we are paying a penalty in "programming" when using this ingenious system. For example, the number 2, which is simply 10 in the real-valued binary systems, is a little more "complex" (in both senses of the word) when using the base $1 - i$.

Actually, calling the system based on $1 - i$ complex is something of an understatement. Figure 22B shows all those numbers in the complex (Gaussian) number plane that are *proper fractions*, that is, numbers in which only *negative* powers of the base appear, such as $0.1 = (1 - i)^{-1}$ or $0.\overline{1} = (1 - i)^{-1} + (1 - i)^{-2} + \cdots = i$. (Note that the periodic fraction $0.\overline{1}$ does *not* equal 1, as it does in the real binary system.)

As is readily apparent from this illustration, the proper fractions occupy a simply connected area with a fractal perimeter, the *twin dragon*. But, in contrast to the Koch flake and the fractal "hexagons" that we encountered in previous sections, the twin dragon's skin is born of a generator consisting of pieces of different lengths: one piece of length $r_1 = 1/\sqrt{2}$ and two pieces at right angles of length $r_2 = r_1^3$ (see Figure 23). The Hausdorff dimension D_H of such fractals is a straightforward generalization of the formula for N equal-length pieces, which

Figure 23   The generator for the twin dragon's skin: Hausdorff dimension 1.52 . . . .

can be written $Nr^{D_H} = 1$. Replacing the unique length r of the generator by N different lengths $r_1, r_2, \ldots r_N$ results in

$$r_1^{D_H} + r_2^{D_H} + \cdots + r_N^{D_H} = 1$$

Of course, for $r_1 = r_2 = r_3 = r_4 = \frac{1}{3}$ we get the old value for the Koch flake: $4(\frac{1}{3})^{D_H} = 1$ or $D_H = \log 4/\log 3$.

To obtain D_H for the twin-dragon generator, we have to solve a cubic equation in $r_1^{D_H}$, namely, $r_1^{D_H} + 2r_1^{3D_H} = 1$. Although there are closed formulas involving radicals for cubic equations, I prefer a more conservative approach: my pocket calculator tells me that $r_1^{D_H} = 0.5897545123 \ldots$ and, with $r_1 = 1/\sqrt{2}$, $D_H = 1.523627 \ldots$

The twin dragon can be cut up into four pieces similar to itself (see Figure 24). Thus, according to our generalization of Euclid's scaling theorem for fractal figures (see pages 13–15), the skin of the mother dragon must contain the skin of one of the four child dragons not 2 times but $2^{D_H} = 2.875 \ldots$ times. (Now, of course, one wishes for the radical solution to see what this irrational ratio could possibly mean.)

Golden and Silver Means and Hyperbolic Chaos

Iteration, as was noted before, is one of the richest sources of self-similarity. Given the proper jump start, the repeated application of some self-same operation, be it geometric, arithmetic, or simply symbolic, leads almost invariably to self-similarity. Take for example the simple rule $F_{n+2} = F_{n+1} + F_n$. Starting with $F_0 = 0$ and $F_1 = 1$, this recursion generates the well-known Fibonacci numbers: 0, 1, 1, 2, 3, 5, 8, 13, 21, . . . What is self-similar about them? Multiplying each number by 1.6 and rounding to the nearest integer, we get 0, 2, 2, 3, 5, 8, 13,

Figure 24 The twin dragon contains four smaller replicas similar to itself, but its fractal skin violates Euclid's scaling law for areas and perimeters of similar figures.

21, 34, . . .—the same sequence, except for a few initial terms (and perhaps later ones).

Taking ratios of successive numbers, we find $F_n/F_{n+1} = 0, 1, 0.5, 0.\bar{6}, 0.6,$ 0.625, 0.615 . . . , 0.619 . . . —numbers that appear to approach some constant. In fact, a little arithmetic shows that these ratios approach the irrational number $r = (\sqrt{5} - 1)/2 = 0.618$ . . . , the famous golden mean that tells us how to subdivide a piece of straight line so that the ratio of the shorter segment to the larger equals the ratio of the larger to the whole. Thus, the nth Fibonacci number should equal, approximately, some constant times γ^{-n}. In fact, the approximation is uncannily close: simply divide γ^{-n} by $\sqrt{5}$, which yields 0.4 . . . , 0.7 . . . ,

1.1..., 1.8..., 3.0..., 4.9..., 8.0...; and round to the nearest integer and—presto—the Fibonacci numbers, even for $n = 0$.

What about the golden mean γ itself? Does it hide any self-similarities? Perhaps they are revealed if we write γ down in the proper number notation: not in "dumb" decimal (0.61803...), not in "bitsy" binary (0.100111...), and not in any other system that elevates some base number beyond its proper station. Rather, let us try a more natural representation, namely, *continued fractions*. Written as a continued fraction, γ becomes $[1, 1, 1, \ldots]$. In general, $[a_0, a_1, a_2, \ldots]$ is shorthand for

$$\cfrac{1}{a_0 + \cfrac{1}{a_1 + \cfrac{1}{a_2 + \cdots}}}$$

which should be banished even as a typist's punishment.

Figure 25 shows a rendering of the periodic continued fraction for the golden mean γ that is geometrically self-similar. But what is self-similar about γ *arithmetically?*

The continued fraction for a given positive irrational number $\alpha < 1$ is calculated as follows: set $x_0 = 1/\alpha$ and apply the iteration $x_{n+1} = 1/\langle x_n \rangle_1$, where the pointed brackets with the subscript 1 mean "take the remainder modulo 1" (for example, $\langle \pi \rangle_1 = 0.14 \ldots$). Then the continued fraction for α is $[\lfloor x_0 \rfloor, \lfloor x_1 \rfloor, \lfloor x_2 \rfloor, \lfloor x_3 \rfloor, \ldots]$, where $\lfloor \; \rfloor$ stands for rounding down to the nearest integer and the left-most term is the integer part of α. Since all terms (except the first) in the continued fraction for the golden mean equal 1, the number γ is a *fixed point* of the iteration $x_{n+1} = 1/\langle x_n \rangle_1$, also called the *hyperbolic map*.

The hyperbolic map is particularly simple to execute if the "condemned" number x_0 is given as a continued fraction: simply move all terms of $[a_0, a_1, a_2, \ldots]$ one place to the left and drop the first term: $[a_1, a_2, \ldots]$. Thus, the golden mean $\gamma = [1, 1, 1, \ldots]$ is indeed a fixed point of the hyperbolic map.

This map is also called the *Gauss map* because Gauss derived many of its properties, including the invariant distributions of x and a_k [Schr 90].

Are there other such precious numbers expressible as periodic continued fractions with periodic length 1? Note that $1/\gamma = \gamma + 1$. If we replace the $+1$ by any other positive integer, we get the *silver means* τ_n defined by $1/\tau_n = \tau_n + n$, which have the continued fractions $\tau_n = [n, n, n, \ldots] = [\bar{n}]$.

The silver means $[\bar{n}]$ play an enormous role in a sheer, limitless wonderland of applications, encompassing curious quasicrystals, (easy) Ising spins, the mode-locking route to chaos, the "multiplication" of rabbits—and some even curiouser games, such as the Fibonacci fleecing, effected by the golden mean.

These numbers, like so many self-similar objects, contain the seeds of chaos. Try iterating the hyperbolic map, starting with a silver mean, on a computer of any finite precision: after a while the result will be utter chaos. As an example,

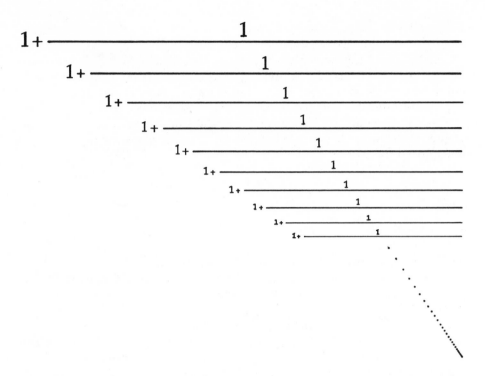

Figure 25 Geometrically self-similar continued fraction for the reciprocal golden mean.

take as a starting value $(\sqrt{13} - 3)/2 = [3, 3, 3, \ldots] = 0.3027756\ldots$, the silver mean τ_3. Just keeping the first decimal digit, the hyperbolic map iterated on a small pocket calculator gives 0.3 eight times, followed by 0.2, 0.8, 0.2, 0.6, 0.4, 0.0, 0.2, and so on—a completely unpredictable sequence with a totally chaotic tail.

Similarly, the successive continued fraction obtained from τ_3 by iterating the hyperbolic map eventually become chaotic: showing only the first term, $[3, \ldots]$ occurs nine times, followed by $[1, \ldots]$ $[4, \ldots]$, $[1, \ldots]$, $[2, \ldots]$, $[10, \ldots]$, $[4, \ldots]$, $[11, \ldots]$, $[90, \ldots]$, $[1, \ldots]$, and so forth. Where does this chaos come from? The continued fraction of any irrational number does not terminate; it has infinitely many terms. But finite machines have finite precision, and no matter how high the precision, the less significant terms, starting with some term, must be indeterminate. Thus, in our example, τ_3 is represented *not* by $[\overline{3}]$, where the line over the 3 stands for infinitely many 3s, but by $[3, 3, 3, 3, 3, 3, 3, 3, 3, 1, 4, 1, 2, 10, 4, 11, 90, 1, \ldots]$. After iterating the hyperbolic map nine times, the random digits have "moved to the front" and take over after that in a characteristic case of chaos.

Winning at Fibonacci Nim

Let us put the golden mean γ through some of its paces right now. Consider the two integer sequences.

$$a_k = \lfloor k/\gamma \rfloor = 1, 3, 4, 6, 8, \ldots$$
$$b_k = \lfloor k/\gamma^2 \rfloor = 2, 5, 7, 10, 13, \ldots$$

called a pair of Beatty sequences. Between them, they "exhaust" the positive integers. Note that $b_k = a_k + k$ and that a_k is always the smallest positive integer not already used up by a_n and b_n for $n < k$.

This property of the Beatty sequences just given leads to an interesting game called *Fibonacci nim* (also named Wythoff's nim): two players alternate taking coins from two piles, always at least one coin, and if a player takes coins from both piles, then he must take the same number of coins from both. The player who takes the last coin(s) wins.

Suppose initially there are 7 and 12 coins in the two piles and I have the first move. To win, I have to leave my opponent a Beatty pair (b_k, a_k), which I can always attain from a non-Beatty pair. The Beatty partner of the smaller[11] number ($7 = b_3$) is $a_3 = 4$; thus I take $12 - 4 = 8$ coins from the larger pile, leaving my opponent the Beatty pair (7, 4), which means that he has now lost, because he cannot obtain another Beatty pair.

Suppose he leaves me the pair (5, 4), which I cannot convert into (7, 4). But notice that the difference between 5 and 4 is 1. Thus, by taking 3 coins from each pile, I can realize the Beatty pair with the difference 1, namely, (2, 1). I leave it to the reader to convince himself that no matter which of the four possible options my opponent now chooses, I can always take the last coin(s) and thereby win. Once you receive a Beatty pair, you cannot, by yourself, recover from it, and you are beaten.

It is interesting to note that there is a simple board game, called "corner the lady" [Gar 89], that is equivalent ("homomorphic") to Fibonacci nim. Take a chessboard and place a "queen" anywhere in the top row or the rightmost column, shown in gray in Figure 26A. Two players alternate moving the queen either "west," "southwest," or "south." To which cells should I move the queen so that, no matter what my opponent does, I have the last move to the starred square and win?

The queen's moves to the west or south correspond, of course, to taking tokens from either of the two piles in the nim game. And the moves to the southwest correspond to taking on *equal* numbers of tokens from both piles. The

---

11. If both numbers are equal, I take all the coins and win immediately.

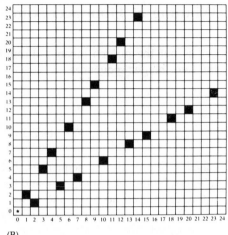

(A)   (B)

Figure 26   (A) The board game "corner the lady" of Rufus P. Isaacs: The initial positions for the queen are shown in gray. The goal is to reach the lower left corner. (B) The safe squares for the lady are shown in dark shading. They are all those squares for which two opposing sides are pierced by one of two straight lines whose slopes equal the golden mean and its reciprocal.

"safe" squares are therefore whose coordinates correspond to our Beatty pairs derived from the golden mean: (1, 2), (3, 5), (4, 7), (6, 10), (8, 13), and so on. These safe squares are black in the large board shown in Figure 26B.

Is there a simple, purely *geometric*, method of finding all the safe squares? There certainly is. Draw a straight line from the lower left corner with slope equal to the golden mean (see Figure 26B). The squares whose east *and* west sides are pierced by the straight line are all the lower safe squares. (The upper safe squares are their images mirrored at the 45° diagonal.)

We also remark in passing that Fibonacci's "multiplying" rabbits have known this strategy since long ago—which is why the nim version of the game is called *Fibonacci* nim.

In his *Liber Abaci*, published in 1202, the Italian mathematician Leonardo da Pisa, better known as Fibonacci (son of Bonacci), considered the question of how many rabbit pairs one would have after *n* breeding seasons, starting with a simple immature pair and assuming the following idealized rules for the growth of their numbers [Fib 1202]:

• Rabbits mature in one season after birth.

• Mature rabbit pairs produce one new pair every breeding season.

• Rabbits never die.

It is easy to see that, with these rules, the number of rabbit pairs F_n in the nth generation must equal the sum of the number of rabbit pairs in the two preceding generations: $F_n = F_{n-1} + F_{n-2}$. Starting with $F_1 = F_2 = 1$ yields the justly famous sequence of Fibonacci numbers: 1, 1, 2, 3, 5, 8, 13, 21, 34, 55, . . . , which appears in innumerable situations.

But the rabbits produce still another number sequence, a binary bit sequence which I have called the *rabbit* sequence [Schr 90]. Consider the two maps $0 \to 1$ ("young rabbits grow old") and $1 \to 10$ ("old rabbits stay old and beget young ones"). Beginning with a single 0, continued iteration gives 1, 10, 101, 10110, . . . , resulting in the infinite rabbit sequence 1011010110110 . . . . Is this sequence self-similar? Naturally it is. Just underline all the "10" pairs— 10 1 10 10 1 10 1 10 · · ·—and read them as 1s, and read the nonunderlined 1s as 0s, and the infinite rabbit sequence reproduces itself: 10110101 . . . !

Where are all the 1s in the rabbit sequence? They occupy the places numbered 1, 3, 4, 6, 8, 9, 11, 12, . . . , which is the first of our golden-mean Beatty sequences, a_k. And the 0s are located at places 2, 5, 7, 10, 13, which is our second Beatty sequence, b_k. So, apparently, the rabbits know the ropes of Fibonacci nim—but they have to go on multiplying. . . .

There is still another curious connection between the rabbit sequence and the Fibonacci numbers, discovered by John Horton Conway [Gar 89, p. 21]. The "rabbit constant," defined by the binary fraction .1 0 1 1 0 1 0 1 1 0 1 1 0 . . . obtained by putting a "decimal" point in front of the rabbit sequence, equals the continued fraction $[2^0, 2^1, 2^1, 2^2, 2^3, 2^5, 2^8, 2^{13}, 2^{21}, 2^{34}, \ldots]$, where the exponents are none other than the Fibonacci numbers.

We leave it to the reader to concoct similarly mean games based on a silver—rather than the golden—mean. What are the corresponding rabbit multiplication rules for $\tau_2 = [\overline{2}] = \sqrt{2} - 1$, say?

Self-Similar Sequences from Square Lattices

Take a square lattice (see Figure 27) and draw a straight cutting line through the origin with a slope equal to the golden mean $\gamma = [\overline{1}] = 0.618 \ldots$. Next, write a 1 every time the inclined line crosses a vertical lattice line and write a 0 when it crosses a horizontal lattice line. The resulting sequence of 1s and 0s is aperiodic, because the golden mean is irrational, yet it has infinite long-range order because the sequence is deduced from an infinite rigid square lattice.

What other properties are there to the infinite sequence, which begins 1011010110110 . . . ? It is self-similar in the following sense. Consider each 1 the beginning of a new "sentence," and abbreviate the sentence 10 by 1 and the other possible sentence, 1, by 0. The result is the original (infinitely long) "novel": 10110101 . . . . In fact, our novel is not that novel after all: it is the same old-young rabbit sequence we have just encountered in the Fibonacci nim

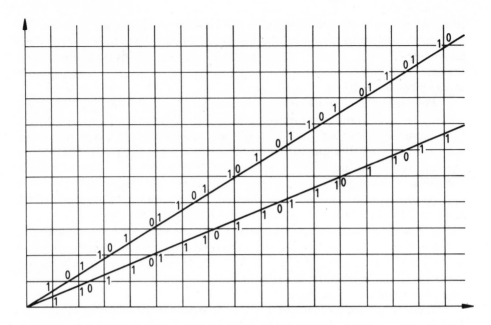

Figure 27   Square lattice and straight line with golden-mean slope γ generates the rabbit sequence 10110 . . . . The lower straight line has the silver-mean slope $\sqrt{2} - 1$ and generates another self-similar binary sequence.

game, only here the rabbits emerge from a square lattice. (I am sorely tempted—but will resist for the moment—to call such a lattice *lettuce.*)

If, instead of abbreviating the golden-mean novel, we want to *write* it in the first place, we can start (like most authors) with nothing (0) and iterate the "rabbit" mapping $0 \rightarrow 1$, $1 \rightarrow 10$.

What if the slope of the cutting line is one of the silver means $\tau_n = [\overline{n}]$? For $n = 2$, we have $\tau_2 = 1/(2 + 1/(2 + \cdots)) = \sqrt{2} - 1$ (see the lower cutting line in Figure 27). It generates the infinite sequence 110110111011011110110110 . . . . It, too, is self-similar, another "scaling" novel. But where do the sentences start or end that reproduce the novel? Consider each 0 and likewise the first 1 of each triple 1 the end of a sentence. Abbreviate the first kind of sentence (110) by 1 and the other sentence (1) by 0. The result? The same old story: 1101101110 . . . .

Conversely, the cutting sequence for τ_2 is generated by the iterated mapping $0 \rightarrow 1$ and $1 \rightarrow 110$.

For a cutting line with a slope corresponding to the next silver mean, $\tau_3 = [\overline{3}] = (\sqrt{13} - 3)/2$, the cutting sequence is 11101110111011110110 . . . . Where are the periods ("full stops") between sentences? And how do we write the novel by iteration starting from scratch (0)?

Self-similar sequences, similar to the ones just exhibited, have recently gained notoriety as generators of one-dimensional *quasiperiodic* "lattices" whose generalizations to three dimensions are good mathematical models for a newly discovered state of matter called *quasicrystals* (see Chapter 13).

Sequences, similar to the one discussed here, have also become prominent in computer graphics—specifically, the digitization of straight lines. For *any* slope m (not just the golden or silver means), the staircase function that best approximates a straight line running through an integer lattice is characterized by a sequence of 0s and 1s, called a *chain code*. Each horizontal step in the staircase is represented by a 1 in the chain code, and each vertical step by 0. Chain codes have the following property: one symbol occurs always in isolation, and the other symbol occurs in runs with at most two different run lengths. If the two run lengths differ, they differ by 1. In fact, for $m \geq 1$, the run lengths are $\lfloor m \rfloor$ and $\lceil m \rceil$. (The "gallows" $\lceil \ \rceil$ stand for rounding up to the nearest integer.) For $m < 1$, the run lengths are $\lfloor 1/m \rfloor$ and $\lceil 1/m \rceil$. The run lengths are different only for noninteger m or $1/m$.

One method of recoding chain codes with different run lengths by two symbols is to assign one symbol to the shorter run length and the other symbol to the longer run length. The result of this recoding is another sequence of symbols of which one occurs in isolation and the other occurs with at most two run lengths that differ by 1. This invariance was discovered by Azriel Rosenfeld in 1973 [RK 82]. But the underlying number-theoretic question, namely, whether a given sequence of integers can be represented by rounding a linear function, was already addressed by one of the Bernoullis [GLL 78].

These results have been applied to efficient image coding and in picture recognition, specifically, for distinguishing straight lines from curved contours [WWM 87]. Present research in this area focuses on computationally efficient algorithms for detecting straight-line segments [KS 87].

A related recoding simply counts the number of places between two successive symbols that occur with different run length. Thus, the chain code 1011010110110 . . . is recoded as 10110101 . . . . I conjecture that this recoding of chain codes corresponds to a left shift in the continued fraction representing the slope m.

John Horton Conway's "Death Bet"

John H. Conway, the prolific British mathematician—now serenely ensconced in Princeton, New Jersey—became widely known, even outside mathematics, through his ingenious game called *life* (see Chapter 17, Cellular Automata). During a captivating talk entitled "Some Crazy Sequences" at AT&T Bell Laboratories in Murray Hill, New Jersey, on July 15, 1988, Conway delivered himself of one

Figure 28 John Horton Conway wearing self-similar horned sphere. (Drawing by Simon Fraser; courtesy of J. H. Conway.)

more direct proof that mathematics is not only great fun but outright fun*ny*. After a few preliminary reminiscences about the Fibonacci numbers and the like, he introduced a sequence, $a(n)$, that began, harmlessly enough, 1, 1, 2, 2, 3, 4, 4, 4, 5, ... The simple iterative law for this sequence is

$$a(n) = a(a(n-1)) + a(n - a(n-1))$$

with $a(1) = a(2) = 1$ for starters. As in the Fibonacci sequence, each new term is the sum of two previous terms.

Numerical evidence suggests that $a(n)/n$ approaches $\frac{1}{2}$ as n becomes large, and Conway challenged the audience to find an n_0 such that for all $n > n_0$ the absolute error $|a(n)/n - \frac{1}{2}|$ is smaller than 0.05. Since he and his wife (also a mathematician) found the sequence rather intractable, he offered $100 for the first solution to reach him. In a barely audible aside (but clearly detectable on the videotape that was made of his talk), he offered a $10,000 bet to the first finder of the *smallest* such n_0.

Precisely 34 days later, on August 18, 1988, Colin Mallows, a most capable colleague at Bell, presented the solution, including a formal proof, to Conway's $10,000 question: $n_0 = 317\,337\,5556$. I have written the solution as a U.S. telephone number, in Indiana, incidentally, because it was suggested that Conway was just kidding, that he knew the solution all along and that—upon dialing (317) 337-5556—his voice would come on the line to reassure the keen caller that he had the right number, all right, but was, unfortunately, a bit too late. (Actually, when dialing n_0, one gets a recorded message to "try again." Try *again*? After all the trouble to get the number in the first place!)

As one might have guessed, the sequence $a(n)$, being generated by a simple recursion, is replete with appealing self-similarities that contain the clue to the problem's solution. These self-similarities were speedily brought to light by Mallows, a statistician and data analyst, employing nothing more sophisticated than straightforward numerical computation and graphic displays [Mal 91].

We leave it to the PC-equipped rapt reader to discover for himself the tip-off self-similarities and other symmetries of Conway's sequence $a(n)$ or the simpler $b(n) = 2a(n) - n$ (to take out the trend). What happens to $b(n)$ for $n = 2^m$, and why? And what is the Hausdorff dimension of the fractal function to which a properly normalized $b(n)$ between n and $2n$ converges as n goes to infinity? Is there a simple *direct* formula for $b(n)$? As a warm-up workout, the reader may want to unravel the run-length law of $a(n)$.

Mallows, the grand winner, and Conway later agreed that Conway had meant to offer *one* thousand dollars instead of ten thousand. So Conway sent another check, for the smaller amount, but Mallows kept the original prize check for framing. (Figure 28 shows John Horton Conway wearing a self-similar head-gear, the aptly named *horned sphere*.)

2
C H A P T E R

Similarity and Dissimilarity

*The Universe is built on a plan the
profound symmetry of which is somehow
present in the inner structure of our
intellect.*

—PAUL VALERY

The concepts of similarity and dissimilarity have long played an important role
in human affairs and nature. We welcome similarity, but are often more powerfully
attracted by the *dis*similar. Mass attracts mass, but electrons are attracted by their
opposite numbers: antielectrons, or *positrons*. In fact, electrons and positrons can
consummate their love in a small bang with an energy of a million electron volts.
(But more often than not, the dissimilar evokes discrimination. For a mild example,
think of the Russian word for *German: nyemets*, meaning "the mute one" or "not
one of us." The reader will have no difficulty finding more partisan instances.)

In the sciences, similarity has both mystified and illuminated. Why are all
electrons similar to each other—in fact, for all we know, *identical*? Here is one
of the great unsolved riddles that Nature likes to tease us with. Once we know
the answer and why the electron and other "elementary" particles have precisely
the masses, charges, and spins they have, we shall know a lot more about the
environment we inhabit (and inhibit). In this chapter, before indulging further in
self-similarity, we shall touch upon some of the uses to which the ideas of similarity
and *dis*similarity have been put in physics, psychophysics, biology, geology
(mountaineering?), and other fields in which *scaling* is a crucial concept.

More Than One Scale

Measurement is usually thought of as an unambiguous, if imprecise, process. A
soccer field has an area of roughly 50 meters times 100 meters. Thus, 5000 square

meters is the area of the field as far as the soccer player is concerned, or the real estate agent who sold the land.

But there is another area associated with a soccer field or any other lawn or meadow: the area important to the little bug that stalks up and down the grass blades. This area, corresponding to the total surface area of all the grass blades, is much larger than the soccer-playing area of the field, perhaps by a factor of 100. This larger area is also the relevant area for the sun's photons that are absorbed by the chlorophyll in the grass to convert carbon dioxide in the air to carbohydrates and oxygen.

Thus, for a soccer field, the question *What is its area?* has at least *two* true answers; the field is characterized by two area *scales* that differ by a very large factor. In other situations, measurements can lead to *many* answers. For example, the boundary between two European countries typically depends on the scale used in its determination. Thus, on a globe of the world, the length of the border between Spain and Andorra (or Austria and Liechtenstein, if shown) is considerably shorter than that determined from a map of Europe, which in turn is shorter than the border length obtained from a map of the Pyrenees (or the Alps).

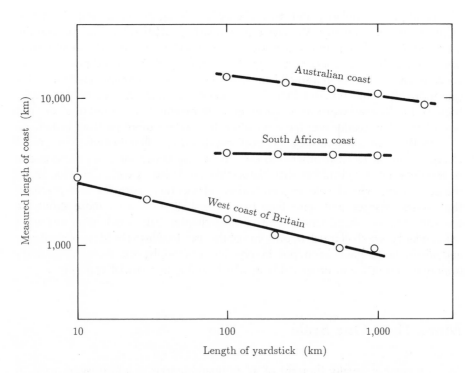

Figure 1   Measured lengths increase as the length of the yardstick is reduced.

Still longer lengths are obtained from more detailed maps, showing just the region in question, and from hiking maps. And actually walking (swimming or climbing) along the border will reveal an even longer length (see Figure 1). Thus, there is no *one* border length—there are *many*. In contemporary parlance, the border, like the fractal Koch curve discussed on pages 7–8 in Chapter 1, is said to have *many length scales*, an important concept in self-similarity and fractals.

In physics, there are numerous phenomena that are said to be "true on all scales," such as the Heisenberg uncertainty relation, to which no exception has been found over vast ranges of the variables involved (such as energy versus time, or momentum versus position). But even when the size ranges are limited, as in galaxy clusters (by the size of the universe) or the magnetic domains in a piece of iron near the transition point to ferromagnetism (by the size of the magnet), the concept *true on all scales* is an important postulate in analyzing otherwise often obscure observations.

To Scale or Not to Scale: A Bit of Biology and Astrophysics

Elephants and hippopotamus have grown
clumsy as well as big, and the elk is of
necessity less graceful than the gazelle.
—D'ARCY THOMPSON

Ironically, Galileo (see Figure 2), who discovered the scaling law for falling objects and thereby inaugurated modern experimental science, was also the one who noticed that some laws of physics (and biology) are *not* unchanged under changes of scale. In reflecting about the strength of bones, he argued that an animal twice as long, wide, and tall will weigh 8 times more. But, he pointed out, bones that are twice as wide have only four times the cross section and can support only *four* times the weight. Thus, to support the full weight, bone width must be scaled up by a factor greater than 2. This deviation from simple similarity introduces a natural scale in the design of animals, land-bound or aquatic: at some roughly predictable size, the bones become larger than the rest of the animal, and scaling (and the hypothetical beast) break down; see the essay by J. B. S. Haldane (1892–1964) *On Being the Right Size* [Hal 28].

Another instance of scaling in biology is the energy dissipation of warm-blooded animals as a function of their weight or mass (see Figure 3). One would naively expect the energy dissipation P as measured by daily caloric consumption to be proportional to the animal's surface area, which, for "similar" animals, is roughly proportional to the two-thirds power of its volume or mass m: $P \sim m^{2/3}$.

GALILEO GALILEI
BORN: 1564
CENSURED: 1616
DIED: 1642
TOMB BUILT: 1737
HONORED WITH
MONUMENT: 1842
EXONERATED: 1982

Figure 2   Galileo Galilei Vindicated [Har 77]. (© 1991 Sidney Harris)

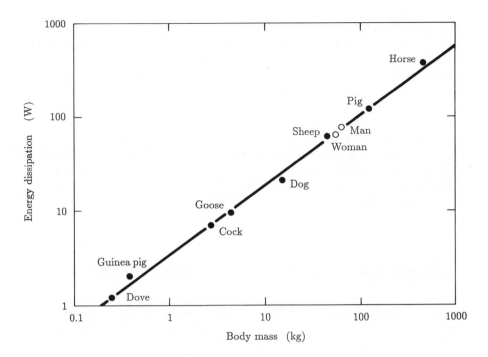

Figure 3   Energy dissipation of warm-blooded animals as a function of their weight.

In fact, the slope in Figure 3 is tantalizingly close to $\frac{2}{3}$, yet there is a small but systematic deviation from the expected slope: larger animals dissipate more power than the relation $P \sim m^{2/3}$ would predict. Actually, the data for a wide variety of species, including *Homo sapiens*, are much better fitted by an exponent of $\frac{3}{4}$. Why? This is a good question that merits further study. Is it that larger animals are less energy-efficient? Are they pushing at the same constraint on size that finally did the dinosaurs in? It seems that *people* are getting taller and bigger by the decade and they should perhaps be careful lest they join the extinct mammoth in oblivion.

A similar scaling failure occurs in photography, commonly referred to as *reciprocity failure*. It was discovered by the German astrophysicist (in fact, the founder of astrophysics) Karl Schwarzschild (1873–1916).[1] Schwarzschild, in

---

1.   Schwarzschild was barely 16 when he published his first papers (on the orbits of double stars). As early as 1899 he developed theories of the curvature of space, and in 1916 he gave the first exact solution of Einstein's general relativity equations, predicting the existence of *black holes* once a star shrinks below the "Schwarzschild radius", a characteristic length at which gravity overpowers all other forces. Thus, gravity limits the mass of both stars and animals.

cataloging the brightness of stars, found that a star half as bright as a reference star required *more* than twice the exposure time to blacken the photographic plate to the same degree as the reference. To achieve a given degree of blackening at low brightness b, the required exposure time t is *not* proportional to the reciprocal of the brightness (as is true at higher brightnesses and shorter exposure times). Rather, Schwarzschild found, $t^p \sim b^{-1}$, where the *Schwarzschild exponent* p is less than unity.

Thus, when it gets dark (or when you stop down the lens of your camera too much), you should expose longer than you might think. In color photography, since the different colors show different reciprocity failures, the color balance may change at low illumination unless corrected by special filters.

In spite of these failures, there is much to be gained by scaling—from fish to physics.

Similarity in Physics: Some Astounding Consequences

In physics, similarity arguments have carried us quite far. But even on an elementary level, similarity reasoning can help a great deal. Think of a physical system whose potential energy U is a homogeneous function of degree k of the spatial coordinates r_m:

$$U(\alpha r_1, \alpha r_2, \ldots) = \alpha^k U(r_1, r_2, \ldots) \tag{1}$$

If we scale all spatial coordinates by a factor α and time by a factor β, then velocities are changed by a factor α/β and kinetic energy changes by α^2/β^2. Now, if α^2/β^2 equals the factor α^k for the potential energy U, then the *Lagrangian* of the physical system is multiplied by a constant factor α^k and the equations of motion are unaltered. The resulting paths of all mass points ("particles") remain similar to the original ones; the only change is a change of scales [LL 76].

Time intervals along the new path scale as $\beta = \alpha^{1-k/2}$. Energies, of course, vary as α^k, and angular momenta, having the same dimension as Planck's quantum of action (energy $\times$ time), are multiplied by $\alpha^{1+k/2}$.

What can we conclude from all this? Let us look at a linear oscillator, whose potential energy is a homogeneous *quadratic* function, that is, a function for which the exponent k in equation 1 equals 2. A pendulum swinging with very small amplitudes is a simple example of a linear oscillator. Does the period of oscillation depend on the amplitude of the oscillation? We could, of course, solve the equation of motion and see that it does not. But that is unnecessary just to answer the question. As we just noted, times scale as $\alpha^{1-k/2}$ which, for $k = 2$, equals α^0. Thus, we see immediately that all times, including the period of oscillation, are unaffected: the natural frequency of a *linear* oscillator, adhering strictly to a quadratic potential energy function, is independent of its amplitude or energy. (In quantum mechanics this fact is reflected in the constant spacing, $h\nu$, between adjacent energy levels.)

What about a *non*linear oscillator with a cubic-law restoring force, that is, a quartic (fourth-power, $k = 4$) homogeneous potential? Now we cannot solve the equation of motion with a simple trigonometric function. But similarity tells us that times must scale as $\alpha^{1-k/2} = \alpha^{-1}$, that is, that frequencies are proportional to α: the higher the energy of the oscillator, the higher its resonance frequency, as might be expected for a spring with increasing stiffness. More precisely, the resonance frequency of such a nonlinear oscillator scales as the fourth root of its energy. (Quantum mechanically, $E_n \sim n^{4/3}$.)

In a *uniform* force field, the potential energy is a homogeneous *linear* function of the spatial coordinates; that is, $k = 1$. As a consequence, times scale as $\alpha^{1-k/2} = \alpha^{1/2}$, as indeed they do, a fact that an early practitioner of scaling, Galileo, discovered a long time ago in Pisa: he had to scale 4 times as many steps on the Leaning Tower to double the fall time (a fine tale, albeit apocryphal).

In Newtonian attraction, the potential energy is *inversely* proportional to distance. Thus, $k = -1$ and, for circular orbits around a massive center, we must expect times to scale as $\alpha^{1-k/2} = \alpha^{3/2}$. In other words, the square of a planetary orbital period is proportional to the cube of its size. And so we have just rediscovered a special case of Kepler's immortal third law of celestial mechanics—without solving a single integral!

Isaac Newton, in his *Principia*, even considered more general "planetary" laws. For a circular orbit of radius r and period τ, he deduced, from an assumed scaling relation $\tau \sim r^n$, the following law for the gravitational potential: $U \sim r^{2-2n}$. The same result follows directly from our similarity principle. For $n = \frac{3}{2}$, we are back to $U \sim r^{-1}$ and the real world of falling apples and orbiting moons. In fact, it was Newton's sudden inspiration[2] that the gravitational pull the earth exerted on an apple was $3600 = 60^2$ times larger than the pull it exerted on the moon (60 times more distant from the earth's center) that led him to formulate his universal law of gravitation: gravitational force must be proportional to the reciprocal of the distance squared.

For $U \sim r^{-2}$, a possible orbit is a logarithmic spiral: $r(\phi) = r_0 e^{\gamma\phi}$ in polar coordinates, a self-similar object! See pages 89–92 in Chapter 3 and Figure 4 for an artistic elaboration of the logarithmic spiral by S. Kim. What does scaling tell us about velocities and timing for *this* motion as a "planet" spirals into (or away from) its sun?

For $U \sim r^{-3}$, a cardioid $r = r_0(1 + \sin \phi(t))$ is a possible motion. What can we say about the angle $\phi(t)$?

Exploiting similarity, we can even prove the *virial theorem*, which relates *average* potential energy $\overline{U}$ to the *average* kinetic energy $\overline{T}$ for bounded motions. Since the kinetic energy T is a homogeneous *quadratic* function of the velocities

---

2.   Intriguingly, the German word for inspiration is *Einfall*, spelled like *ein Fall* (a fall).

Figure 4   Logarithmic spiral to infinity [Kim 81].

and the potential energy U is a homogeneous function of degree k of the spatial coordinates, it follows (almost) immediately from equation 1 that $2\overline{T} = k\overline{U}$.

For the linear oscillator ($k = 2$) we recover the well-known equality between average kinetic and potential energies: $\overline{T} = \overline{U}$.

Similarity in Concert Halls, Microwaves, and Hydrodynamics

Similarity transformations have been particularly fruitful in hydrodynamics and other difficult fields. Already in the nineteenth century, Sophus Lie (1842–1899) and later George David Birkhoff (1884–1944) had looked for transformation groups that leave given partial differential equations, and thus their solutions, invariant. Such solutions are called *similarity solutions*.

Suppose the following limit of a solution $\phi(x, y)$ exists:

$$\lim_{\varepsilon \to 0} \varepsilon^a \phi(\varepsilon^b x, \varepsilon^c y) = \Phi(x, y) \tag{2}$$

Then the similarity solution, $\Phi(x, y)$, obeys the following scaling law:

$$\Phi(x, y) = \lambda^a \Phi(\lambda^b x, \lambda^c y)$$

which follows immediately from equation 2 and is, in fact, a generalization of equation 1.

But scaling is not always *that* easy. A good method of designing concert halls and opera houses, for example, is to build *scale models* first and study sound transmission in *them*, instead of in the finished hall.[3] Linear dimensions, wavelengths, and frequencies scale easily. In a one-tenth scale model, for example, all linear dimensions are one-tenth of those in the real hall. Thus, since sound diffraction from hard surfaces depends only on the ratio of the wavelength to the linear dimensions of the scattering surfaces, wavelengths should likewise be scaled down by a factor of 10. For a fixed sound velocity ($c = 343$ m/s in dry air at room temperature), this means that frequencies should be scaled *up* by a factor of 10. Travel times are, of course, 10 times shorter in the scale model. Thus, times scale inversely with respect to frequencies—a good thing, because frequencies are measured in reciprocal seconds. But sound absorption (by people, for example) is more difficult to scale. Nevertheless, special materials (known as "instant people") have been invented that mimic human absorption at upscale frequencies.

Absorption, friction, and other energy loss mechanisms generally cause scaling difficulties. For example, a microwave cavity (a hollow metallic resonator) scaled down in size by a factor of 10 has its resonance frequencies scaled up by the same factor of 10. However, the minimum *bandwidths* of its resonant modes (as determined by skin-effect losses) will go up by a factor of $10^{3/2} \approx 32$, because the small penetration depth of electromagnetic fields in conductors (the finite penetration depth causes the skin effect) is proportional to $f^{-1/2}$, not f^{-1}. (The *relative* bandwidth of an electromagnetic cavity resonance is given approximately by the ratio of two volumes: the inner surface area of the cavity times the skin depth—i.e., the volume where the energy losses occur—divided by the total

---

3. But the alternative approach, *first* building the full-scale hall and *then* the scale models, has also been tried—with disastrous consequences requiring expensive "remodeling" (a euphemism if there ever was one).

volume of the cavity, or the volume where the energy is stored. At least *some* things about microwaves seem safe and simple.

Another instance where friction causes endless scaling problems is in ship-building model basins. A battleship scaled down in size by a factor of 50 experiences rather ill-mannered drag forces because drag is caused by viscous boundary layers, which, like the skin effect, follow a different scale.

How *much* size can affect scaling is illustrated perhaps best by the following eyewitness observation. A large ocean liner, "steaming" into New York harbor during a tugboat strike, had to shut off its engines *miles* ahead of its berth and then drift with the slowing tide up the Hudson River to arrive exactly at its pier just as the tide began to turn. This is no mean feat, because stopping distances of ocean liners, without external assist, are rather larger than even those of titanic trucks on glare ice.

By contrast, stopping distances on microscopic scales in anything but frictionless fluids (superfluids) are *so* short that any particles suspended in a liquid seem devoid of inertia. This was, in fact, the experience of the present investigator while observing hydrodynamic streaming in a model of the inner ear under a microscope: the moment the sound was turned off, the streaming motion stopped instantly, as if everything were massless. Such is "life at low Reynolds numbers" as described in an engaging article of that name by Edward Purcell [Pur 77]. Reynolds numbers are only one kind in a long list of dimensionless numbers that reflect the importance and difficulty of scaling in hydrodynamics [McG 71].

Scaling in Psychology

Whereas measurement in classical physics is a well-understood process, relating an observed quantity to a well-defined unit, the situation in psychology was not so clear-cut until the physiologist E. H. Weber (1795–1878)—brother of the physicist Wilhelm Weber (1804–1891), collaborator of C. F. Gauss— made careful studies of the sensations of sound and touch, thereby laying the foundations of a new science, the science of sensations. According to Weber's law, an increase in stimulus necessary to elicit a just noticeable increase in sensation is not a fixed quantity, but depends on the ratio of increase to the original stimulus. Later, the physicist and philosopher G. T. Fechner (1801–1887) restated Weber's law (as the Weber-Fechner law) and specified its domain of validity.[4] Modern psychologists, and particularly S. S. Stevens, have succeeded in introducing measurement methods into psychology that are nearly as unambiguous

---

4. Fechner also fathomed experimental aesthetics by measuring which shapes and dimensions are most pleasing. He may have been the first to conduct a public opinion poll (to discover which of two Holbein paintings was preferred by most people).

as objective measurements in physics [Ste 69]. The new discipline has therefore rightly earned the designation *psychophysics*, of which psychoacoustics is a special branch, as is psychovisual research.

One of Steven's great contributions was the introduction of *ratio scales* for subjective variables (like loudness and brightness) and the discovery of simple *power-law* relations between these subjective variables and corresponding physical quantities (like energy flux or intensity).

For example, for a sound to double in loudness L, its intensity I has to be multiplied by a factor of 10; this is true over much of the intensity range that the human ear can perceive without pain (a range exceeding 12 orders of magnitude at mid-range audio frequencies). Thus, because $\log_{10} 2 \approx 0.3$, we have the following power law for loudness as a function of acoustic intensity:

$$L \sim I^{0.3} \tag{3}$$

Someone who has not participated in a psychoacoustic scaling test might object that "loudness doubling" is not a well-defined concept. But surprisingly, the random scatter encountered in such tests is remarkably small even between different listeners.

The exponents found in psychophysical power laws, such as the value 0.3 in equation 3, are not universal but are specific to the sense modality studied (subjective brightness, perceived weight, or apparent length, for example) and have been analyzed in great detail by psychophysicists.[5] One important research question concerns the *transitivity* of these exponents when comparing loudness with weight and weight with brightness, for example, and what it might reveal about brain functions.

If we replace the sound intensity I in equation 3 by the sound pressure p, then, because intensity is proportional to pressure *squared*, we have

$$L \sim p^{0.6}$$

Interestingly, the exponent 0.6 can be derived from an exponent of 0.5 found at a more fundamental level, the Fourier-like "critical" frequency-band decomposition of sounds in the inner ear. The exponent 0.5, in turn, turns our attention in the direction of statistical analysis and uncertainty, resulting in the following simplified model of loudness perception. If loudness were perceived as the mean rate of nerve pulses traveling along the acoustic nerve up to higher auditory centers in the brain, and if these pulses were a modulated Poisson process whose mean rate was proportional to the sound pressure p, then the uncertainty of the number of pulses in a given time interval (100 ms, say) would be proportional to $p^{0.5}$. Since many ratio scales in psychophysics are found to be directly related

---

5. Thus, *universality*, so beloved by physicists, is lacking in psychophysics.

to perceptual uncertainties ("just noticeable differences"), the observed power law for subjective loudness *versus* physical intensity would then indeed be predicted by such a statistical model of neural firing rates.

In reality, loudness perception is more complicated, but the observed power laws and their exponents have yielded important clues and steered researchers in the right direction.

Acousticians, Alchemy, and Concert Halls

Concert halls are built to transmit pleasing sounds from performing musicians to attentive listeners (while keeping everybody dry and comfortable at the same time). Thus, nothing seems more apropos for an acoustical scientist than to measure the "frequency response" of a concert hall between the stage and various points in the audience area. Here *frequency response* means the effectiveness with which various frequencies (musical pitches) are transmitted between two distant points in the hall. Figure 5 shows a typical sample of such a frequency response on a logarithmic scale versus frequency.

The many ups and downs of such a response, even over narrow frequency intervals, are immediately apparent, as they were to Edward C. Wente (1889–

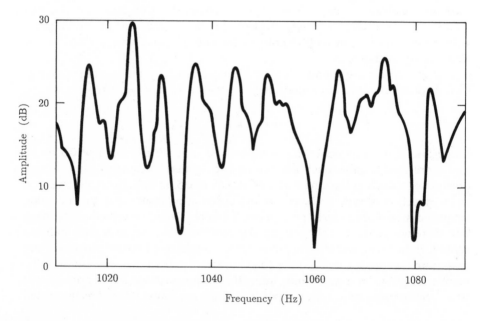

Figure 5   Sound transmission between two points in a concert hall as a function of frequency. Note large statistical fluctuations.

1972), inventor of the condenser microphone, who first published such a response in 1935. As a communications expert, Wente wondered (in print) why people could actually enjoy music in concert halls, given such large response fluctuations—fluctuations that exceed even those of the cheapest loudspeaker. The answer is that, when listening to speech or music, the ear unconsciously "switches" to a *short*-time spectral analysis, which, in accordance with Heisenberg's uncertainty principle, does not resolve fluctuations on such a fine frequency scale as shown in Figure 5.

Oblivious to the profound perceptual insignificance of these high-resolution frequency responses, acousticians around the world kept measuring them with great abandon. Worse, before too long, various supposedly objective criteria for acoustical quality were concocted from these responses. Of course, every time a new concert hall "came on line" (was inaugurated, to use a more archaic term), the extant criterion had to be modified to make the new hall's response characteristics conform with its perceived musical quality. These "scientific" attempts have been compared, quite appropriately I think, to tea-leaf reading and alchemy (although one should not unduly belittle the latter).

When the author, then a young student at Göttingen, heard about these activities—still going strong in 1954—he thought that maybe these frequency responses were nothing but *noise* in the frequency domain (technically: the modulus of a complex Gaussian process, resulting from the random interference of many overlapping room resonances). If this was indeed so, then practically all frequency responses in large halls should be *similar* to each other, characterized by a *single scaling parameter*: the reverberation time of the hall.

This turned out to be the case, as further measurements soon revealed. For example, the average frequency spacing (in hertz) of the response maxima was found to be about 4 divided by the reverberation time (in seconds), in excellent agreement with the theoretical prediction [Schr 54, or, in the most recent English translation, Schr 87]. Thus, people have been measuring nothing but reverberation time all these years, and in a very complicated and roundabout way at that. Ironically, the "new" criteria that had been distilled from frequency responses were supposed to *supplement* reverberation time, which had been found wanting in its predictive power of acoustical quality. Thus, a little insight and a good similarity argument liberated a lot of manpower from a useless pursuit.

High-resolution frequency responses became important later in solving the problem of stability of public-address systems, for which acoustic feedback ("howling") always threatens to become a problem. By shifting all frequency components of a speech signal by the average spacing between the room's response maxima and minima (a few hertz, according to the aforementioned statistical theory), the stability can be considerably increased—a howling success [Schr 64].

More generally (and perhaps more important), the theory of randomly interfering coherent waves has assumed a central role in the analysis of hologram laser speckles and electromagnetic multipath propagation; think of mobile cellular telephones in cars and the cordless handset at home, a marvelous invention, especially when it comes with the matchless sound quality of a good "corded" phone.

Preference and Dissimilarity: Concert Halls Revisited

In this section we shall make a brief call on a problem in psychological scaling with which the author, although not a psychologist himself, is somewhat familiar: the ranking of concert hall acoustics.

A deeply entrenched procedure for getting a reading on the acoustical quality of a concert hall (or opera house) is to collect comments from listeners, musicians, conductors, and music critics. These subjective ratings are then correlated with various architectural and physical characteristics of the hall (such as width of the listening area, reverberation time, and frequency response). From these correlations a mathematical formula is then constructed for predicting acoustic quality on the basis of measurable, objective parameters; see, for example, Beranek's book *Music, Acoustics and Architecture* [Ber 62].

Typical responses elicited from German-speaking music lovers to characterize the acoustics of concert halls include such lovely locutions as *glasklar, jämmerlich, krankhaft, ruinös, unheimlich*, and—last but not least—*wunderbar*. To translate these high-sounding words into basic English would be sheer waste because they are not only ill defined but nearly meaningless in any language.

At Lincoln Center for the Performing Arts in New York City, Philharmonic Hall (now Avery Fisher Hall) had been designed on the basis of the aforementioned approach, and thus it is unsurprising that it required a major acoustic rescue effort. When the author was confronted with this, he and his colleagues recognized that as a first order of business, more reliable methods for the subjective (and objective) evaluation of concert halls had to be developed.

New objective measurement techniques [ASSW 66] revealed that the overhead acoustic panels ("clouds"; see Figure 6) did not reflect low-frequency components (especially from the cellos) with sufficient strength into the main audience area [SASW 66]. This was partly the result of poor *scaling*: to properly reflect musical notes of different wavelengths from an acoustic panel (not a panel of listeners), the panel's geometric dimensions must be at least comparable in size with the longest wavelength present in the sound. In actual fact, they were much too small, a failure that evoked both seering sarcasm and much mirth.[6]

---

6. The maestro of the acclaimed Cleveland Orchestra, the none-too-reticent George Szell, was so enraged by the whole debacle that he dubbed the panels "schwangere Frösche mit beleuchtetem Bauchnabel" (pregnant frogs with illuminated navels—on account of their double-duty function as lighting fixtures). A contemporary *New Yorker* cartoon showed two ladies walking in the foyer under the Lipchitz sculpture (vaguely reminiscent of suspended acoustic panels) and remarking, "No wonder the acoustics is so bad in there; it's all hanging out here!"

Figure 6 The acoustic panels ("clouds") in Philharmonic Hall, Lincoln Center for the Performing Arts, New York.

To put the subjective evaluation of concert hall quality on a firmer basis, the author suggested eschewing the use of any ill-defined epithets, such as those quoted in this section, and restricting listeners' responses strictly to an expression of acoustic *preference* between two halls or a degree of *dissimilarity* that they perceived. To preserve individual differences in musical preference, these responses were not simply averaged but were analyzed by modern *multidimensional scaling* algorithms [SGS 74]. With a sufficient number of responses—even just binary responses as in the case of two-valued preference judgments—these algorithms are capable of constructing a well-defined Euclidean space, usually of two or three dimensions, in which Euclidean distances are closely proportional to the perceived dissimilarities or differences in preference.

Figure 7 shows an example of a so-constructed *preference space*. The different symbols T_1, Q_3, and so forth, represent different concert halls and recording

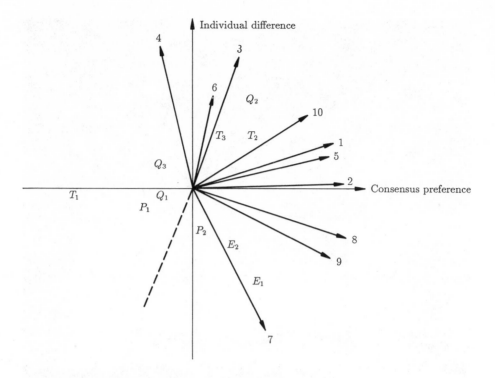

Figure 7 Preference space for concert halls.

locations in these halls (e.g., Q_3 is the third location in concert hall Q). The numbered arrows are unit vectors representing the 10 listeners who participated in this preference test. (The fact that some arrows appear shorter than others indicates that they have nonnegligible components in the third dimension used in the analysis, which is not shown here.)

For each pair of the 10 test conditions, for example T_3 and E_2, each of the 10 listeners states which of the two conditions he prefers. The accumulated preference data (a total of 450 paired comparisons) is subjected to multidimensional scaling by a metric linear factor analysis [Sla 60].

A computer program that implements this factor analysis iteratively changes the position in the preference space of each test condition, for example T_1, and the direction of each listener vector in a three-dimensional Euclidean space until the normal projections for all 10 test conditions on each of the 10 listener vectors agree, as closely as possible, with the preference data. Thus, Figure 7 tells us, for example, that listener 3 prefers test condition T_1 least and Q_2 most.

For an exact representation, a 10-dimensional Euclidean space would generally be required. In fact, almost 90 percent of the total variance in the data is accounted for by the two first dimensions of the preference space.

Since almost all listener's arrows in Figure 7 point into the right half plane, the abscissa could be labeled "consensus preference." Indeed, if some architectural modification of a hall would move the position in the preference space for a given location in that hall to the right, *all* listeners, except one (listener 4), would respond with a higher preference score for that seat.

By contrast, about half the arrows point into the upper half plane and about half into the lower. Thus, the ordinate strongly reflects "individual differences" in musical tastes—a personal dimension that should be honored in the design of spaces for the enjoyment of music.

Subjective tests in which judgments of *dissimilarity* were elicited from the listeners gave very similar results. Thus, our confidence in *multi*dimensional scaling methods for constructing perceptual spaces was further strengthened. The success is doubtless due to two ingredients (or, rather, omissions):

1 Avoiding the use of ill-defined terms or empty words to describe acoustic quality

2 Not forcing musical tastes into a one-dimensional Procrustean bed (Procrustes would have loved the very thought of one-dimensional cots)

In conclusion, we should mention one "technical detail" without which the cited results could not have been achieved. Comparisons between the (often subtle) acoustic differences prevailing in different halls are notoriously difficult. Listening experiences separated by days or weeks, based on different musical programs, executed (good word!) by different orchestras, are highly unreliable. Thus, much would be gained by the possibility of *instantaneous* comparisons between different halls. This requirement was realized in the aforementioned study by an ingenious method that allowed faithful reproduction of music recorded in different halls at different times, using a tape recording of Mozart's *Jupiter* Symphony (and representatives of other musical styles) played by the English Chamber Orchestra in an anechoic environment and kindly made available to the author.

Figure 8A shows three crucial collaborators in this project. The reproduction method [Schr 70] is based on the fact that most people have just two ears. By properly preprocessing the two audio signals from the dummy head shown in Figure 8A, it is possible to transfer these two signals, via two loudspeakers, to the eardrums of a listener seated at some distance in front of the loudspeakers (see Figure 8B). The preprocessing, a kind of inverse filtering, compensates for the *cross talk* from each of the loudspeakers to the "wrong" ear. Since the sound transfer matrix between loudspeakers and ears is nonsingular for proper loud-speaker placement in an anechoic listening room, a physically realizable inverse exists, which is then incorporated into appropriate cross-talk compensation filters. The transfer matrix and its inverse depend upon the geometry and sound diffraction around the listener's head, which is measured for a "standard" head

(A)

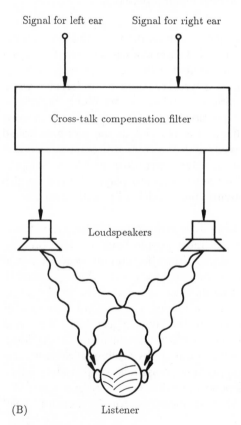

(B)   Listener

Figure 8   (A) Three crucial collaborators in concert hall measurements. (B) Listening to loudspeakers with compensated cross-talk between the two speakers.

shape. (For once, success depends not so much on a head's inner workings but on its outer shape.)

To test this method of sound reproduction, sound waves arriving laterally from an angle of 90° were simulated using two loudspeakers located at $\pm 22.5°$. The simulation turned out to be so realistic that many listeners turned their heads 90° to locate a third (nonexistent) sound source. (Of course, when they turn their heads, the effect disappears because the geometry is changed.)

When this reproduction method was first applied to concert halls, the effect was truly overwhelming: at a flick of a switch the listener could "transport" himself from the Vienna Musikvereinssaal, say, to the Amsterdam Concertgebouw and back again. These instantaneous comparisons finally made reliable quality judgments possible that had eluded acousticians for so long.

The preference coordinates obtained in these tests were correlated with various physical parameters, revealing a consistent absence of strong *laterally* traveling sound waves in the less preferred halls. Thus, good acoustics—given proper reverberation time, frequency balance, and absence of disturbing echoes—is mediated by the presence of strong lateral sound waves that give rise to a preferred "stereophonic" sound. In old-style, narrow and high halls, such lateral sound is naturally provided by the architecture. By contrast, in many a modern, fan-shaped hall with a low ceiling, a "monophonic" sound, arriving in the symmetry plane through the listener's head, predominates, giving rise to an undesirable sensation of detachment from the music.

This, in a nutshell, is the reason why many modern halls project such a poor musical sound. The stimulating role that the concepts of (dis)similarity and multidimensional scaling [She 62] have played in the clarification process can hardly be overemphasized.

To increase the amount of laterally traveling sound in a modern hall, highly efficient sound-scattering surfaces have been invented recently [Schr 90]. These "reflection phase gratings," to use the physicist's term, are based on number-theoretic principles (primitive polynomials over finite fields, quadratic residues, and discrete logarithms) and have the remarkable property of scattering nearly equal acoustic (or radar) intensities into all directions. Such broadly scattering surfaces are now also being introduced into recording studios, churches, and even individual living rooms—and who knows where else.

These sound-dispersing structures should also find good use in noise abatement, because a dispersed (and therefore weakened) noise is more easily "masked" (rendered inaudible) by other sounds.

3

C H A P T E R

Self-Similarity—Discrete, Continuous, Strict, and Otherwise

Big fleas have little fleas upon their backs to bite them,
and little fleas have lesser fleas, and so ad infinitum

—SWIFT,
Poems II.651 (1733)

We say that an object—a geometric figure, for example—is invariant with scaling, or *self-similar*, for short, if it is reproduced by magnifying some portion of it.

Self-similarity comes in many different shapes and forms. Some of it continuous, some discrete; some is accurate over many powers of 10 and some over less than a factor of 10. We find examples of self-similarity in our daily surroundings or deeply hidden in the behavior of physical or biological systems. Some self-similarity is fully deterministic, some is only probabilistic. A few cases of self-similarity are mathematically exact; however, most instances in the real world are only asymptotically self-similar or just approximately so. Cantor sets and Weierstrass functions (see pages 96–98 and Chapter 7) are two well-known delegates from mathematics at this reunion. Brownian motion represents both the physical sciences and probabilistic self-similarity. And Bach's tempered 12-tone scale shows the importance of self-similarity in music.

A well-known example of discrete, albeit limited, self-similarity is a *set* of Chinese boxes or Russian dolls—the (usually wooden) kind where a large doll (discreetly) hides a *similar* smaller one inside its "body" and the smaller doll hides a similar third one and so forth for two or three more "generations." If we had a doubly infinite number of dolls, both ever smaller and ever larger dolls, then, provided the scaling ratio of doll sizes between successive generations was constant, we would have a set with *exact* discrete self-similarity. But such a set could, of course, exist only in our imagination; real dolls have finite measurements: they must be both larger than single atoms and smaller than the full universe. (For Charles Addams's vision of incipient self-similarity, see his posthumously published cartoon given in Figure 1.)

Another example of discrete, if severely limited, self-similarity can be discovered on some product labels. Think of a beer bottle that shows the same beer bottle on its label, which shows the same beer bottle on *its* label.

Or take a look at the cover of Paul Halmos's book *Naive Set Theory* [Hal 74], which shows the cover of the book, which shows the cover of the book, which shows the cover of the book without showing the cover. Of course, printing costs (not to mention other constraints) put an early and abrupt end to this progression of ever smaller images. A cheaper way to get many more scaled-down images is to stand between two almost parallel mirrors such as those found in clothing stores (see Figure 2). But of course, the high-order reflections are

Figure 1   Self-similarity à la Addams. (Drawing by Chas. Addams; © 1987 The New Yorker Magazine, Inc.)

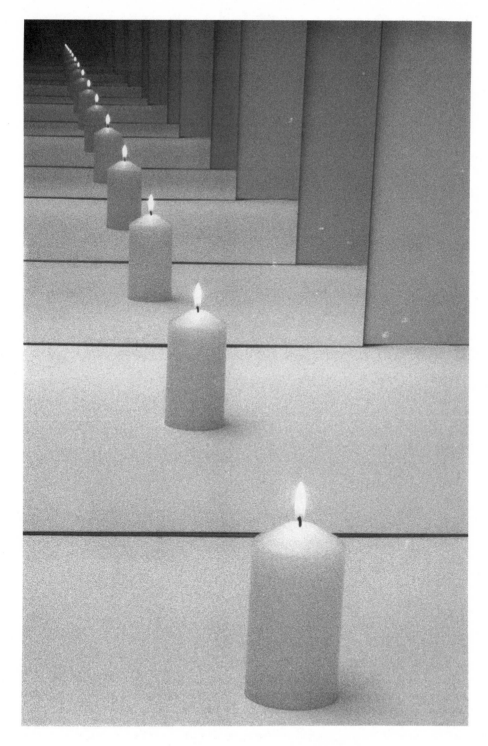

Figure 2   A long row of candles: self-similarity induced by parallel mirrors.

Figure 3   Castel del Monte, an emperor's early attempt at self-similarity.

distorted and weakened, because no real mirror is perfectly planar (or 100 percent reflective).

The result of an early attempt at self-similarity in architecture is the monumental Castel del Monte (Figure 3), designed and built by Holy Roman Emperor—also King of Sicily, Germany, and Jerusalem—Frederick II (1194–1250), the great falconer, rare mathematician (among medieval emperors, anyhow), and last but not least, irrepressible castle builder. The basic shape of the castle is a regular octagon, fortified by eight mighty towers, again shaped like regular octagons. (These towers were devised for the easy release and retrieval of hunting falcons.)

Ironically, Frederick himself was born in a tent, hastily erected in the market square of the little town of Iesi in the Italian Marches near Ancona on the Adriatic Sea.[1] Frederick, who of course wrote and illustrated the book on falconry [Fri 1240], was graced with one of the most stupendous minds of his time (hence

---

1.   Frederick's birth in a tent may have been intentional rather than unexpected because his mother, the Norman queen Constance, had taken "forever" to become expectant with the sorely awaited future sovereign, the prospective Hohenstaufen ruler. Thus, to forestall any suspicion of double-dealing, the precious child was delivered in the most public of places, with no double walls and no false bottoms to conceal a surrogate mother. As an adult, Frederick revisited Iesi, which he called his "Bethlehem" (he was born during Christmas).

his Latin epithet *stupor mundi*). He grew up in Palermo and Apulia, speaking and reading mostly Arabic, Greek, and Latin. He later introduced—by way of poetry—the Italian *volgare* spoken by the people at the imperial court, thereby strengthening its linguistic links with Italy and elevating the vulgar language to official status. Frederick was also a friend of Fibonacci (Leonardo Pisano), furnishing the latter with algebraic problems whose solutions became part of the history of mathematics.

In music, too, self-similarity abounds. Adjacent notes of a well-tempered scale are in a constant frequency ratio ($2^{1/12}$ for an octave divided into 12 semitones). Thus, because of the inverse proportionality between resonant frequency and the length of a tube, clusters of organ pipes exhibit self-similarity. Less pleasing (and more dangerous) but self-similar nevertheless is the perspective of railroad tracks and ties receding to a distant vanishing point (in Figure 10, on page 93, this self-similarity is exploited in a proof of the formula for the sum of an infinite geometric sequence).

Other examples of discrete self-similarity are "log-periodic" antennas (see Figure 4), which cover a wide spectrum of wavelengths in many discrete steps. Note that both the lengths of the adjacent dipoles and their spacings are scaled by the same similarity factor. Thus, except for end effects, these antennas cover a wide range of wavelengths with nearly the same sensitivity and spatial resolution. TV antennas adorning our roofs are but poor cousins of the antenna shown in the illustration; but they, too, must capture many different channels with equal gain and clarity.[2]

Another kind of wideband "antenna," albeit for sound, is the *basilar membrane*, the frequency analyzer in our inner ears. Different frequencies excite different places along the basilar membrane. This resonating membrane therefore effects a mapping from frequency to place. To cover the enormous frequency range of human hearing, from 20 Hz to 20,000 Hz, without unduly compressing the space available for the important low and middle frequencies, the ear must map frequencies on a *logarithmic* scale. In fact, above about 600 Hz, constant frequency *ratios* correspond to constant shifts in the locations of the resonances along the basilar membrane. In the frequency range from 600 Hz to 20,000 Hz, frequency ratios and places (i.e., the locations of the resonance) scale almost perfectly, the scaling factor being 5 mm along the basilar membrane per octave.

There is another good reason for this logarithmic mapping of frequency to place. It means that the *relative* change with place of the parameters (mass density, stiffness) controlling resonance frequency is constant along the basilar membrane. The basilar membrane therefore behaves like an exponential acoustic horn, such as the horn in a woofer loudspeaker, for example, thus minimizing the reflection

---

2. Unfortunately, the abovementioned end effects that limit exact self-similarity often manifest themselves at the expense of public television channels occupying the less sensitive band edge (such as channel 13 for the VHF band in the United States).

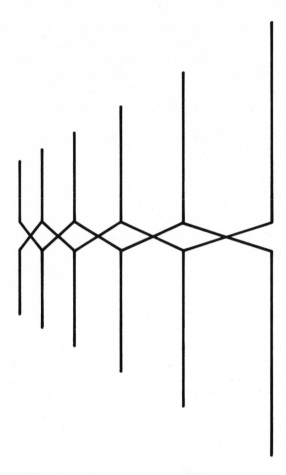

Figure 4   Self-similar TV antenna.

of acoustic energy for a given total length and frequency range. This must have been an important design consideration in the evolution of our ears to a highly sensitive acoustic receptor. (The healthy ear is almost able to detect the Brownian motion of the air molecules!)

For the physicist, the fact that the mechanical parameters change relatively slowly along the length of the basilar membrane—indeed, as slowly as possible for a given total length and frequency range—means that he can analyze wave motion on the basilar membrane in terms of the convenient Wentzel-Kramers-Brillouin approximation [ZLP 76]. This useful method was invented by Joseph Liouville (1809–1882) and rediscovered by the three named authors, who introduced it into quantum mechanics.

Below 600 Hz, the mapping between place and frequency itself (not frequency ratio) is linear. Otherwise, the five octaves below 600 Hz would be given the same space on the basilar membrane as the five octaves above 600 Hz. Since the density of auditory neurons along the basilar membrane is approximately constant, the neural representation of the high frequencies would thus suffer relative to the low frequencies.

The response of the basilar membrane shows another interesting scaling behavior that obtains for its entire frequency range: local wave *velocity* is proportional to local resonance *frequency*. The constant of proportionality is a characteristic length that equals about 1 mm. As a consequence the *delays* of acoustic signals along the basilar membrane are *inversely* proportional to place of detection. This scaling behavior leads to a simple integrable mathematical model of the basilar membrane [Schr 73].

Hierarchical structures, such as phylogenetic trees, for example, often show self-similarity, as do mathematical Cayley trees (also called *Bethe lattices* in physics). A Cayley tree is defined as a graph without loops in which each node has the same number of branches (namely, two, in Figure 5). The self-similarity of such graphs is not necessarily manifest in their geometric representation, but is seen

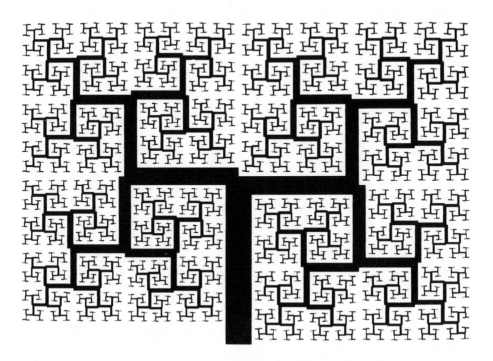

Figure 5   Artistic Cayley tree with 1:2 branching ratio [Man 83].

in their connectivity or topology. Bethe lattices, although highly unphysical. often afford the only exactly solvable models in difficult situations such as Anderson localization and percolation (important concepts in contemporary physics; see Chapter 16).

Figure 6 shows another example, a binary-code tree. It is interesting to note that if we define the distance between two "leaves" (endpoints) by how many generations we have to go back to find a common ancestor, then the space so generated is *ultrametric*. For phylogenetic trees, this means that either the three distances between three existing species are all equal or two are equal and the third is smaller. In other words, all triangles in such a space are either equilateral or short-base isosceles, with interesting consequences in Hensel codes and error-free computing [Schr 90].

If one can identify an ultrametric space in a given problem, there is usually a hierarchical structure lurking behind it that holds the key to better understanding; such structures are found in problems from taxonomy to statistical physics and

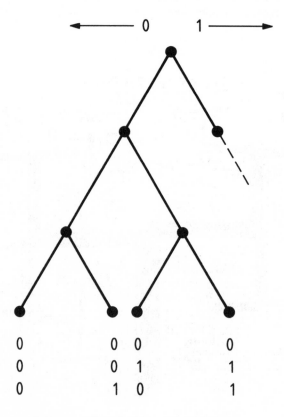

Figure 6   Binary-code tree: a self-similar hierarchical structure.

optimization theory. For an excellent overview of ultrametricity for the mathematical amateur, see the paper by Ramal, Toulouse, and Virasoro [RTV 86].

The Logarithmic Spiral, Cutting Knives, and Wideband Antennas

A charmingly simple example of a self-similar object, with a practical application
to boot, is the *logarithmic spiral*—well known from high school mathematics. In
polar coordinates (r, ϕ) we have $r(\phi) = r_0 \exp(k\phi)$, where $r_0 > 0$ and k are
constants. Scaling the radius vector r, that is, the size of the spiral, by a factor
s results in the same spiral rotated by a constant angle $(\log s)/k$. Since the angle
ϕ is defined only modulo 2π, scaling factors equal to $s = \exp(2\pi m k)$, where m
is an integer, leave the infinite spiral invariant—that is, the logarithmic spiral is
self-similar, with a similarity factor $s = \exp(2\pi|k|)$. If we disregard rotations, the
logarithmic spiral is self-similar for *any* real scaling factor.

The self-similarity of the logarithmic spiral has several interesting consequences and useful applications. For one, the direction of the tangent to the spiral
depends only on the angle ϕ and not on which branch of the spiral one considers
(see Figure 7). This follows directly from the scaling invariance but can also be
verified, of course, by a more circuitous calculation. Furthermore, since the rotated
logarithmic spiral is similar to itself, the angle β between the radius vector and

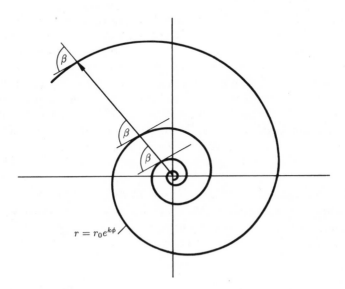

Figure 7   Logarithmic spiral: a *smooth* self-similar curve.

the tangent at any point must be the same for the entire spiral. A little thought (or a small sketch) shows that cot β must equal $(dr/d\phi)/r = d(\log r)/d\phi = k$. So *that* is the geometric meaning of k: it is the cotangent of the constant angle between the radius vector and the tangent at any point. (For $k = 0$, β equals $\pi/2$ and the spiral has degenerated into a circle.)

The just-stated property of the logarithmic spiral has an interesting application. Suppose you had to design a cutting tool with a rotating knife in the form of a disk. Which form should the knife have so that the cutting angle is constant, independent of the angle of rotation of the knife? Why, the perimeter of the knife should follow a logarithmic spiral! Since the perimeter of the knife is necessarily a single-valued function of the rotation angle, it must, of course, "jump back," by an amount $r_0 \exp(2\pi|k|)$, at some angle.

But things can be even more exciting. As we discovered, the logarithmic spiral is self-similar with *arbitrary* scale factors if we disregard rotation. This means that, if we do not care about rotations, the logarithmic spiral has *no* length scale—contrary to the impression that its mathematical formula, $r(\phi) = r_0 \exp(k\phi)$, imparts because the factor r_0 seems to imply a length scale. However, r_0 can be absorbed into a rotation, as can be seen by writing $r(\phi) = \exp\{k[\phi + (\log r_0)/k]\}$, where $(\log r_0)/k$ is just a constant angle.

Once we have something that is scale-free, we should find any number of useful applications. People are forever baffled by problems engendered by changes in scale or size.[3] Suppose we could construct scale-free footwear, shoes that fit all sizes—what a boon! (But to whom?) For stockings, of course, limited scale-free-ness has in fact been achieved: stretch hose that fit all sizes (or perhaps none).

Another field where scale-free-ness is urgently desired is the design of transmitting and receiving antennas for communication systems that have to cover a wide range of wavelengths, such as the log-periodic antenna shown in Figure 4, which is self-similar at a set of discrete wavelengths. If only such antennas were equally efficient at a *continuous* range of wavelengths! Well, suppose we use circularly polarized waves; then a rotation of the antenna would have no effect on the antenna's gain and directivity. And if we gave the antenna the form of a logarithmic spiral (ideally, of thin superconducting wire), then it would work equally well for all wavelengths within any desired range [CM 90]. Antennas exploiting this enticing principle do in fact exist. They look like tapered bedsprings.

But nature, too, has exploited the self-similarity of the logarithmic spiral. In the chambered nautilus (see Figure 8), each chamber is an upscaled replica of the preceding chamber with a constant scale factor. As a result, the nautilus will grow along a logarithmic spiral. And not just nature: Even artists have been

---

3.   Think of the *two* holes the thoughtful Sir Isaac (Newton) is said to have cut in his door to accommodate his two different-size pets.

Figure 8   The chambered snail *Nautilus* follows a logarithmic spiral in its self-similar design. (Photo by Edward Weston; © 1981 Center for Creative Photography, Arizona Board of Regents.)

inspired by spirals. Color Plate 4 shows an infinity of logarithmic spirals in the colors of the rainbow infinitely intertwined.

A logarithmic spiral with a specific value of the scale factor occurs in the following geometric problem. Take a rectangle with sides $a > b$ and cut off a square of side b from one side of the rectangle (see Figure 9). From the remaining rectangle cut off a square of side $a - b$, as shown in Figure 9. For the construction to work as shown, $a - b$ must be smaller than b; that is, a must be smaller than $2b$. At the second stage of construction, the inequality is $2a > 3b$, and at the $(2n)$th stage, the condition that cutting off a square will result in a rectangle whose longer side is the shorter side of the preceding rectangle is $F_{2n-1}a > F_{2n}b$, where the F_k are the Fibonacci numbers. Similarly, $F_{2n}a < F_{2n+1}b$ must hold. For the construction to carry through for arbitrarily large n, the side ratio b/a of the initial rectangle must equal the limit as $n \to \infty$ of F_{2n-1}/F_{2n}, that is, the golden mean $\gamma = (\sqrt{5} - 1)/2$.

The result of this construction is a limitless spiral of ever smaller squares with a scaling factor equal to the golden mean. The logarithmic spiral with $k = -(\pi/2) \log \gamma$, passing through successive cutoff points, is also shown in the illustration. The vanishing point of the squares and origin of the spiral is given by the common intersection of the diagonals of the rectangles.

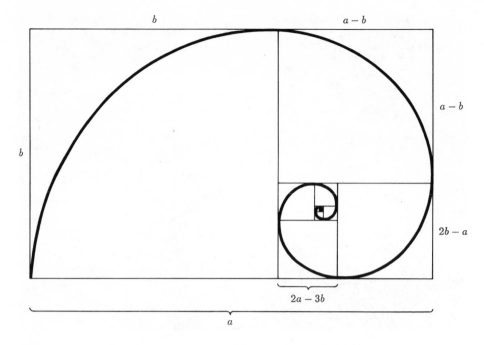

Figure 9   A self-similar succession of golden-mean rectangles.

There is a charming connection between our amputating of squares from residual rectangles and continued fractions. Observe that the continued fraction for the golden mean is $1/(a_1 + 1/(a_2 + \cdots))$, usually abbreviated as $[a_1, a_2, \ldots]$, with $a_1 = a_2 = \cdots = 1$. Thus $\gamma = [1, 1, \ldots]$. Suppose we want to be able to cut off *two* squares at each stage so that the longer side of the remaining rectangle equals the shorter side of the preceding rectangle. What is the appropriate side ratio b/a of the initial (and all subsequent) rectangles? A little experimentation will show that b/a must equal $\sqrt{2} - 1$, which has the continued fraction expansion $[2, 2, 2, \ldots]$. In general, the nth term a_n in the continued fraction of b/a tells us how many squares we must cut off at the nth stage. Thus, self-similar cascades and logarithmic spirals emerge for initial rectangles whose side ratio b/a equals a periodic continued fraction with period length 1. The resulting irrational numbers $[n, n, \ldots]$ with $n > 1$, which I have called *silver means*, also play a role in the construction of quasiperiodic lattices (see Chapter 14) and the modeling of quasicrystals (see Chapter 13).

By the way, the logarithmic spiral, like the infinite straight line, is a specimen of a self-similar object that is *smooth*, in stark contrast to the fractals that we usually associate with self-similarity such as rocky coasts, Brownian motion, and other nondifferentiable functions.

Some Simple Cases of Self-Similarity

One of the simplest self-similar entities is a two-sided infinite geometric sequence, for example,

$$s_n = \ldots, \frac{1}{8}, \frac{1}{4}, \frac{1}{2}, 1, 2, 4, 8, \ldots$$

Multiplying each member of this sequence by a factor of 2 produces

$$2s_n = \ldots, \frac{1}{4}, \frac{1}{2}, 1, 2, 4, 8, 16, \ldots$$

which is of course the same sequence. The factor of 2 is called the similarity factor or *scaling factor*. Obviously, together with 2 all integer powers of 2 are also scaling factors, including $\frac{1}{4}$, 8, and $\frac{1}{128}$. Of all those scaling factors, we shall call the smallest factor whose magnitude exceeds 1 the *primitive* scaling factor. In our example, 2 is the primitive scaling factor; 4 is not primitive. (And $\frac{1}{2}$, although capable of generating all scaling factors, is excluded by our definition because it is not greater than 1.)

In many practical applications, the self-similar sequence will be only one-sided; for example,

$$1, r, r^2, r^3, \ldots$$

This self-similar sequence is illustrated geometrically in Figure 10, where it is

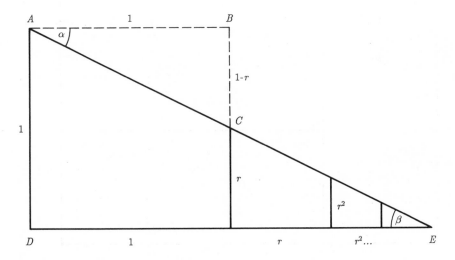

Figure 10   Proof by self-similarity [KB 88].

exploited in a lovely "look-see" proof, due to Benjamin Klein and Irl Bivens [KB 88], of the formula for the sum of geometric series:

$$1 + r + r^2 + r^3 + \cdots = \frac{1}{1-r} \qquad (|r| < 1) \tag{1}$$

Note that the two right triangles ABC and EDA are similar because the angle α equals β. Thus the ratios of corresponding sides must be the *same*. Specifically, the side ratio $\overline{DE}/\overline{DA} = 1 + r + r^2 + r^3 + \cdots$ must equal $\overline{BA}/\overline{BC} = 1/(1 - r)$. End of proof. (But how would the figure have to be drawn for $r < 0$?)

Another geometric proof of equation 1 relying on self-similarity, due to Warren Page, is shown in Figure 11 [Pag 81]. Starting with the unit square, we cut off a rectangle with side $0 < q < 1$. From the remaining rectangle we cut off a cascade of rectangles, each having an area smaller by a factor $1 - q$ than

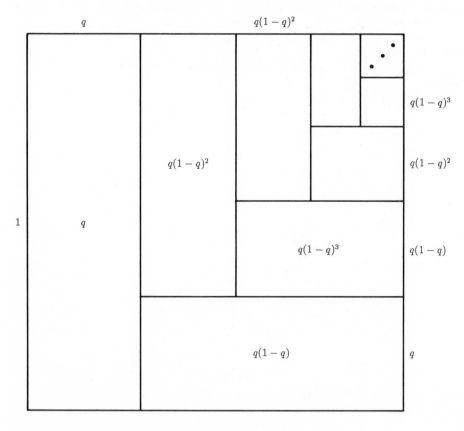

Figure 11   Another proof by self-similarity [Pag 81].

the preceding rectangle. Since all the rectangles together cover the original unit square, we have $q + q(1 - q) + q(1 - q)^2 + \cdots = 1$. Setting $1 - q = r$, we obtain equation 1.

Most readers, I am sure, are familiar with three-way light bulbs and benefit from the three different levels of brightness they offer: low, medium, and bright. Of course, most commercial bulbs accomplish this with just *two* filaments, consuming separately x watts and y watts, respectively; and $x + y$ watts when both are turned on together. For typical values, say, $x = 50$ watts and $y = 100$ watts, the bright condition (150 watts) unfortunately is not much brighter than the medium one (100 watts). It would be preferable if the three wattages formed a self-similar geometric progression; that is, for $y > x$, $(x + y)/y$ should equal y/x. The solution involves the golden mean $\gamma = (\sqrt{5} - 1)/2 = 0.618$. Indeed, for $x/y = \gamma$, $y/(x + y)$ will likewise equal γ.

Similarly, with three filaments, one can realize five different wattages that form a self-similar sequence if the third filament has a wattage $z = y/\gamma^2$. With $y = 100$ watts, for example, the self-similar wattages, rounded to the nearest integer watt, are $x = 62, y = 100, x + y = 162, z = 262$, and $x + y = 424$ watts.

Of course, the reader reading by the light of a multifilament bulb is not really interested in self-similar physical wattages; he is concerned with subjective *brightnesses*. Fortunately, over a fairly large brightness range, brightness B follows a simple power law, $B \sim W^\alpha$, as a function of wattage W. Since power laws are themselves self-similar (see Chapter 4), our choice of filaments is still the proper one even if we desire equal *brightness* ratios.

An amusing auditory paradox, based on a finite self-similar sequence of musical notes, was devised by Roger Shepard [She 64]. The paradoxical sound consists of a superposition of 12 notes with each note an octave higher in frequency than its lower-frequency neighbor. Beginning with 10 Hz, the other 11 frequencies in the composite sound are then 20, 40, 80, 160, 320, 640, 1280, 2560, 5120, 10,240, and 20,480 Hz. Increasing all 12 frequencies by a semitone (about 6 percent) will yield a sound with frequency components at 10.6, 21.2, . . . , 10,489, and 21,698 Hz, which will of course sound a bit higher in pitch (in fact, a semitone higher) because all frequencies have been increased by a semitone.

Increasing the frequencies by another semitone will result in a sound still higher in pitch. Repeating the process a few more times will give rise to further increases in pitch. But after 12 increases in pitch, the sound will be indistinguishable from the original sound! (The 10-Hz component present in the original stimulus and the extra component at 40,960 Hz are inaudible.)

By supplying a sufficient number of (inaudible)low-frequency components, Shepard was able to generate a succession of sounds whose pitches increase forever! With a personal computer, connected to a digital-to-analog converter, such sounds are now easy to generate and I encourage interested readers to subject themselves to this weird perceptual paradox. Figure 12 shows a self-similar waveform that has a constant pitch when all frequencies are doubled.

Figure 12   Paradoxical self-similar waveform. Frequency doubling leads to same pitch.

Weierstrass Functions and a Musical Paradox

The ever-ascending tones of Shepard are closely related to the nondifferentiable functions introduced by Karl Weierstrass (1815–1897) as examples of *continuous* functions that are nowhere differentiable. Such functions seemed to defy common sense and were hotly debated in the nineteenth century. A Weierstrass function is defined by

$$w(t) := \sum_{k=1}^{\infty} \alpha^k \cos (\beta^k t) \qquad \alpha \text{ real, } \beta \text{ odd}$$

Weierstrass showed that, for $\alpha\beta > 1 + 3\pi/2$, $w(t)$ is a continuous but nowhere differentiable function, such as the fractal Koch flake and Hilbert's space-filling curve, which we first encountered in Chapter 1. Like Cantor sets, nondifferentiable functions are a rich mine of paradoxes, such as Shepard's ever-ascending tones.

Another example of a musical chord patterned after a Weierstrass function can have the following weird property. If recorded on magnetic tape and replayed at *twice* the recording speed, the chord will not sound an octave higher in pitch, as every well-behaved recorded sound would, but a semitone *lower* [Ris 71, Ris 75, Schr 86]. How is this possible? Let us construct a finite-sum approximation to a Weierstrass function (we can omit the factors α^k that are needed to make the infinite series converge):

$$w_K(t) = \sum_{k=1}^{K} \cos (\beta^k t)$$

If we scale the time dimension t by a factor β, we obtain

$$w_K(\beta t) = \sum_{k=1}^{K} \cos (\beta^{k+1}t) = \sum_{k=2}^{K+1} \cos (\beta^k t)$$

That is, except for end effects, $w_K(\beta t)$ equals the unscaled function $w_K(t)$. Thus, $w_K(t)$ is approximately self-similar (see Figure 12, where $K = 7$ and $\beta = 2$).[4] Obviously, in the limit as $k \to \infty$, such a function cannot have a finite nonzero derivative anywhere, because derivatives change with scaling.

Now suppose we select $\beta = 2^{13/12}$ to give

$$w_K(t) = \sum_{k} \cos (2^{k(13/12)}t)$$

and make $w_K(t)$ audible as a sound. Then the frequencies $2^{k(13/12)}/2\pi$ have to cover only the audio frequency range (10 Hz to 20,000 Hz). Recording $w_K(t)$ on magnetic tape and playing it back at twice the speed produces

$$w_K(2t) = \sum_{k} \cos (2^{k(13/12)+1}t) = \sum_{k'} \cos (2^{k'(13/12)} \cdot 2^{-1/12}t)$$

where $k' = k + 1$. Now, if these summations cover the entire audio range, then, as far as the human ear is concerned,

$$w_K(2t) = w_K(2^{-1/12}t)$$

Thus, a doubling of the tape speed will produce a sound with a pitch lowered by a factor of $2^{1/12}$. In musical terms, the chord will sound one semitone *lower* rather than an octave higher. Such are the paradoxes engendered by fractals!

It is easy to program a personal computer to produce a $w_k(t)$ with 11 components comprising the frequencies from 10.0 Hz to 18,245.6 Hz. By doubling the playback speed, the sixth component, for example, will change in frequency from 427.15 Hz to 854.3 Hz. But in comparing the two chords, the human auditory system will identify the doubled sixth component at 854.3 Hz with the *nearest* component of the original chord, namely, the seventh at 905.1 Hz. Since 854.3 Hz is a semitone lower than 905.1 Hz and the same argument can be made for all frequency-doubled components, a pitch lowered by one semitone will be perceived.

---

4.   It is interesting to note that the concept of self-similarity entered mathematics at two independent points, Cantor sets and Weierstrass functions, at about the same time and for similar reasons: to elucidate two of the foundations of mathematics, numbers and functions. Even earlier, however, Leibniz had used the concept of self-similarity ("worlds within worlds") in his *Monadology* [Lei 1714] and in his definition of a straight line.

A frequency ratio $\beta = 2^{14/12}$ will produce a lowering by a full note upon tape speed doubling, and $\beta^{15/12}$ will lower the perceived pitch by three semitones, and so on. However, for $\beta = 2^{24/12} = 4$, the percept will be ambiguous, because when the numbers 1, 4, 16, 64, ... are doubled, the resulting sequence (2, 8, 32, ...) could be considered either one octave lower *or* one octave higher.

All physical objects that are "self-similar" have limited self-similarity—just as there are no perfectly periodic functions, in the mathematical sense, in the real world: most oscillations have a beginning and an end (with the possible exception of our universe,[5] if it is closed and begins a new life cycle after every "big crunch"; see the admirably equationless bestseller by Stephen Hawking, *A Brief History of Time* [Haw 88]). Nevertheless, self-similarity is a useful abstraction, just as periodicity is one of the most useful concepts in the sciences, any finite extent notwithstanding.

A mathematical object that is self-similar except for an "end effect" is the sinc function

$$\text{sinc } x := \frac{\sin \pi x}{\pi x}$$

which describes the wave diffraction pattern of a rectangular slit and plays an important role in interpolating sampled functions in discrete-time digital systems. (Here the slit is a rectangular "window" through which the function's spectrum is seen.)

By applying the trigonometric identity

$$\sin 2x = 2 \sin x \cos x$$

to the sinc function, we obtain

$$\text{sinc } x = \cos\left(\frac{\pi x}{2}\right) \frac{\sin (\pi x/2)}{\pi x/2} = \cos\left(\frac{\pi x}{2}\right) \text{sinc}\left(\frac{x}{2}\right)$$

Thus, except for the factor $\cos (\pi x/2)$, the sinc function is self-similar with a similarity factor of 2.

By repeating the factoring process, we obtain Euler's famous infinite product

$$\text{sinc } x = \cos\left(\frac{\pi x}{2}\right) \cos\left(\frac{\pi x}{4}\right) \cos\left(\frac{\pi x}{8}\right) \cdots$$

---

5.   A *universe is something that happens once in a while*—according to some of the latest physical phantasies.

The sinc function has zeros for all positive and negative integers x. The first factor in the Euler product produces the zeros of sinc x at all odd values of x. The second factor gives the zeros for x equal to twice an odd number. The third factor creates the zeros at x equal to 4 times an odd number, and so forth, giving the required zeros at all integer values of x except at $x = 0$. (Note that each nonzero integer n can be written, uniquely, in the form $n = 2^m k$, where k is an odd integer and $m \geq 0$.)

In many applications, self-similarity is only approximate; there may be statistical or deterministic "perturbations." Thus, the sequence (based on another well-known infinite product, namely, for $2/\pi$)

$$s_n = \prod_{k=1}^{n} f_k - \frac{2}{\pi}$$

with the recursion

$$f_{k+1} = \left(\frac{1 + f_k}{2} \right)^{1/2} \qquad f_0 = 0$$

is self-similar only in the limit as $n \to \infty$. For $n = 1, 2, 3, 4, \ldots$, we obtain

$$s_n = 0.070482,\ 0.01662,\ 0.004109,\ 0.0001024, \ldots$$

Its terms approach a constant scaling factor of 4. Such asymptotic self-similarities are often encountered in recursive computations.

In subsequent chapters we shall meet the other deviation from strict self-similarity: *statistical* self-similarity, in which the *statistical* laws that govern the object exhibit a scaling invariance. The object itself may change upon scaling, but its probabilistic aspects remain the same.

More Self-Similarity in Music: The Tempered Scales of Bach

The ancient Greeks, with their abundance of string instruments, discovered that dividing a string into two equal parts resulted in a pleasant musical interval, now called the *octave*. The corresponding physical frequency ratio is $2:1$.

"Chopping off" one-third of a string produced another pleasant interval, the *perfect fifth*, with a frequency ratio of $3:2$.

The Pythagoreans asked themselves whether an integral number of octaves could be constructed from the fifth alone by repeated application of the simple frequency ratio 3/2. In mathematical notation, they asked for a solution of the equation

$$\left(\frac{3}{2}\right)^n = 2^m$$

in positive integers n and m. But the fundamental theorem of number theory tells us that no positive power of 3 can equal a power of 2, that is, that the equation $3^n = 2^k$ has no integer solutions for $n > 0$.

However, the Greeks were not discouraged and, by trial and error, found an excellent approximate solution:

$$\left(\frac{3}{2}\right)^{12} \approx 2^7$$

which is based on the near equality of $3^{1/19}$ and $2^{1/12}$.

A systematic way of finding such near-coincidences is to write the ratio of the logarithms of the two integer bases (2 and 3) as a continued fraction:

$$\frac{\log 2}{\log 3} = [1, 1, 1, 2, 2, \ldots]$$

where the bracket notation is a convenient way to write the continued fraction

$$\cfrac{1}{1 + \cfrac{1}{1 + \cfrac{1}{1 + \cfrac{1}{2 + \cfrac{1}{2 + \cdots}}}}}$$

Continued fractions generally yield good rational approximations to irrational numbers; for example, $\pi \approx \frac{355}{113}$. This excellent approximation to π using not very large integers was known already to the ancient Chinese.

Breaking the foregoing continued fraction off after the fifth term (as shown) yields the excellent approximating fraction for the musical fifth:

$$\frac{\log 2}{\log 3} \approx \frac{12}{19}$$

from which follows $(\frac{3}{2})^{12} \approx 2^7$.

Another important fact here is that the exponents 12 and 7 are coprime, so that repeated application of the perfect fifth modulo the octave (the "circle of fifths") will not be close to a previously generated frequency until the twelfth step. These 12 different frequencies within an octave are all approximate powers of the basic frequency ratio $1:2^{1/12}$, the *semitone*. Thus, there is always some value of k for which

$$\left(\frac{3}{2}\right)^k \approx 2^{r/12} \qquad r = 1, 2, 3, \ldots$$

The solution of this approximate equation is $k \equiv r/7 \equiv 7r \bmod 12$. One-third of the octave, or $2^{1/3}$ ($r = 4$, frequency ratio ≈ 1.260), for example, is equivalent (modulo the octave) to $k = 4$ fifths (frequency ratio ≈ 1.266).

The third part of the octave, is also close to the pure major third (frequency ratio $5/4$). This is the lucky result of another, *independent*, number-theoretic near-coincidence, $5^3 \approx 2^7$, relating the next prime number above 3, namely 5, with the smallest prime, 2.

The fifth itself is approximated by seven semitone intervals with an accuracy of 0.1 percent: $2^{7/12} \approx 1.4983$. The resulting shortfall from the exact value 1.5 is called the *Pythagorean comma*. It is interesting to note that not only do seven semitones make one fifth, but, modulo the octave, seven fifths make one semitone.[6] This coincidence results from still another number-theoretic fluke, namely, that 7 is its own inverse modulo 12: to wit, $7 \cdot 7 = 49 \equiv 1 \bmod 12$.

To ensure that fixed-note instruments, such as the piano, can be played in many different musical keys, the frequencies of the different keys should be selected from the same basic set of frequencies. This led to the development of Bach's tempered scales, based on the semitone with a frequency ratio of $2^{1/12}$. A musical instrument tuned according to the tempered scale thus has frequencies approaching the following multiples of the lowest note:

$$1, 2^{1/12}, 2^{2/12}, 2^{3/12}, 2^{4/12}, 2^{5/12}, \ldots$$

up to some highest note.

Thus we see that the frequencies of a well-tempered instrument form a self-similar sequence, with the similarity factor $2^{1/12}$. If all these notes were sounded *simultaneously*, the instrument would produce an acoustic output (to put it mildly) that approximates a self-similar Weierstrass function. (Actual tuning of pianos differs from exact self-similarity, the tuning being somewhat "stretched" to minimize the beating of overtones, which are not precisely harmonic, as a result of the finite bending stiffness of the strings.)

For an excellent introduction to the science of musical sound see the book of that title by John R. Pierce [Pie 83].

---

6. Of course, seven fifths can also make one "semi*conscious*."

The Excellent Relations between the Primes 3, 5, and 7

John R. Pierce of communications satellite fame asked himself a few years ago whether one could not replace the frequency ratio 2:1 of the octave by the frequency ratio 3:1, which he called the *tritave*, and design a self-similar (equal-tempered) scale that matches frequency ratios constructed from the *next* two prime numbers, namely, 5 and 7 [MPRR 88]. In other words, is there some integral root of 3, $3^{1/N}$, such that $\frac{5}{3}$ and $\frac{7}{5}$ are well approximated by integer powers of $3^{1/N}$—in analogy to the good approximations $\frac{3}{2} \approx 2^{7/12}$ and $\frac{5}{3} \approx 2^{9/12}$ for Bach's well-tempered scale?

To answer this question in a systematic way, we have to expand the ratios log 3/log 5 and log 3/log 7 into continued fractions. This yields for the first ratio

$$\frac{\log 3}{\log 5} = [1, 2, 6, \ldots]$$

Breaking the continued fraction off after the 6 gives the following rational approximation:

$$\frac{\log 3}{\log 5} \approx \frac{13}{19}$$

or $3^{1/13} \approx 5^{1/19}$. This means that the basic frequency ratio $3^{1/13} = 1.088 \ldots$ is a good "semitone" for constructing a musical scale that closely matches, modulo the tritave, notes that are generated from the frequency ratio 5:3. In fact, $3^{6/13}$ equals $\frac{5}{3}$ within 0.4 percent!

But what about the frequency ratio 7:3? Continued fraction expansion of log 3/log 7 gives the excellent approximation $3^{1/13} \approx 7^{1/23}$. *Again* $3^{1/13}$ emerges as the preferred basic interval to construct the well-tempered Pierce scale— another number-theoretic fluke! In fact, the match between $\frac{7}{3}$ and a power of $3^{1/13}$, namely, $3^{4/13}$, is uncannily close: the difference is but 0.16 percent. The resulting frequency ratios again form a self-similar sequence:

$$1, 3^{1/13}, 3^{2/13}, 3^{3/13}, \ldots$$

But while the number theory of this new musical scale may be nearly perfect, its compositional value is a matter of taste and open to debate.[7]

---

7.   This patient listener, who has served as a subject in musical tests involving the Pierce scale, showed a stubborn preference for compositions written in well-established traditional scales.

4

C H A P T E R

Power Laws: Endless Sources of Self-Similarity

What really interests me is whether God had any choice in the creation of the world.

—ALBERT EINSTEIN

Similarity not only reigns in plane geometry, as in Einstein's triangles (see Chapter 1), but underlies much of algebra too. Think of a homogeneous power function such as

$$f(x) = cx^\alpha$$

where c and α are constants. For example, for $\alpha = 1$ we get the special case $f(x) = cx$, which, for $c < 0$, describes the restoring force of a linear spring; for $\alpha = -2$ (and c still negative) we get Newton's law of gravitational attraction $f(x) = cx^{-2}$. Such simple power laws, which abound in nature, are in fact self-similar: if x is rescaled (multiplied by a constant), then $f(x)$ is still proportional to x^α, albeit with a different constant of proportionality. As we shall see in the rest of this recitation, power laws, with integer or fractional exponents, are one of the most fertile fields and abundant sources of self-similarity.

The Sizes of Cities and Meteorites

Many objects that come in different sizes have a self-similar power-law distribution of their relative abundance over large size ranges. This is true for objects that have grown, like cities, as well as for objects that have been shattered, like crushed stones [Mek 90]. The only prerequisite for a self-similar law to prevail in a given size range is the absence of an inherent size scale. (Of course, no earthly "city" has fewer than 1 inhabitant or more than 1 billion, and no stone on earth is smaller than an atom or larger than a continent.)

One of the best behaved shattering mechanisms occurs not on earth but in outer space: the mean frequency with which different kinds of interplanetary debris (shooting stars or meteors) slam into the earth's atmosphere is inversely proportional to the squared diameter of the projectile, and this is true over 10 orders of magnitude (see Figure 1). Whereas the space shuttle is hit at a rate of one particle every 30 microseconds (10^{12} particles per year) with a diameter under 1 micrometer (μm), meteorites of the size that created the Arizona crater, with a diameter of 100 meters or more, are expected (thank heavens!) only once every 10^4 years. And the next "shooting star" of the size that hit Sudbury, Ontario, with an "astronomical" 10,000-meter diameter, should not rock the earth for another 10^8 years.

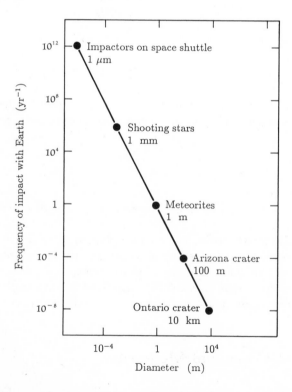

Figure 1   Frequency of meteor collisions with earth in relation to particle diameter. (After E. Schoemaker, U.S. Geological Survey.)

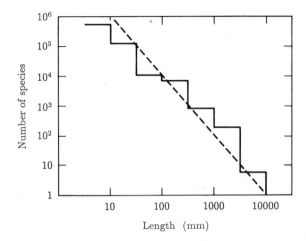

Figure 2 Number of species of terrestrial animals in relation to individuals' lengths [May 88].

Luis Alvarez, whom we already encountered chasing submarines during World War II (see pages 33–34 in Chapter 1), and his associates used the data in Figure 1 to help support his theory of the sudden disappearance of the dinosaurs 65 million years ago. According to Alvarez, the impact of a large meteorite kicked up a lot of sunlight-blocking dust, thereby depriving the dinosaurs of the greenery necessary for their survival [AAMA 82].

Speaking of dinosaurs, one is reminded of the sizes of animals and their distribution. Figure 2 shows the estimated distribution of the number of species of terrestrial animals as a function of their length. Again we find a power law with an exponent of −2 over four orders of magnitude, from 1 millimeter to 10 meters [May 88].

A Fifth Force of Attraction

One of the early laws of physics that shows self-similarity resulted from Galileo's observation that large stones and small stones dropped from the Leaning Tower of Pisa fall with (nearly) equal speed.[1] In fact, ignoring aerodynamic drag, the

---

1. Actually, Galileo used rolling balls on inclined plains, but the Leaning Tower story has taken on a life of its own.

falling time t is simply proportional to the square root of the dropping height h, independent of the stone's mass: $t \sim h^{1/2}$, a law that is independent of scale—or nearly so, because when one goes to astronomical heights (or even up a tall mountain), the gravitational pull of the earth is diminished. Thus, there *is* a natural length scale that limits the scale invariance or self-similarity of Galileo's $t \sim h^{1/2}$: the radius of the earth. Self-similarity is only *approximate*, a situation that we shall encounter again and again: self-similarity reigns supreme, but only over bounded domains.

In fact, even Newton's just mentioned universal law of gravitation—in full, $f(x) = GMx^{-2}$, where G is the gravitational constant and $f(x)$ is the attractive force of a mass M at distance x—is being called into question these days (even before the advent of a theory of quantum gravity, which will fuse Newton's G and Planck's h). A reexamination of the old attraction data and careful new force measurements seem to have revealed a *nonscaling* correction to Newton's law, called the "fifth force" by some imaginative minds. (The other four forces are gravity, electromagnetism, and the weak and strong nuclear interactions.) The fifth force, whatever its origin and if it is real, appears to depend on distance x as $\exp(-x/x_c) \cdot x^{-2}$, which is *not* a homogeneous power law and therefore not self-similar: it contains a characteristic cutoff length, x_c, as it must because the exponential function calls for a dimensionless argument.

What is the significance of x_c, whose order of magnitude is 100 m? Forces in modern physics are mediated by *particles*, with a rest mass equal to h/cx_c, where h is again Planck's constant and c is the velocity of light. For forces with an infinite range ($x_c = \infty$), as for electromagnetic fields, the rest mass is zero, which is believed to be the case for the electromagnetic particle, commonly called the photon. In fact, efforts to establish an upper limit for the rest mass of the photon focus on the *range* of the electromagnetic field. And the same is true for the rest mass of the neutrino—with potentially enormous consequences for the total weight (and the final fate) of our universe if the neutrino's rest mass should turn out to be nonzero.

The importance of scaling (or rather *nonscaling*) with distance was never as dramatically demonstrated as when the Japanese physicist Hideki Yukawa (1907–1981), in the 1930s, concluded from the finite cutoff length of nuclear forces ($x_c \approx 10^{-14}$ m) that a particle, called the *meson*, must exist with a mass of approximately 240 electron masses. A bit later, such a particle was indeed found (in the cosmic showers of particles that "rain" down on earth), but it turned out to be just a heavy brother to the common electron, now called the *muon* (with a mass 207 times that of the electron).[2] Thus, the search for Yukawa's hypothetical meson continued until it *was* found, weighing in at 270 electron masses and baptized pi-meson or *pion*.

---

2. "Who ordered *that*?" as Isidor Rabi (1898–1987) asked, in an often quoted question, when this completely "unnecessary" particle was first brought to his attention.

Now what particle goes with the fifth force? With $x_c \approx 100$ m, its mass would have to be less than 10^{-13} of the electron's mass, which is already extremely small. Perhaps the apparent range of the fifth force is the result of two (or more) new forces that almost cancel each other. Or maybe there is no new force at all, as mass scaling and the latest experimental results would suggest [TKFFHKMM 89, BBFT 89, JER 90].

Free of Natural Scales

As we said, homogeneous functions have an interesting scaling property: they reproduce themselves upon rescaling. This scaling invariance can shed light into some of the darker corners of physics, biology, and other sciences, and even illuminate our appreciation of music.

Scaling invariance results from the fact that homogeneous power laws lack natural scales; they do not harbor a characteristic unit (such as a unit length, a unit time, or a unit mass). Such laws are therefore also said to be *scale-free* or, somewhat paradoxically, "true on all scales." Of course, this is strictly true only for our mathematical models. A *real* spring will *not* expand linearly on all scales; it will eventually break, at some characteristic dilation length. And even Newton's law of gravitation, once properly quantized, will no doubt sprout a characteristic length.[3]

This concept of something happening *on all scales* (another much liked locution) is one of our central themes. In fact, it is said (and Mandelbrot was perhaps the first to say it emphatically enough) that for mountainscapes to be interesting they must have features (cliffs, crevices, peaks, and valleys) *on many length scales*. And music, written in any scale, to be appealing, had better have pitch changes on many frequency scales and rhythm changes on more than one time scale. This is in fact how the Bachs (J. S. et al.) composed their music, although they never said so.

Bach Composing on All Scales

When Johann Sebastian Bach (1685–1750) composed his *Brandenburg* Concertos he was, unwittingly no doubt, using homogeneous power functions in the selection of his notes [VC 78]. The power spectrum (the squared magnitude of the

---

3.   Heisenberg heralded the appearance of a new constant of nature, a characteristic *length*, in the basic laws of physics more than 50 years ago, but it still has not happened. (The Planck length, $(G\hbar/c^3)^{1/2} \approx 10^{-35}$ m, which governed the "big bang" that may have created our universe, is but a derived entity.)

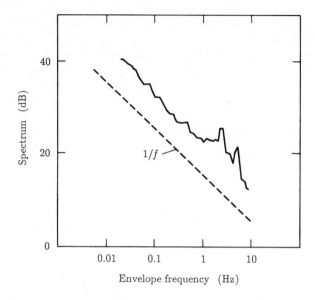

Figure 3   Spectrum of amplitude variations for Bach's First Brandenburg Concerto [VC 78].

Fourier transform) $f(x)$ of the relative frequency intervals x between successive notes can be approximated over a large range by a homogeneous power function with an exponent of -1:

$$f(x) = cx^{-1}$$

which is also called a *hyperbolic law* (because its plot looks like a hyperbola). Taking logarithms yields

$$\log f(x) = \text{const} - \log x$$

where x is in semitones. Thus, on a doubly logarithmic plot, $\log f$ versus $\log x$, the data follow a straight line with a slope of $-45°$.

Not only does the spectrum of the frequency intervals follow a homogeneous power law, but the spectrum of the *amplitudes* (instantaneous "loudness") of Bach's music follows a homogeneous power law with the same exponent (see Figure 3). The amplitude of the music is obtained by temporal smoothing[4] of

---

4.   The smoothing, which could falsify the result, can be circumvented by taking Hilbert transforms and computing the so-called *Hilbert envelope*. The Hilbert envelope of a function is defined as the envelope of the *family* of functions generated by phase-shifting the Fourier transform of the given function through all angles from 0 to 2π.

the magnitude of the sound pressure recorded near the orchestra. Note that the time scale for these amplitude or envelope variations extends to 100 s, corresponding to 0.01 Hz.

Why is it that Bach should have chosen the simple hyperbolic power law when composing his music? Well, first one has to say that he (and countless other composers) did nothing of the kind. Composers compose to create interesting music. So the real question should perhaps be, Why does (at least some) interesting music have hyperbolic spectra for frequency intervals and amplitudes?

Birkhoff's Aesthetic Theory

A partial answer may come from the *"theory of aesthetic value"* propounded by the American mathematician George David Birkhoff (1884–1944). Birkhoff's theory, in a nutshell, says that for a work of art to be pleasing and interesting it should neither be too regular and predictable nor pack too many surprises. Translated to mathematical functions, this might be interpreted as meaning that the power spectrum of the function should behave neither like a boring "brown" noise, with a frequency dependence f^{-2}, nor like an unpredictable white noise, with a frequency dependence of f^0.

In a white-noise process, every value of the process (e.g., the successive frequencies of a melody) is completely independent of its past—it is a total surprise (see Figure 4A). By contrast, in "brown music" (a term derived from Brownian motion), only the *increments* are independent of the past, giving rise to a rather boring tune (see Figure 4B). Apparently, what most listeners like best, and not only in Bach's time, is music in which the succession of notes is neither too predictable nor too surprising—in other words, a spectrum that varies according to f^α, with the exponent α between 0 and -2. As Richard Voss discovered, the exponents found in most music are right near the middle of this range: $\alpha = -1$, giving rise to the hyperbolic power law f^{-1} (see Figure 4C) [VC 78]. Or, as Balthazaar van der Pol once said of Bach's music, "It is great because it is inevitable [implying $\alpha < 0$] and yet surprising [$\alpha > -2$]." (I found this quotation in Marc Kac's captivating autobiography, *Enigmas of Chance* [Kac 85].)

Figure 5B shows a sample of a noise waveform with hyperbolic power spectrum, f^{-1}. Such time functions are also called *pink* noise, because they are intermediate between brown(ian) (f^{-2}) and white (f^0) (see Figure 5C and 5A, respectively). Since the power spectrum of any noise that obeys a homogeneous power law (f^α) is self-similar, the underlying waveform must likewise be self-similar. In fact, if the frequency axis of the power spectrum is scaled by a factor r, then, by the law of Fourier reciprocity, the time axis of the corresponding

Figure 4 (A) "White" music produced from independent notes; (B) "brown" music produced from notes with independent increments in frequency; and (C) "pink" music—frequencies and durations of notes are determined by $1/f$ (pink) noise [VC 78].

waveform is scaled by $1/r$. Of course, in the case of noise (and other probabilistic phenomena), the self-similarity is only *statistical*; a magnified excerpt is not an exact, deterministic replica of the unscaled waveform.

Also, to preserve power when rescaling frequencies, amplitudes should be adjusted by a factor $r^{-\alpha/2}$. Strictly speaking, such stochastic processes are therefore

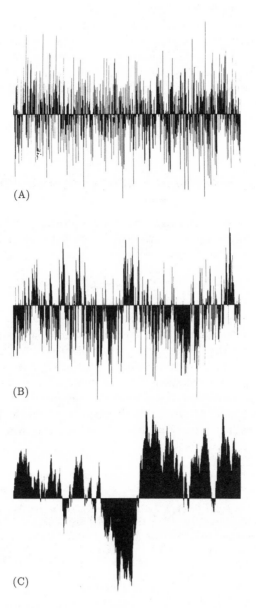

Figure 5 Sample of (A) white noise with f^0 power spectrum; (B) "pink" noise with $1/f$ power spectrum; and (C) "brown" noise with $1/f^2$ power spectrum.

self-affine—that is, they have more than one scaling factor: r for frequencies (or, equivalently, $1/r$ for times) and $r^{-\alpha/2}$ for amplitudes.

In fact, a pink (or white, or brown) noise is the very paradigm of a statistically self-similar process. Phenomena whose power spectra are homogeneous power[5] functions lack inherent time and frequency scales; they are *scale-free*. There is no characteristic time or frequency—whatever happens in one time or frequency range happens on *all* time or frequency scales. If such noises are recorded on magnetic tape and played back at various speeds, they sound the same—unlike a human voice, which sounds like the cartoon character Donald Duck when played at twice the tape speed. There are even self-similar tones that go *down* in pitch when the tape speed is doubled (see pages 96–98 in Chapter 3). We shall further explore noises of different colors in Chapter 5.

Heisenberg's Hyperbolic Uncertainty Principle

Hyperbolic laws are so widespread that they are often not even recognized as such, especially when they are written as a product that equals a constant. A case in point is Heisenberg's famous uncertainty relation of quantum mechanics:

$$\Delta q \cdot \Delta p \geq \hbar$$

where q and p are two (quantum mechanically) "canonical conjugate" variables such as position and momentum or energy and time, and where $\hbar$ is Planck's constant (divided by 2π).[6] The uncertainty principle says that the smaller the error (in the sense of a statistical standard deviation) in one variable, the greater the error in the conjugate variable. Thus, if someone wants to determine, in a *Gedanken* (thought) *experiment*, the position of, say, an electron very accurately, he has to use photons (light particles) of very short wavelengths, that is, very high momentum, which, after bouncing off the poor electron, leave it in a state

---

5. Note the double duty that the noun *power* is serving here: power as an exponent, as in *third power*; power as a force, as in *third-rate power* or *nuclear power* (both physical and political)

6. "What *is* this thing called 'h bar'?"—a whispered question overheard at a recent physics conference (between two metallurgists?). It seems that Planck's "quantum of action" needs more time (or energy) to penetrate universal consciousness. In fact, sad to say, this can even be said of some professional physicists. This patient chair of the Göttingen General Physics Colloquium was once witness at a talk in which the speaker (quite matter-of-factly) mentioned that a (camera) shutter chopping up a stream of Mössbauer gamma quanta in time would (of course) widen their energy spectrum. Whereupon one of the *experts* in the field (of Mössbauer spectroscopy!) objected that the shutter, moving only laterally to the quanta, could not *possibly* alter their energy. Well, as Richard Feynman said in 1967, "I think it is safe to say that no one understands quantum mechanics!"

of highly uncertain momentum. This statement is often phrased more qualitatively as "any observation disturbs the system to be observed."

But, however phrased, the uncertainty principle is nothing but a consequence of the well-known reciprocal scaling relationship predicted by Fourier transformation theory for a pair of Fourier variables. (However, the fact that Fourier transformation—and the Hilbert spaces—are the proper domains for quantum systems is anything but trivial; it is one of the deeper insights into the makings of nature so far afforded the human mind.)

The greatest possible accuracy in p and q is achieved if both variables are distributed according to Gauss's normal law of probability, in which case the equal sign in the preceding inequality holds:

$$\Delta q = \frac{\hbar}{\Delta p}$$

Although this minimal-uncertainty relation is not usually thought of as a hyperbolic law, once we have written it as such, we can ask, What is its range of validity? The answer is perhaps one of the more awesome in all of physics: although tested and retested over vast ranges of energy, time, position, and momentum, never has the sligh*test* violation of $\Delta q \geq \hbar/\Delta p$ been found. There is no doubt in the mind of physicists that uncertainty, like relativity, is of an absolutely fundamental nature that admits no exceptions. Theories may come and go, but "h bar" will always be with us.

One of the fundamental consequences of uncertainty is the very size of atoms (which, without it, would collapse to an infinitesimal point). In fact, we can calculate the radius of the lightest atom, hydrogen, directly from the uncertainty relation: the potential energy U of the atom is proportional to the reciprocal of the radius. Thus, making the radius smaller increases the magnitude of the potential energy. By the virial theorem (see pages 66–68 in Chapter 2), this engenders a proportional increase in the atom's kinetic energy T, which means that the atom would have to *increase* in size. The minimum or *Bohr radius*, r_{Bohr}, is given by the uncertainty relation with the equal sign, $\Delta x \cdot \Delta p = \hbar$: we simply identify r_{Bohr} with Δx and the momentum corresponding to the kinetic energy with Δp. Using the exact relations for these two energies ($U = e^2/r4\pi\varepsilon_0$ and $T = p^2/2m_e$, where e and m_e are the charge and mass of the electron and ε_0 is the permittivity of the vacuum) yields, in one fell swoop, without the usual extensive calculations, the correct value for the Bohr radius. This radius, also called the *atomic unit of length*, equals, in terms of four fundamental physical constants,

$$r_{Bohr} = \frac{4\pi\varepsilon_0\hbar^2}{m_e e^2}$$

or $5.3 \cdot 10^{-11}$ m. Thus, the diameter of the hydrogen atom, $2r_{Bohr}$, equals approximately 10^{-10} m, or 1 angstrom (Å).

The immense range of validity of the uncertainty relation is impressively illustrated by the large *coherence length* of laser light, which is of great importance in holography and refined tests of relativity. With a relative energy uncertainty $\Delta E/E = 10^{-7}$, the "length uncertainty" Δx (i.e., the coherence length) for a helium-neon laser with a wavelength $\lambda = 630$ nm becomes $\Delta x = \lambda E/\Delta E \cdot 2\pi$, or about 1 m—a *macroscopic* length more than 10 powers of 10 larger than the Bohr radius derived from the same principle.

In Mössbauer spectroscopy, we deal with even smaller relative energy uncertainties (10^{-14} or less), yet Heisenberg's principle still holds.

While the coherence time (time uncertainty) Δt for laser light, say, 10^{-9} s, is perhaps not very long, *neutron* spectroscopy makes energy measurements with an incredibly small uncertainty of $\Delta E = 2 \cdot 10^{-36}$ watt-seconds (W · s) possible, because of the macroscopic coherence time for neutron waves of 50 seconds! Again, the hyperbolic uncertainty relation is still firmly entrenched, spanning more than 10 orders of magnitude.

Perhaps the most astounding consequence of uncertainty is the excommunication of nothingness, innocently called *vacuum*, from our worldview. A classical vacuum contains neither matter nor energy. But *zero* energy would be a precise value, and that is forbidden by uncertainty. Thus, the modern, quantum mechanical vacuum has finite energy fluctuations as dictated by Heisenberg's prescription—just as the finite size of the smallest atom in its state of lowest energy is prescribed by the same law.

The reality of vacuum fluctuations is now an integral part of quantum physics, with numerous consequences that are testable with great precision, such as the hyperfine structure of atomic spectra. There are even creditable theories of the origin of our universe as a vacuum fluctuation run amok [Haw 88]. As Thomas Cranmer, archbishop of Canterbury, wrote as early as 1550, "Naturall reason abhorreth vacuum."[7]

For all we know, the range of validity of the uncertainty relation is unlimited. On the other hand, for any homogeneous power law representing energy as a function of frequency (called a *spectrum*), there must, of course, be an upper or a lower limit (or both) beyond which the homogeneous power law cannot hold. White noise, for example, has a flat power spectrum only up to some, possibly very high, frequency. And pink noise must have an upper ("ultraviolet") *and* a lower ("infrared") transition frequency, possibly far apart, beyond which the hyperbolic law breaks down because otherwise the total power (the integral over the spectrum) would be infinite. (Of course, physicists use such laws anyhow—

---

7.   This authentic citation, recalled in the *Oxford English Dictionary* [Bur 87], is not to be confused with a more modern quote, circulated by John Robinson Pierce on the occasion of the transistor's birth (which he so baptized in 1948): "Nature abhors vacuum tubes."

since they scale so well—and then they complain of ultraviolet and infrared "catastrophes.")

Fractional Exponents

Power laws are not restricted to integer exponents as in white, pink, and brown noises. In fact, *fractional* exponents abound in nature. After all, self-similarity prevails for integer and noninteger exponents alike. And not infrequently a fractional exponent contains an important clue to the solution of an intricate puzzle. Often such exponents seem to be the same in rather different situations (such as melting or magnetism, for instance), providing a hint of similar underlying "universal" mechanisms.

A simple example of the appearance in nature of a self-similar law with a fractional exponent is the relation between the radiation density ρ_r and matter density ρ_m in the expanding universe which prevailed shortly after its creation:

$$\rho_r \sim \rho_m^{4/3}$$

From this simple relation Alpher and Herman were able to calculate, back in 1948, the present radiation density of the universe [AH 48]. Given other, albeit defective, data, they predicted a blackbody radiation—a remnant from the big bang that gave birth to the universe and is now bathing it (like a baby?) in a "warm" background—corresponding to a temperature of about 5 kelvins (degrees centigrade above absolute zero). Later Arno Penzias and Robert Wilson, while "tuning" microwave antennas, found this cosmic background radiation with a temperature of 2.7 degrees, and received the 1978 Nobel Prize for their discovery of this ancient "footprint" of the early universe.

In subsequent chapters we shall encounter other simple power laws with fractional exponents showing scaling invariances with far-reaching consequences in a wide variety of real-world situations, from the floods of the Nile to the gambler's ruin and the distribution of galaxies in the universe. In fact, in a surprising number of instances, complicated functions of two or more variables exhibit simple power-law behavior near "critical points." Thus, the function of two variables $f(x, y)$ can very often be written in the generic form

$$f(x, y) = x^\alpha g(y/x^\beta)$$

in which $f(x, y)$, has been replaced by a function of only one variable, g. For any range of the variables over which g is relatively constant, $f(x, y)$ is then approximated by a simple power law in x.

This kind of representation, in terms of power laws and their exponents, is enormously fruitful in the analysis of critical phenomena from percolation (see Chapter 15) to ferromagnetism and superconductivity.

The Peculiar Distribution of the First Digit

Power laws, or relations of the form $f(x) \sim x^\alpha$, lead to skewed nonuniform distributions of the first (leftmost) digit when the self-similar data are listed numerically. For the exponent $\alpha = -1$, the probability p_m that the most significant digit equals $m > 0$ is given by

$$p_m = \log_b \left(1 + \frac{1}{m} \right)$$

where b is the base of the number system used [Pin 62, Rai 76]. For decimal data, these probabilities are approximately $p_1 = 0.301$, $p_2 = 0.176$, $p_3 = 0.125$, $\ldots, p_9 = 0.046$. Note that $p_2 + p_3$ equals p_1, as it must because the digits 2 and 3 cover the same *relative* data range as the digit 1.

These probabilities p_m, which favor the digit 1 as the leftmost digit, are obtained by integrating x^α, with $\alpha = -1$, from m to $m + 1$. For other values of α this integration yields

$$p_m = \frac{m^{\alpha+1} - (m+1)^{\alpha+1}}{1 - b^{\alpha+1}} \qquad \alpha \neq -1$$

where b is again the base of the number system in which the self-similar data are expressed. For example, for $b = 10$ and $\alpha = -2$, $p_1 = 0.\overline{5}$, $p_2 = 0.\overline{185}$, $p_3 = 0.0\overline{925}, \ldots, p_9 = 0.\overline{012345679}$.

Real-world data are of course never *exactly* scale-invariant, if only because of "end effects." No living village has fewer than 1 inhabitant or more than 100 million, say—except the proverbial "global village." For recent results on leading digits, see the paper by Diaconis [Dia 77].

There are of course plenty of nonscaling "data," such as telephone numbers and the digits on automobile license plates, for which our skewed distributions of digits do not hold.

Skewed distributions of *numbers* (rather than digits) are also known from continued fractions. According to Carl Friedrich Gauss (1777–1855), for most irrational numbers, the asymptotic probability that a given term a_n in the continued fraction expansion equals k is given by the following expression:

$$\text{prob } (a_n = k) = \log_2 \frac{(k+1)^2}{k(k+2)} \qquad n \to \infty$$

The exceptional irrational numbers, such as the golden mean $\gamma = (\sqrt{5} - 1)/2$, for which $a_n = 1$ for all n, have measure zero.

Again, much as in the case of self-similar power laws, the number 1 has the greatest probability: prob $(a_n = 1) \approx 0.415$.

In Chapters 5 and 6 we shall further explore the connection between power laws and statistics.

The Diameter Exponents of Trees, Rivers, Arteries, and Lungs

Consider a tree trunk of diameter d bifurcating into two main branches with diameters d_1 and d_2. Is there any consistent relationship between these diameters as one moves up the tree to branches bifurcating into subbranches and subbranches bifurcating into twigs and so on up to the leaf-bearing stems?

Leonardo da Vinci argued that for the sap to be able to flow unimpeded up the tree, the combined cross-sectional areas of the two main branches must equal that of the trunk [RI 57]. In other words, Leonardo believed that $d^2 = d_1^2 + d_2^2$. This claim has stood the test of time and is now enshrined in the "pipe model" of biological tree design [Zim 78]. The pipe model rests on the mental image of the sap being carried up the tree from roots to leaves by many nonbranching vessels ("pipes"), which occupy a fixed proportion of the cross section of each branch.

The same relationship, namely,

$$d^\Delta = d_1^\Delta + d_2^\Delta \tag{1}$$

with $\Delta = 2$, holds for the confluence of two rivers, where d, d_1, and d_2 are the river widths. In fact, the width d of a river is found proportional to the square root of the quantity of water Q transported by the river: $d \sim Q^{0.5}$ [Leo 62]. But the depth t of a river typically varies only as $Q^{0.4}$. The resulting slack is taken up by an increase in the water's velocity v, which is found to be proportional to $Q^{0.1}$. In other words, a river having two tributaries of equal size and thus carrying twice the volume of water per second is typically 1.4 times wider but only 1.3 times deeper than one of its tributaries. But its water velocity is about 1.1 times higher than that of the tributaries. Of course, 1.1 times 1.3 times 1.4 equals very nearly 2, as it should if no water is lost or is added at the confluence.

As Mandelbrot has pointed out, it is impossible to estimate the scale of a map if all river widths are shown to scale. And if the meanders of the river are also self-similar, the course of the river, too, contains no clue to the map's scale.

By contrast, converging or bifurcating roads, which, unlike rivers, possess no depth, should have widths that scale according to equation 1 with an exponent $\Delta = 1$ provided that the traffic flow in cars per lane and minute is the same on all roads. Here the traffic lanes play the same role as the sap pipes in the pipe model for trees.

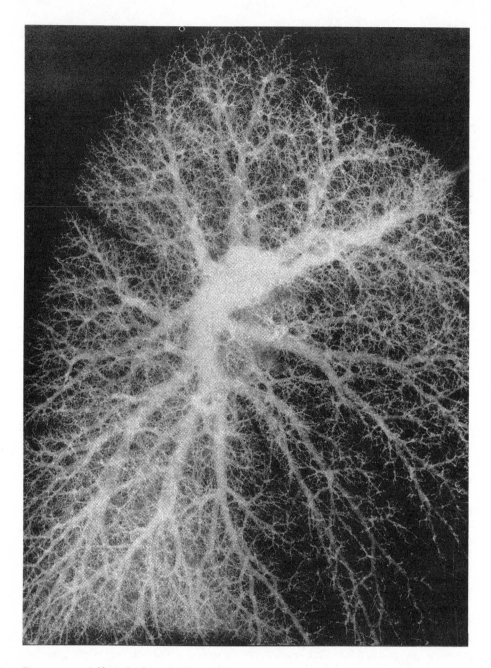

Figure 6   Self-similar bronchial tree [Com 66].

Arteries and veins in mammalian vascular systems, too, have been found to obey the scaling law (equation 1) over a range of 20 bifurcations between heart and capillaries. Estimates of the exponent Δ give values near 2.7 [ST 71]. This is a reasonable value for biological evolution to have attained, given the requirement that arteries and veins should come close to every point of the body that needs nourishment and waste disposal. But the ideal value $\Delta = 3$ for this purpose is, of course, unattainable, because a space-filling vascular system leaves too little tissue for other tasks.

By contrast, the bronchi of the lung do attain a scaling exponent very close to $\Delta = 3$, the value for a fractal that fills three-dimensional space. In fact, the bronchial tree is nearly self-similar over 15 successive bifurcations (see Figure 6 [Com 66]).

The exponent $\Delta = 3$ can be derived from assuming that the geometry of the bronchial tree is determined by the least possible resistance to airflow in the entire bronchial system [Tho 61]. This implies a fixed branching ratio of $d/d_1 = d/d_2 = 2^{1/3}$. With equation 1, the exponent Δ must therefore equal 3 [Wil 67].

However, Mandelbrot has a much more convincing argument, which does not require the branching ratio d/d_1 to be encoded genetically [Man 83]. Rather, Mandelbrot assumes a simple self-similar growth process during the prenatal stage of lung developments: "The growth starts with a bud, which grows into a pipe, which forms two buds, each of which behaves as above."

Iteration of these rules results in a self-similar tree structure for the lung. Thus, the empirically observed self-similarity is obtained not because it is optimum but as a result of the shortest growth-governing program: each step repeats the previous one on a smaller scale. The lung's geometry is therefore fully determined by two parameters: the width/length ratio of the branches and the diameter exponent Δ. In this model, the value $\Delta \approx 3$ simply results from the fact that a large number of bifurcations should be nearly space-filling without crowding each other out.

5

N oises: White, Pink, Brown, and Black

And the rain was upon the earth
forty days and forty nights.

—GENESIS 7.12

Earnings momentum and visibility should
continue to propel the market to new highs.

—E. F. HUTTON, the Wall Street brokerage
firm, from a forecast issued on October 19, 1987,
moments before the stock market plunged

Among the many domains where self-similar power laws flourish, statistics ranks very high. Especially, the *power spectra* (squared magnitude of the Fourier transform) of statistical time series, often known as *noises*, seem addicted to simple, homogeneous power laws in the form $f^{-\beta}$ as functions of frequency f. Prominent among these is *white noise*, with a spectral exponent $\beta = 0$. Thus, the power spectrum of white noise is independent of frequency. But white noise, that is, a noise with a constant or flat power spectrum, is a convenient fiction—a little white lie. Just like white light (whence the name *white* noise), the spectrum of white noise is flat only over some finite frequency range. Nevertheless, white spectra provide a supremely practical paradigm, modeling untold processes across a wide spectrum of disciplines. The increments of Brownian motion and numerous other *innovation processes*, the learned name for a succession of surprises, belong to this class. Electronic and photonic shot noises, thermal noise, and many a hiss from man or beast aspire to membership in the white noise "sonority."

If we integrate a white noise over time, we get a "brown" noise, such as the projection of a Brownian motion onto one spatial dimension. Brown noise has a power spectrum that is proportional to f^{-2} over an extended frequency range. Some of the paradoxical consequences of such processes, such as the gambler's bad fortune, and the "dusty" consistency of their isosets (sets of constant capital), will be discussed in Chapter 6. However, white and brown noises are far from exhausting the spectral possibilities: between white and brown there is *pink* noise with an f^{-1} spectrum. And beyond brown, *black* noise lurks, with a power spectrum proportional to $f^{-\beta}$ with $\beta > 2$. Figure 5 on page 111 in Chapter 4 showed waveforms of white, pink, and brown noise; Figure 1 shows a waveform of black noise with $\beta = 3$.

As it turns out, both pink and black noises are widespread. Pink processes make their appearance in many physical situations and have surprising aesthetic implications in music and other arts.

Black-noise phenomena govern natural and unnatural catastrophes like floods, droughts, bear markets, and various outrageous outages, such as those of electrical power. Because of their black spectra, such disasters often come in clusters. Indeed, "Wyse men sayth . . . that one myshap fortuneth never alone"; so says A. Barclay in his translation of *The Ship of Fools* [Bar 1509].

All of these phenomena share an important trait: their power spectra are *homogeneous power* functions of the form $f^{-\beta}$ over some respectable range of frequencies, with the exponent β running the gamut from 0 to 4.

Such homogeneous spectra, and the space or time records from which they result, exhibit a simple scaling invariance: if such a process is compressed by a constant scale factor s, the corresponding Fourier spectrum is expanded[1] by the reciprocal factor $1/s$. However, changing the frequency scale by any constant factor does not change the frequency *dependence* for power-law spectra; they keep their form. This can be nicely demonstrated acoustically: when such processes (properly time-scaled to fall into the audio frequency range) are recorded on magnetic tape and played back at a higher or lower tape speed, they do not sound higher or lower in "pitch"; apart from a change in volume, they sound the same! Thus, such spectra are self-similar and the underlying processes are statistically self-similar or self-affine.

Pink Noise

Pink noise, also called $1/f$ noise, has equal power in octave frequency bands or any constant intervals on a *logarithmic* frequency scale. This is a desirable attribute in many applications. For example, pink noise is a favorite test signal in hearing research and acoustics in general because it approximates many naturally occurring noises. Pink noise also has the approximate property of exciting equal-length portions of the basilar membrane in our inner ears to equal-amplitude vibrations, thus stimulating a constant density of the acoustic nerve endings that report

---

1. This is a fundamental property of the Fourier transform, which underlies the uncertainty principle (see pages 112–114) and many other facts of physics that don't wear such a nice label.

Figure 1   Waveform of "black" noise with $\beta = 3$.

sounds to the brain (see pages 85–86 in Chapter 3). Pink noise is therefore the *psychoacoustic* equivalent of white noise.

Pink noise is also encountered in a wide variety of physical systems, including semiconductor devices. One possible reason for the ubiquity of $1/f$ noises is their genesis through parallel relaxation processes, which abound in nature [Agu 76]. In a relaxation process (think of electrons trapped inside the walls of a potential well in a semiconductor), the trapped particle enters an excited state, where it remains for an exponentially distributed time interval with relaxation time τ. The power spectrum—that is, the squared magnitude of the Fourier transform—$P_\tau(f)$ of such a process is the familiar Lorentz resonance line centered at 0 frequency (a first-order lowpass filter response, to the electrical engineer):

$$P_\tau(f) = \frac{4\tau P_0}{1 + (2\pi f \tau)^2} \tag{1}$$

Here the total power P_0 of the relaxation process—that is, the integral over all positive frequencies $P_\tau(f)$—is independent of τ.

In many physical, chemical, or biological systems there is not just *one* relaxation time but there are a whole spectrum of relaxation times τ that depend on the energy barriers E which keep the structure temporarily trapped in the excited state. The relation between relaxation time τ and energy barrier E is the famous law named after Svante August Arrhenius (1859–1927):

$$\tau = \tau_0 e^{E/kT} \tag{2}$$

where T is the absolute temperature and k is Boltzmann's constant. Suppose these energies are distributed uniformly in the interval $[E_1, E_2]$; then the distribution of the relaxation time $p(\tau)$ can be obtained from equation 2 by applying the elementary rules for transforming probabilities, yielding a hyperbolic distribution for τ

$$p(\tau) = \frac{kT}{E_2 - E_1} \cdot \frac{1}{\tau} \qquad \tau_1 \leq \tau \leq \tau_2 \tag{3}$$

where $\tau_{1,2} = \tau_0 \cdot \exp(E_{1,2}/kT)$

Superposition of many independent relaxation processes with power spectra given by equation 1 and relaxation times distributed according to equation 3 yields

$$P(f) = \frac{2kTP_0}{\pi(E_2 - E_1)f} [\arctan(2\pi f\tau_2) - \arctan(2\pi f\tau_1)]$$

where the difference inside the brackets, in spite of its somewhat forbidding appearance, is roughly constant in the frequency interval

$$\frac{1}{\pi\tau_2} < f < \frac{1}{4\pi\tau_1} \tag{4}$$

The main point now is that the frequency interval in expression (4) could be, and in numerous situations *is*, very wide. Suppose, for example, that the barrier energies span a range as narrow as $7kT$; then $\tau_2/\tau_1 \approx 10^3$. The corresponding frequencies for which the $1/f$ law $P(f) \sim f^{-1}$ holds within an accuracy of 3 decibels (dB) range over a factor greater than 1200.

Distributions of relaxation times over wide ranges of values have been observed in many physical and biological phenomena [Man 83]. Thus, the electrical voltage on a Leyden jar, an early storage capacitor for electricity, does not decay exponentially with time, with a single relaxation time. Rather, the charge decays *hyperbolically*, implying a wide range of relaxation times [Koh 1854]. The ubiquitous electret microphone, a kind of latter-day Leyden jar, shows a kindred decay of its internal charge [Ses 80].

Similarly, the recovery times of neurons after firing were found by Jerry Lettvin to stretch from fractions of milliseconds to hours and days. And when Wilhelm Weber, following a suggestion by Carl Friedrich Gauss, measured the lengthening of elastic silk threads used in his apparatus, he found that an applied load would give rise not only to an immediate stretching but to a long-lasting aftereffect, a continual further lengthening that followed a hyperbolic law with elapsed time [Koh 1847].

Hyperbolic decay with time can even be observed in concert halls with insufficient sound diffusion. As a consequence, sound decay in such halls cannot be characterized by a single reverberation time even at a single frequency [Schr 70]. It seems that wherever we look or listen, we see or hear that exponential behavior is much less common than commonly supposed.

We shall later encounter still another mechanism for hyperbolic behavior that gives wide-range $1/f$ spectra: *intermittency*, stemming from *tangent bifurcation* in the logistic parabola and other iterated nonlinear mappings. A generator for generic $1/f$ noise in chaotic Hamiltonian systems, hostage to a self-similar hierarchy of "cantori," was recently proposed by Geisel [GZR 87].

While brown noise is easy to generate (just keep summing independent random numbers), pink noise is a bit more difficult to produce. A relatively simple method of generating pink or $1/f$ noise on a computer is to add several relaxation (first-order lowpass) processes with power spectra like equation 1 and relaxation times τ that form a self-similar progression with a similarity factor equal to 10 (or less for a better approximation). In this manner, just three relaxation times will cover a frequency range of nearly three powers of 10 (see Figure 2).

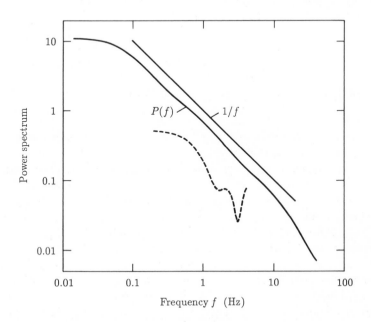

Figure 2 Pink noise from relaxation processes. Solid curve: superposition of three relaxation processes. Dashed curve: Superposition of values on three dice.

A relaxation process with discrete time samples x_n can be generated with the help of a computer's internal random number generator for producing independent random samples r_n and using these (zero mean) samples in the recursion relation

$$x_{n+1} = \rho x_n + \sqrt{1 - \rho^2}\, r_n \qquad x_0 = 0$$

Here ρ is the desired correlation coefficient between adjacent samples. It is related to the relaxation time τ by the equation $\rho = \exp(-1/\tau)$. Thus, for a set of relaxation times that increase by a factor of 10 ($\tau = 1, 10, 100, \ldots$), the correlation coefficients are obtained by taking successive tenth roots (e.g., $\rho = 0.37, 0.90, 0.99, \ldots$).

For less accuracy, three dice, instead of a computer, suffice as random number generators: the first die is rolled for every new sample of the pink noise, the second die is "updated" only every other time, and the third die is thrown only every fourth time. This ingeneously dicey idea is due to Richard Voss [Gar 78]. The sum of the dots on all three dice then forms a random variable with a mean of 10.5 and a variance (noise power) of 8.75 that is a rough approximation to a pink noise over a limited frequency range.

In this parlor-game approach to pink noise, the different relaxation times are mimicked by different persistence times of the dice (increasing by a factor of 2 in our example). However, the three-dice method (the dashed curve in Figure 2) does not approach a straight $1/f$ slope nearly as well as three relaxation processes (the solid curve in Figure 2).

Self-Similar Trends on the Stock Market

One of the neighborhoods where power-law noises dominate the scene, and chaos reigns the charts, is Wall Street, U.S.A. At stock and commodity exchanges, self-similarity weighs in on many scales. This is perhaps best illustrated by my once mistaking a chart of *minute-by-minute* stock averages (see Figure 3) for *day-by-day* fluctuations. I would not have been surprised by self-similarity between daily, weekly, and monthly prices, but I never suspected the same kind of fluctuations to prevail right down to 30-second intervals, which is what the "minute-by-minute" charts actually show.

Of course, once in a while there is an uncharacteristic jump in the data, as in October 1987 (see Figure 4), when the computers that handle programmed trading went wild (through no fault of theirs, of course). Technically, such price jumps are also known as "innovation processes"—some *innovation*, and certainly no consolation to the unlucky trader waiting impatiently for his broker to answer

Figure 3 Minute-by-minute stock averages look much like daily averages because stock averages are a self-affine process.

the phone! But after the "novelty" has worn off, price fluctuations resume their habitual course.[2]

The trends and fluctuations of stock prices have been analyzed in great detail in terms of information-theoretic concepts, such as cross entropy and mutual information. In fact, Claude Shannon, the father of communication theory (as he called it), is reported to have become quite rich after he began to apply his theory to the stock market. Now, market analysis is a firmly entrenched branch of information theory, as are other economic applications of entropic principles. But the emergence of programmed trading, executed by fast and soulless machines governed by instant feedback, will necessitate much rethinking of the rules by all concerned: the exchange board, the numerical analyst, and the hapless investor.

---

2. Speech signals, which are highly redundant (no matter what the semantic content), can also be reduced to innovation sequences (by linear prediction from past sample values). These "prediction residuals" can then be quantized and encoded by a single binary information bit for every four samples of the speech signal, or 0.25 bit per sample, without loss of quality [SA 85]. Ordinarily, speech signals require 8 bits per sample, and compact disks use 16 bits per channel.

Figure 4   Drop of stock market in October 1987.

In a first approximation to stock averages, one assumes the actual prices to be generated by *independent increments*. The resulting price "noise" has a power spectrum that is proportional to the inverse square of the frequency. Such random runs are now often called brown noises, in an allusion to Brownian motion, that jittery dance of floating dust specks seen in a microscope by the Scottish botanist Brown. (In Brownian motion, the innovation process consists of the independent kicks given the suspended particles by the molecules of the liquid in which they float.)

Another, purer paradigm of brown noise is the fluctuating capital of a gambler, for which independent rolls of dice constitute the innovation process. Suppose the probability of winning a dollar is p for every roll (and that of losing a dollar is $1 - p$). What optimum strategy does information theory teach? Answer: Unless your chances of winning exceed 50 percent, don't play! (Another case of science most profound confirming common sense.)

But what to do if $p \geq 0.5$? This is not as unrealistic as it may seem. The gambler could have side information—legal or otherwise—of the mechanical statistics of the roulette wheel. Here information theory provides another useful answer: Bet, but don't bet *all* your money! To maximize your capital growth, bet the fraction $2p - 1$ of your present capital [Kel 56]. Then the logarithmic growth rate of your capital will attain its highest value, given by Shannon's *information capacity* $C(p) = 1 - H(p)$ of the binary symmetric channel with error probability p. Here $H(p)$ is the entropy function $H(p) = -[p \log p + (1 - p) \log (1 - p)]$. Thus, for $p = 0.6$, say, the gambler should invest 20 percent of his current capital at every roll. Taking logarithms to base 2, $H(p)$ equals 0.97 bits per roll and $2^C = 1.02$. Thus, the well-informed gambler can expect an average gain of 2 percent *per roll*—which is a lot better than the taxable 2 percent interest per *year* that a major European bank recently offered the author, without blushing.

The more timid player who bets only 5 percent of his capital will gain an average of less than 1 percent per roll. After 200 rolls, his gain will be only one-tenth that of the optimum player who properly exploits information theory.

By contrast, the greedy gambler who always bets half of his current capital will, on average, *lose* 3.5 percent of his money per roll, or practically all of it (99.9 percent) during a 200-roll evening. And the reckless player who bets all his current money on every roll will, of course, be cleaned out completely after surviving an average of two rolls. (See also pages 150–152 in Chapter 6.)

Brownian motion contains several subtle statistical self-similarities, and we shall return to the Brownian theme, including more on gambling and the topic of constructing interesting topographies from noise, in the following sections and in Chapter 6.

Black Noises and Nile Floods

If the gambler thinks brown noise is bad enough, expose him to processes with power spectra proportional to $f^{-\beta}$ with $\beta > 2$, which we have called *black* noises. A diffusion process with independent increments Δx diverges but does so only with the square root of elapsed time t; that is, the root mean square distance is proportional to the square root of time:

$$\langle x^2 \rangle^{1/2} \sim t^{1/2}$$

To characterize black processes, we need a new measure of divergence. This was provided by Harold Edwin Hurst (1900–1978) [HBS 65] and Mandelbrot [Man 83].

The quantity in question is the *rescaled range R/S*, which is essentially the range $R(\Delta t)$ of the data over a time interval Δt (after subtracting any linear trend) divided by the sample standard deviation $S(\Delta t)$. For a white Gaussian noise, the ratio R/S tends to a constant for large Δt. In a sense, both R and S measure the range of the data, but R "looks" at the data linearly and S is based on the squared data. For some processes this yields no new information and R/S is asymptotically constant, that is, proportional to Δt^0. But this is not so for numerous geophysical records such as floods and a host of other inhospitable data.

For a Brownian function (power spectrum proportional to f^{-2}), R/S is proportional to $\Delta t^{0.5}$, reflecting the long-range dependence, or *persistence*, hiding behind brown processes. The water-flow statistic of the river Rhine (at the Swiss-French-German triple point near Basel, anyhow) converges on a similar long-term behavior with $R/S \approx \Delta t^{0.55}$ (see Figure 5A).

But other rivers are not as mild and tame as the Rhine [MW 69]. The water-level minima of the river Nile, for instance, taken between the years 622 and 1469 (the drought-dreading Egyptians must have been very patient papyrus

Figure 5   (A) Water-flow statistics of the Rhine; (B) water-level statistics of the Nile [MW 69].

keepers!) show a dependence $R/S \sim \Delta t^{0.9}$ (see Figure 5B), an exponent that reflects a high degree of persistence, of which the Bible has taken due note in the heart- (and coat-) rending Joseph story (Genesis 41).

The *Hurst exponent*, defined by $H := \log (R/S)/\log (\Delta t)$, is a convenient measure of the persistence of a statistical phenomenon. For white noise, which has no persistence, $H = -0.5$; for brown noise, which does have persistence, $H = 0.5$.

Interestingly, there is a simple relation between the Hurst exponent H and the spectral exponent β: $\beta = 2H + 1$. Thus, the Nile noise has a power spectrum proportional to $f^{-\beta} = f^{-2.8}$, implying, like the large Hurst exponent 0.9, a long-range persistence that requires unusually high barriers, such as the Aswân High Dam, to contain damage and rein in the floods.

Warning: World Warming

Processes with pronounced persistence pose perplexing puzzles—and are prone to frequent misinterpretation. Again and again one hears alarmists crying wolf when confronted with seemingly threatening data, but impartial analysis may reveal nothing more threatening than a statistical artifact.

Let us look at a Nile-like noise, with a power spectrum that decays as f^{-3} for large frequencies f. Of course, for small frequencies, the spectrum would diverge, implying an infinite energy. But no matter how catastrophic, *real* calamities are finite. If nothing else, finite observation times T would limit observable excesses. Thus, a realistic power spectrum $P(f)$ with asymptotic f^{-3} dependence, obtained from data collected over a time period T, might look like this:

$$P(f) = \frac{T^4 f}{1 + T^4 f^4} \quad (f > 0)$$

This spectrum is plotted in Figure 6 for an observation period $T = 1$ (say, 1 year).

But now suppose observations are extended over *two* years. The "new" power spectrum, as seen by the extended observations, is shown by a dashed line in Figure 6. Thus, just extending the observations from 1 year to 2 years has added all the doom-ordaining power shown in gray in the plot.

For an actual example, let us look at a study of the annual population variation of a large number of terrestrial animals over a period of 50 years. Ecologists found that fluctuations over 20 years are roughly twice as large as those observed over 2-year spans—in spite of the fact that the populations appeared to be relatively stable over half a century [RP 88].

Thus, before drawing doomsday conclusions from the exceedingly warm 1988 summer in the continental United States, one should remember Hurst and his exponent, and the strong dependence of extremes on the length of observation. There may well be a "greenhouse" effect of global warming in the air, but confirmation requires a lot of patience (but perhaps not much more carbon dioxide [Fis 90, Wei 90]).

On the other hand, minimum viable populations of endangered species must be considerably larger than current estimates, based on time-limited data, would indicate [Law 88]. (For further paradoxes resulting from power-law statistics, see Chapter 6.)

Fractional Integration: A Modern Tool

Brownian motion is obtained from summing independent increments. Summing (or integrating) the increments changes the spectrum from f^0, for innovation

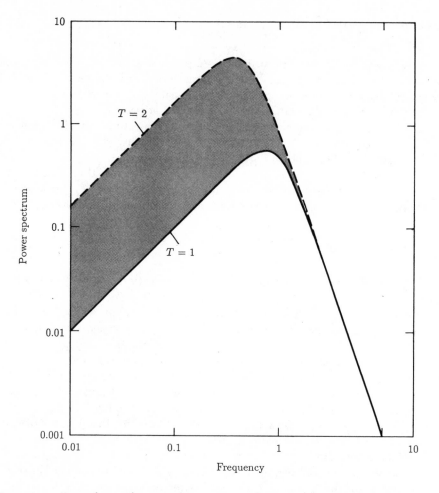

Figure 6 Dependence of measured spectrum on observation time T.

processes, to f^{-2} for the integrated process. Can we not get f^{-1} processes from integration? Yes, but we must first reinvent *fractional* integration, and before that, we must *define* it.[3]

Since integration multiplies a power spectrum by f^{-2}, let us define a *half integration* as an operator that multiplies the power spectrum by f^{-1}. In the conjugate Fourier variable (time or space, say), the operation is a convolution

---

3. Amazingly, as already noted, Leibniz thought of fractional differentiation and integration 300 years ago, right after the invention of calculus proper. (The incredible Leibniz also seems to have invented the grooves under the head of a nail to improve its cohesion with the material into which it has been hammered.)

integral whose kernel is the inverse Fourier transform of $|f|^{-1/2} \cdot \exp{(i\phi(f))}$ with appropriate phase $\phi(f)$.

More generally, we can define ν-fractional integration by an operator that multiplies the power spectrum by $f^{-2\nu}$. The corresponding convolution kernel is proportional to $t^{\nu-1}$ [Erd 54].

Fractional integration and differentiation have been useful tools in quantum mechanics and other fields for some time. Now they may also serve in the automated assembly of fractal landscapes and other self-similar structures. While the required numerical convolution in the time or space domain may consume much computer time and memory space, the alternative—Fourier synthesis from the prescribed spectrum—imposes potentially unrealistic periodicities on the resulting fractal.

Brownian Mountains

How can we generalize the Brownian function, $B(t)$, to a function of *two* variables, $B(t_1, t_2)$? In other words, we want to erect a Brownian *mountain*, $B(t_1, t_2)$, over the (t_1, t_2) plane in such a way that any cut of the mountain with a plane perpendicular to the (t_1, t_2) plane will be a typical Brownian function $B(t)$. Interestingly, as seen in the next section, the answer is connected with computer tomography—and, equally unlikely, with imaging by rotating cylinder lenses.

We begin by recalling that $B(t)$ has a power spectrum proportional to f^{-2}. Its amplitude spectrum is therefore proportional to $|f|^{-1}$. For the two-dimensional case, we need an amplitude spectrum which is proportional to $|\mathbf{f}|^{-1}$, where $\mathbf{f}$ is the frequency *vector* (f_1, f_2) with components f_1 (corresponding to the "time" variable t_1) and f_2 (corresponding to t_2). The length of the frequency vector is $|\mathbf{f}| = (f_1^2 + f_2^2)^{1/2}$.

Thus, one method of constructing a Brownian mountain $B(\mathbf{t})$, where $\mathbf{t}$ is the "time" vector (t_1, t_2), is to take sufficiently many independent, identically distributed random samples on a square lattice in the frequency plane, to multiply them by $|\mathbf{f}|^{-1}$, and to Fourier-transform the product into the time plane. For mountains in our world, the time plane will, of course, be a spatial plain, measured in square kilometers.

An alternative method of constructing $B(\mathbf{t})$ is to start with independent samples in the *time* plane and perform an operation on them that is equivalent to multiplying by $|\mathbf{f}|^{-1}$ in the frequency plane. This equivalent operation is a *convolution* with the inverse Fourier transform of the function $|\mathbf{f}|^{-1}$, which, in two dimensions, happens to be the function $|\mathbf{t}|^{-1}$ (within a constant factor that is of no interest). In other words, the two functions $|\mathbf{f}|^{-1}$ and $|\mathbf{t}|^{-1}$ form a *Fourier pair* for the *two*-dimensional Fourier transform. (Another, better known case where a function is similar to its Fourier transform is, of course, the Gaussian distribution function. In fact, for the Gaussian function this similarity holds in any number of dimensions.)

Thus, if we use $W(\mathbf{t})$ to refer to a two-dimensional array of independent, identically distributed random samples in the time plane, a Brownian mountain is given by

$$B(\mathbf{t}) = W(\mathbf{t}) * |\mathbf{t}|^{-1} \tag{5}$$

where $*$ stands for a convolution integral. Figure 7 shows a mountainscape generated in this fashion. Other methods for generating fractal landscapes and other self-affine fractals are described by Richard Voss [Vos 88] and Dietmar Saupe in *The Science of Fractal Images* [PS 88].

Radon Transform and Computer Tomography

B. Julesz has shown that imaging an object in the $\mathbf{t}$ plane with a rotating cylinder lens, combined with time averaging, is equivalent to convolving with the function $|\mathbf{t}|^{-1}$ [JSS 69]. The same is true for tomographic imaging (using x rays or some

Figure 7   Computer-generated Brownian mountain [Man 83].

other shadow-casting radiation) [SHJ 88]. The transformation in equation 5, which describes these"fuzzy" imaging methods, is also called the *Radon transform*. Figure 8 shows the image of a letter A obtained with a rotating cylinder lens and time averaging.

To recreate a sharp image $W(t)$ from the "blurred image" $B(t)$, we need to execute an *inverse* Radon transform on $B(t)$. And we already know how to do this! We Fourier-transform the $B(t)$ information (preferably on a computer; hence the term *computer* tomography), multiply the result by $|f|$ to cancel the factor $|f|^{-1}$, and Fourier-transform back into the t plane to yield the sharp image $W(t)$. Computer tomography is that simple—if we shun dull and deadly detail!

Fresh and Tired Mountains

A Brownian mountain with an $|f|^{-2}$ power spectrum has a surface whose Hausdorff dimension is $D = 2.5$ [Vos 85], which makes it look rather rugged, like a geologically very young, ragged mountain (see Figure 9A) whose jagged peaks have had no time to erode.

If we want smoother mountains, like the North American Rockies (except the Grand Tetons), the power spectrum must fall off faster than $|f|^{-2}$. A good earthlike mountainscape is obtained by multiplying independent samples in the frequency plane by a factor $|f|^{-\gamma}$ with $\gamma > 1$. Fourier transformation then yields

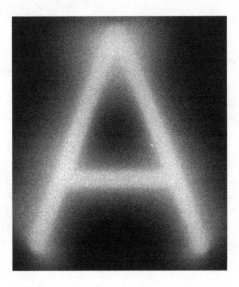

Figure 8   Image of letter A obtained with rotating cylinder lens.

(A)

(B)

Figure 9   (A) Brownian mountain with fractal surface dimension 2.5, (B) mountain with surface dimension 2.1 [Vos 88].

Figure 10   Coastline with fractal dimension 1.33 [Man 83].

mountains with a fractal surface dimension $D = 3.5 - \gamma$. For the spectral exponent $\beta = 2\gamma$, we thus have $\beta = 7 - 2D$.

Figure 9B shows a mountainscape with $D_M = 2.1$ having a power spectrum proportional to $|\mathbf{f}|^{-2.8}$.

Many mountain lakes (see the white areas in Figure 9A) have coastlines with Hausdorff dimensions equal to $D_c = D - 1$, which equals $D_c = 1.5$ in the Brownian case ($\beta = 2$) and $D_c = 1.1$ for $\beta = 2.8$ (Figure 9B). This is somewhat less than the fractal dimension of the west coast of Britain ($D_c \approx 1.25$).

A coastline with $D_c \approx 1.33$ generated in this manner is shown in Figure 10.

In general, for a fractal embedded in E Euclidean dimensions, the fractal dimension D is given by

$$D = E + \frac{3 - \beta}{2}$$

There is also a connection between D and the previously introduced Hurst exponent H (see pages 129–130). With $\beta = 2H + 1$, we have

$$D = E + 1 - H$$

Since for many shapes in nature (mountains and clouds, for instance) $D \approx E + 0.2$, Hurst exponents near 0.8 are an excellent general choice for generating such natural designs. Today, the use of computer-generated fractal landscapes in movie animation is pervasive.

6

Brownian Motion, Gambling Losses, and Intergalactic Voids: Random Fractals par Excellence

As far as the laws of mathematics refer to reality, they are not certain; and as far as they are certain, they do not refer to reality.

—ALBERT EINSTEIN

Some of the most fertile fields for fractals are fluctuating phenomena. In fact, nature abounds with self-similar structures that are statistical in nature, covering many different disciplines: from the distribution of galaxies in astronomy to cloud formation, climate, and the weather in meteorology; from polymerization and rusting in chemistry to the design, in biology, of our lungs and vascular systems and the growth patterns of many plants; from "fingering" in oil exploration, the branching and drainage basins of river systems, and the occurrence of floods in geophysics to physics proper, where we encounter fractals and statistical self-similarity in Brownian motion, fracture surfaces, soap bubbles, coagulation, percolation, diffusion-limited aggregation, and dielectric breakdown—such as lightning and Lichtenberg figures—not to mention the energy valleys in *spin glasses* and, last but not least, *turbulence*.

Cosmic strings, too, those wispy threads thought to have been created during the birth of the universe and potentially responsible for the clumping of galaxies, are statistically self-similar as the universe expands [Vil 87].

In view of this encompassing scope of statistical self-similarity, it is perhaps not surprising that fractals have even invaded the art industry. Following Mandelbrot's pioneering work and his specific suggestions, mountains and other "backdrops" in videos, movies, and images in general are increasingly being generated by computers, programmed to produce self-similar structures with the desired pleasing statistics—pleasing, it is said, because fractals have interesting features *on many size scales*.

In this chapter we shall first explore the random fractal par excellence: Brownian motion and certain games of chance.

The Brownian Beast Tamed

Brownian motion, the paradigm of random fractals, was first observed in the nineteenth century by the Scottish botanist Robert Brown (1773–1858) and properly described by him in 1827 as a *physical*[1] phenomenon.

Thus, the stage was set for mathematical physics to step in, and it was none other than Albert Einstein, in 1905, and Marian Smoluchowski (1872–1917), a bit later, who first shed light into a very murky situation [Pai 82]. Interestingly, when Einstein began thinking about random thermal motions of macroscopic objects, he was not even sure that there was such a thing as Brownian motion.[2] But he felt that there *should* be macroscopic manifestations of molecular motion and that their observation would confirm the molecular theory of heat, thereby proving the existence of finite-size atoms. This is what Jean Baptiste Perrin (1870–1942) proceeded to do, and in 1926 he was awarded the Nobel Prize in physics for his work on Brownian motion. In fact, proper observation of the motion under the microscope and an application of the law of large numbers allowed Perrin, following Einstein's suggestion, to "count" the number of molecules in a given volume.

But the headaches that Brownian motion had caused were far from over. Now the puzzling questions arose in *mathematics*, owing to the nondifferentiability of the motion, until laid to rest by Norbert Wiener (1874–1964) and others. So, to mathematical physicists, Brownian motion is now known as a Wiener process—really nothing new or novel, considering that Weierstrass functions had been available for some time as a good mathematical model for nondifferentiable continuous functions. In fact, it was one of the great minds of the nineteenth century, Ludwig Boltzmann (1844–1906), who felt that there were physical

---

1. This is not the place to retell the hilarious tales stimulated by that wiggly motion seen under the microscope, the fanciful interpretations ranging from living molecules—endowed with their own free will—to the outright supranatural. Suffice it to say that, when Brown had the liquid boiled, frozen, and reheated, the little specks still wiggled as madly as many a modern disco crowd.

2. In his first paper on the subject Einstein wrote: "It is possible that the motions discussed here are identical with the so-called Brownian molecular motion; the references accessible to me on the latter subject are so imprecise, however, that I could not form an opinion about this,"[Ein 05].

problems that are best described by nondifferentiable functions and that one could have invented such functions from the proper consideration of physical problems.[3]

Brownian Motion as a Fractal

Figure 1A shows a typical way of portraying the Brownian motion of a dust particle, say, as seen under the microscope. However, this portrayal is exceedingly misleading: Does the particle really move in straight lines between vertices? No! Does it move in curved lines then? No again! Then how *does* the particle move from point A to point B in Figure 1A?

Let us photograph the particle's motion at a shutter speed 100 times faster so that we see the particle 100 times between A and B. The result, magnified 10 diameters, is shown in Figure 1B: the straight line connecting A with B has metamorphosed into 100 straight-line segments, each, on average, about as long as the segments in Figure 1A (really 10 times shorter, because Figure 1B is magnified tenfold).

Does the particle move in a straight line from C to D in Figure 1B? Once more the answer is no. Looking again 100 times more often and magnifying 10 diameters will result again in a picture *statistically similar* to Figure 1B. That is why we call the Brownian motion *statistically self-similar*.[4] Every time we make our spatial resolution 10 times higher, we get 100 times as many pieces. In general, if we increase the spatial resolution by a factor of $1/r$, we get $N(r) \sim 1/r^2$ more pieces to cover. Hence, the Hausdorff dimension (see pages 9–10 in Chapter 1) of Brownian motion is given by

$$D_H = \frac{\log N(r)}{\log (1/r)} = 2 \tag{1}$$

which happens to be an integer. With $D_H = 2$, Brownian motion in two dimensions could be plane-filling, but it is not; there is much self-overlap (see Figure 1A and B). In fact, for a Brownian motion in two dimensions (think of enzymes wandering around on the surface of a cell), the probability of returning to the neighborhood of a given location, no matter how narrowly defined, is 1. By contrast, for Brownian motion in three dimensions, the embedding dimension 3

---

3.   In a letter dated Vienna, January 15, 1898, to Felix Klein (1849–1925).

4.   More accurately, geometric figures whose parts can be brought into correspondence with the whole by scaling different directions by different factors are called *self-affine*. Thus, Brownian motion is statistically self-affine, with a scaling factor r, say, in the spatial dimensions and a scaling factor r^2 in the time dimension.

(A)

(B)

Figure 1 (A) Brownian motion. (B) The segment *AB* of the Brownian motion in part A sampled 100 times more frequently and magnified 10 times.

being larger than the Hausdorff dimension of the motion ($D_H = 2$), the return probability is smaller than 1.[5]

As we remarked earlier, every self-similar process in the real world has to have a largest and a smallest scale; scaling up or down cannot go on forever. But for Brownian motion, the range of lengths over which self-similarity prevails covers many powers of 10: from the size of the vessel containing the liquid, 0.1 m, say, to the mean free path between molecular impacts, which, for small test *particles*, could be as small as 10^{-9} m. In many situations we are willing to call an object self-similar if it scales over a range of just 10 to 1 or even less, in perhaps as few as three discrete steps. By contrast, Brownian motion scales over a range of 10^8 to 1, covered by a *continuum* of intermediate scales.

Brownian motion is as close as we get in physics to a *nondifferentiable* function. And, as Boltzmann remarked so soberly in his previously cited letter to Klein (of *Klein bottle* fame), if [Weierstrass] had not already thought of such functions (in his attempt to show the world how totally counterintuitive something innocently called a *function* could be) then physicists, or botanists, would have had to invent the strange mathematical beast themselves.

How Many Molecules?

The physical law behind the scaling behavior that leads to equation 1 is the *diffusion equation*

$$\overline{x^2} = 2Dt \tag{2}$$

where $\overline{x^2}$ is the mean square displacement of a Brownian particle in time t. Equation 2 is simply an expression of the mathematical fact that if independent random lengths (whose average is zero and whose distribution has a second moment) are added, then the total distance is obtained by adding up the *squared* individual lengths and then taking the square root. We have called the constant of proportionality $2D$ in equation 2, following Einstein's original nomenclature.

The so-called *diffusion constant* D must be related to the "microscopic" variables (really, the variables the botanist can*not* see under the microscope), namely, the mean free path λ and the average time between collisions, τ. A "little thinking" will reveal that the relationship is as follows:

$$2D = \frac{\lambda^2}{\tau} \tag{3}$$

---

5.   This is believed to be an important reason why Mother Nature lets many crucial life-sustaining chemical reactions occur on *surfaces* rather than in three-dimensional space.

or, introducing the thermal velocity $v \approx \lambda/\tau$ of the observed particle,

$$2D = \overline{v^2}\tau \tag{4}$$

where $\overline{v^2}$ is the mean square velocity. Now, the thermodynamic equipartition of energy tells us that

$$\overline{v^2} = 3\,\frac{kT}{m} \tag{5}$$

where k is Boltzmann's constant, T is the absolute temperature, and m is the mass of the particle.

The average time between collisions is given by

$$\tau = \frac{4}{\overline{v}nF} \tag{6}$$

where n is the *number of buffeting molecules* per unit volume and F is the cross section of the buffeted particle (which, like m, can be measured macroscopically).

Now, measuring $\overline{x^2}$ and putting equations 2 to 6 together, allows us to determine the number of molecules n in a given volume. If the result is finite, as it will be, then there must be a finite number of molecules in the liquid. And since the total weight is also known, measuring Brownian motion under the microscope allowed Perrin to measure the weight of an individual molecule. To have foreseen this possibility of establishing the finite, nonvanishing reality of atoms and molecules is one of Einstein's greatest contributions to our understanding of the physical world in which we live—almost on a par with his redefinitions of time and space in special and general relativity.

The Spectrum of Brownian Motion

What is the *power spectrum* of a Brownian function $B(t)$, which we define as the projection of Brownian motion onto one spatial dimension as a function of time? Brownian motion is generated by independent increments (the individual impacts of the buffeting molecules), which have a flat ("white") power spectrum. The sum or integral of the increments therefore has a power spectrum that is proportional to f^{-2}. Noises having such spectra are now called *brown* noises (see Chapter 5)—an allusion also to the fact that brown light has a stronger admixture of red light (low optical frequencies) than white light.

The Gambler's Ruin, Random Walks, and Information Theory

Other instances of brown noise are stock market prices and gains or losses from other *jeux d'hasard* and, generally, random walks subject to independent increments. Let us look at the infamous paradigm called the "gambler's ruin." Each time a coin is thrown and heads come up (with probability p), the player wins a dollar, and he loses a dollar when tails shows. His capital as a function of time (number of throws) is a Brownian process with fixed increments, also called a Markov-Wiener process, after A. A. Markov (1903–1922) and Wiener.

Let the player start with an initial capital K (K as in Karl Marx and Das Kapital). After the first trial (apt word!) his capital is either $K + 1$ (with probability p) or $K - 1$ (with probability $q = 1 - p$). Calling his probability of ultimate ruin q_K, we have the following difference equation:

$$q_K = pq_{K+1} + qq_{K-1} \qquad 0 < K < B \qquad (7)$$

where B is the capital of the bank, $q_0 = 1$, and $q_B = 0$.

Such difference equations can be solved by the generating-function (or z-transform) *ansatz* $q_K = z^K$, which yields a quadratic equation in z: $z = pz^2 + q$, with the two solutions (for $p \neq q$) $z = 1$ and $z = q/p$. Thus, for $p \neq q$, with the particular solutions $q_K = 1$ and $q_K = (q/p)^K$, the general solution is

$$q_K = a + b\left(\frac{q}{p}\right)^K$$

or, with the "boundary conditions" $q_0 = 1$ and $q_B = 0$,

$$q_K = \frac{(q/p)^B - (q/p)^K}{(q/p)^B - 1} \qquad p \neq q \qquad (8)$$

For $p = q$, we apply l'Hospital's rule to the limit as $p \to 0.5$, yielding for the probability of ultimate ruin the simple result

$$q_K = 1 - \frac{K}{B}$$

which satisfies the necessary symmetry for $p = 0.5$, namely, $q_K + q_{B-K} = 1$. Of course, for the poor gambler whose capital K is small compared with that of the bank B, ultimate ruin is almost certain:[6] $q_K \approx 1$.

Counterintuition Runs Rampant in Random Runs

One of the several counterintuitive facts of the fair-coin tossing game is that the expected number of times equals 1 that a player will increase his capital by any given amount G before the first return to his initial capital—*independent of how large* the gain G is! More tangibly, in a $1 per bet fair-coin tossing game, the player will reach $1 million on the average once before he has incurred any loss (i.e., dropped below his initial capital). The only consolation for the bank is that the *expected* gain is still 0. (As we saw in the preceding section, for the game without a time limit, the probability that the player will lose his entire capital of $K < B$ dollars is $1 - K/B$, and the probability that the bank will go broke is K/B).

A similarly counterintuitive result concerns the *duration* D_K of the fair game, obtained by solving a difference equation similar to equation 7:

$$D_K = K(B - K)$$

In other words, if the player has just a single dollar and the bank $1 million, the expected duration of the game is 999,999 tosses! As we shall see in what follows, these mind-boggling conclusions are related to the fact that returns to a given position in the unconstrained ($B = \infty$) fair game, while occurring with probability 1, are Cantor sets with fractal dimension $\frac{1}{2}$.

The counterintuitive results just described are symptomatic for unbiased random walks, expressed most forcefully by the so-called *arcsine law*. Couched in the language of a discrete random walk of a diffusing particle in one spatial dimension, the arcsine law says: The probability $p_{2n}(2k)$ that in the time interval from 0 to $2n$ the particle will spend $2k$ time units on the positive side is given by

$$p_{2n}(2k) = \binom{2k}{k}\binom{2n - 2k}{n - k} 2^{-2n}$$

$$\approx \frac{1}{\pi[k(n - k)]^{1/2}} \tag{9}$$

---

6. Surprisingly, a certain statistics *professor*, who shall remain nameless, was not able to derive the formula for the probability of ruin for the finite-duration game which he had posed as an exercise for the *students* at the beginning of the summer semester in 1948. During the entire semester his recurrent refrain was "we'll tackle that one next week." But he never did tackle it.

[Fel 68], a distribution which has most of its weight near $k = 0$ and $k = n$ (just like the amplitude distribution of a sine wave oscillating between its extreme values, a fact on which time-averaging holography of vibrating bodies is based).

The arcsine law, so named because the integral of equation 9 is the arcsine function, has many curious consequences. For example, for a fair game, the probability that in 20 tosses each player will lead 10 times (which seems fair enough) is only about 6 percent. But the probability that one of the players will lead for *all* 20 tosses (how unfair!) is greater than 35 percent! In other words, in more than one-third of the cases the lead never changes!

More Food for Fair Thought

Another "unfair" result of fair-coin tossing is as follows. Consider $2n$ tosses, half of which turn out heads (and half tails). And let $2k$ again measure how often the accumulated number of occurrences of heads is greater than the accumulated number of occurrences of tails. Then the number of possibilities $N_{2n}(2k)$ for this outcome is given by the "Catalan number:"

$$N_{2n}(2k) = \binom{2n}{n} \cdot \frac{1}{n+1}$$

independent of k.[7]

This kind of counterintuitiveness has led to numerous false conclusions in the history of science, and statistics in particular. In 1876 Sir Francis Galton (1822–1911), inventor of the Galton board (a kind of fakir's bed for bouncing balls), tested some data on plants furnished him by the even more famous Charles Darwin (1809–1882). There were 15 treated plants and 15 untreated specimens (the control group). In rank-ordering the data, Galton saw that the treated plants were ahead of the untreated plants with the same rank in 13 out of 15 cases. Galton concluded, understandably, that the treatment was effective. But assuming perfect randomness in the data (30 measurements from the *same* pool of plants), the probability of Galton's observation is $\frac{3}{16}$. In other words, in 3 out of 16 cases a perfectly ineffectual treatment appears very effective. How many bad answers and bogus inferences have been drawn from this single source of statistical misdemeanor alone?

---

7. Catalan numbers, although perhaps not widely known, are truly ubiquitous. For example, N_{2n} is the number of ways $2n$ people seated around a table can shake hands in n pairs without their arms crossing. For many other applications of Catalan numbers, see the paper by Eggleton and Guy [EG 88].

The St. Petersburg Paradox

Games of chance have led to more than one paradox, most often related to the counterintuitive aspects of random walks and their inherent fractal nature.

Around 1700, Nicolas Bernoulli (1687–1759), nephew of Jakob (1654–1705) and Johann (1677–1748), introduced a curious game of chance with infinite mean winnings (*mean* certainly from the bank's point of view). The game was analyzed by still another Bernoulli,[8] Daniel (1700–1782), in the *Commentarii* of the St. Petersburg Academy—depository of much of Euler's writings.

Suppose a coin has a probability $p > \frac{1}{2}$ of coming up heads. The player flips the coin until heads comes up for the first time. If it takes n flips, the player wins 2^{n-1} dollars. What are his expected winnings W if the game can continue indefinitely? The answer is simple enough:

$$W = 2^0 p + 2^1 (1 - p)p + 2^2 (1 - p)^2 p + \cdots \qquad (10)$$

or

$$W = \frac{p}{1 - 2(1 - p)}$$

For $p = 0.55$, for example, the expected gain is $W = 5.5$ dollars, and for $p = 0.51$ it is $W = 25.5$ dollars.

What happens for a "fair" coin? For $p = \frac{1}{2}$, the geometric series (equation 10) does not converge, and the expected winnings become infinite! Thus, a fair fee for the game would be an infinite ante, or so a bickering banker could reason.

But a prudent player, quite apart from being "temporarily out of" infinite capital, would consider the infinite *mean* winnings promised by equation 10 quite unfair and would prefer to base his ante on the finite *median* winnings, namely, a single dollar. Banker and player are at odds—much more so than even they would have surmised!

How can mean and median be so different? The answer is, of course, that for $p = \frac{1}{2}$, the mean does not even exist, and what does not exist lacks the ability of being different.

The divergent mean winnings may be reminiscent of the infinite length of a fractal curve, and indeed we can tame the St. Petersburg paradox by introducing fractals and Hausdorff dimensions, as we shall see in subsequent sections.

---

8.   A parallel between the abundant Bernoullis and the shingle of a typical American law firm is drawn in Figure 2.

"WHICH BERNOULLI DO YOU WISH TO SEE—
'HYDRODYNAMICS' BERNOULLI, 'CALCULUS' BERNOULLI,
'GEODESIC' BERNOULLI, 'LARGE NUMBERS'
BERNOULLI OR 'PROBABILITY' BERNOULLI?"

Figure 2   Bernoulli, Bernoulli, Bernoulli & Company [Har 77]. (© 1991 by Sidney Harris)

Shannon's Outguessing Machine

Not all games of chance are fair, perhaps least of all those which (who) proclaim fairness the loudest ("I am not a crook"). But some games make no pretensions of fair play; in fact, *un*fairness is their very reason for being—such as Claude Shannon's engaging "outguessing machine" [Sha 53].

Shannon's entrapping contraption initially makes random heads-tails choices against a human contender. But once the machine has experienced its first win, it begins to analyze the opponent's "strategy" to a depth of two throws. Does he or she change after losing a throw? Does the player keep on choosing tails if tails has brought two previous wins? Or does the gambler get chary and head for heads next? For most people such strategies are mostly subconscious, but the machine assumes the human to act like a second-order Markov process and uncovers the underlying transition probabilities without fail. Exploiting these, the machine always wins over the long haul, except against its creator. Shannon, keeping track of his machine's inner states, can beat it 6 times out of 10. Of course, anyone could win 5 out of 10 throws on average by playing random (perhaps by flipping a true coin). But this is precisely what people, deprived of proper props, are incapable of doing, as Shannon's machine has demonstrated again and again by beating a wide variety of human would-be winners. Specifically, man's mind appears to abhor long strings of like outcomes—as occur perfectly naturally in truly random sequences.

Of course, the machine can have bad luck too, especially in its initial guessing phase. I once wanted to show off the machine's prowess to a foreign friend (mathematician Fritz Hirzebruch) visiting Bell Laboratories. As luck would have it, Hirzebruch won 13 times in a row before his first loss. But thereafter the machine took off with a vengeance, overtaking the renowned mathematician on throw 31 (i.e., the machine won 16 out of the next 18 throws!) and never fell behind again—in spite of the fact that Hirzebruch had been told (in general terms) how the machine worked.

The Classical Mechanics of Roulette and Shannon's Channel Capacity

While Shannon's outguessing machine is intrinsically (and intentionally) unfair, some ostensibly fair games of chance can be subverted into unfairness. Great fortunes (and some unfortunate lives) have been lost through gambling at roulette while banking on a bewildering mélange of "infallible" stratagems. In the end the bank always wins, because, for an "even" bet, its odds are 19 to 18 (or even 20 to 18 at tables with a *double* zero, prevalent in American casinos).

Yet this inveterate observer of the gambling scene thought that simple mechanics—classical, not quantum—allied with fast, real-time computation could moderate the odds in his favor. Clad in a brand-new tuxedo and armed with an array of stopwatches hidden in his pockets, he set sail for some of the world's best known casinos: Baden-Baden in the Black Forest (parent to Monte Carlo), Baden near the Vienna Woods, and Évian-les-Bains on Lake Geneva. These field trips, undertaken in the early 1960s, confirmed what I had first found on a full-size regulation roulette wheel lodged in my basement: at the time the average croupier cries *rien ne vas plus* (no more bets, *nichts geht mehr*), the final resting position of the ball is not totally unpredictable; its probability distribution is anything but exactly uniform. In fact, the modulation depth of the probability amplitude around the wheel is typically 10 percent. Thus, instead of losing on average 19 out of 37 even bets, one could *win* about 20 out of 37. For some casinos the winning chances are even greater, as, for example, in Évian, where an easy-going croupier at one table spun the wheel so slowly that the ball, once having left the upper rim, sailed down into its final slot without further hopping to and fro.

To reap the promised harvest, one first has to determine the friction coefficients of the ball and the wheel (usually quite small) and enter these parameters into a small portable computer. (A friend and I had built ourselves a special-purpose *analog* computer for this task. Now, a quarter century later, one would of course place one's trust in a *digital* chip to dislodge the bank's chips and steer them in the right direction [Bas 90].) The computer, through two pushbuttons, gets timing information signaling two successive passes of the ball and the zero of the wheel past a preselected mark on the rim. This fixes speeds and relative positions of wheel and ball. (Friction is determined in a prior measurement.) The most likely outcome computed from these data is communicated via a modified hearing aid to the prospective winner.

How does one maximize the expected rate of increase of one's capital knowing that the chances of winning, p, are better than even ($p > 0.5$)? How much of one's capital should one risk for each spin of the wheel? The answer comes from *information theory*, in fact, one of its earliest applications to gambling. In a landmark paper that we already mentioned on pages 128–129 in Chapter 5, John L. Kelly, Jr. [Kel 56], proved that, in order to maximize the rate of increase of one's capital, one should bet a fraction $2p - 1$ of the current capital. And of course, for $p \leq 0.5$ one should abstain completely and seek another pastime. The expected (exponential) growth of one's capital is then given by the factor $2^{C(p)}$, where $C(p)$ is Shannon's channel capacity of a *binary symmetric channel* with error probability p:

$$C(p) = 1 + p \log_2 p + (1 - p) \log_2 (1 - p)$$

Thus, for $p = 0.55$, for example, one should risk $2p - 1$ or 10 percent of one's capital for every spin and expect a rate of enrichment of $2^{C(0.55)} = 0.005$ or 0.5 percent *with each spin*. (This means that, during a weekend of 138 spins, one's holdings would be doubled—less tips and taxes, of course.)

This application of information theory was the first instance (and is still the only one, as far as I know) in which a benefit can be reaped without the elaborate coding that is necessary to realize the error-free transmission promised by Shannon's theory.

How much did I win? Well, once I knew that classical mechanics (and my analog computer) worked, I lost all interest in the project. Also, I knew that casinos can banish anyone who wins "too much" from their hallowed halls without having to state any reason. Finally, once the management learns the mechanics behind the winning method, it can simply instruct its croupiers to spin their wheels a bit faster and to call *rien ne va plus* a little earlier. End of dream, Eden lost.

The Clustering of Poverty and Galaxies

Certain perfectly simple statistical rules generate random collections of points, called *point processes*, that exhibit unexpectedly large voids with a statistical self-similar structure. The distribution of galaxies in the universe is a good case in point: the voids between the largest clusters of galaxies are only a few times smaller than the entire universe itself. And the voids between galaxies are as large as large clusters of galaxies, and so on. A similar structure also governs games of chance between successive ruins and numerous other events infested by holes. Let us look at a fair game of chance, with probability $p = 0.5$ of either winning or losing a yen. The current capital, $K(t)$, of the player as the game progresses may look somewhat like Figure 3A: the capital has a tendency to drift to large positive or negative values but eventually returns to 0. (The ruined player, who has lost all, is allowed to continue on credit.) Once the capital has reached 0, the probability that it will do so several times in a short time span

Figure 3   (A) Gambler's capital as a function of time in an honest game of chance. (B) Fluctuations of the gambler's capital over very long times.

is, for obvious reasons, very high. In other words, the zeros of the capital function $K(t)$ form clusters.

What else can we say about the function $K(t)$? Other than the sizes of the basic steps in time ($\Delta t = 1$) and money ($\Delta K = 1$), the problem has no further scales. We therefore expect the player's ruin to exhibit self-similarity and self-affinity. Indeed, if we plot $K(t)$ over a longer time span and scale t and K appropriately, the new plot (see Figure 3B) will look much like the old one. The proper scaling factor for K is the square root of the scaling factor for t, just as in Brownian motion.

The number of expected zeros, N_0, in the time interval t also scales with the square root of t [Fel 68]: $N_0 \sim t^{1/2}$. Waiting 4 times as long will increase the number of ruins only by a factor 2 (sounds good!).

What is the probability $p(z)$ of observing a given distance z between successive zeros? Since the problem, given infinite capital resources, contains no long-range cutoff, $p(z)$ must approach a self-similar power law, for long games:

$$p(z) \approx \text{const} \cdot z^{\alpha} \qquad 1 \ll z \leq t$$

Since N_0 is proportional to $t^{1/2}$, so is the mean gap length $\bar{z} = t/N_0$. With $\bar{z} \approx \text{const} \cdot t^{\alpha+2} \sim t^{1/2}$, we have $\alpha = -3/2$ and, asymptotically,

$$p(z) \approx \text{const} \cdot z^{-3/2} \qquad 1 \ll z \leq t$$

From this we obtain the cumulative distribution of the lengths of the zero-free voids exceeding length z:

$$P(z) := \sum_{k=z}^{t} p(k) \approx \text{const} \cdot z^{-1/2} \qquad (11)$$

This distribution has a very long tail that drops off to zero very slowly with increasing gap size. Figure 4 shows experimental results (obtained with a pocket calculator) that confirm equation 11 over five orders of magnitude.

Figure 5 shows the "voids within voids within voids" structure of the zero-free regions. Roughly one-half of every "cluster of zeros" is actually devoid of zeros! Going to the limit of a continuous time scale, the zeros form a rather thin dust indeed, a Cantor set with Hausdorff dimension equal to the negative of the exponent in the cumulative distribution equation (equation 11): $D_H = 0.5$. This value can also be derived from the value $D_H = 2.5$ for the surface of Brownian mountains (pages 134–136 in Chapter 5). A vertical cut through such a mountain produces a Brownian "profile" with $D_H = 1.5$, which corresponds to our $K(t)$. A further cut with the line $K = 0$ produces our zero set (i.e., the set of t values for which $K(t) = 0$) with $D_H = 0.5$. In general, lowering the *topological* dimension of a fractal by 1 by forming a zero set also reduces the *fractal* dimension by 1. For example, Brownian mountains with a surface $D_H = 2.2$ generate coastlines

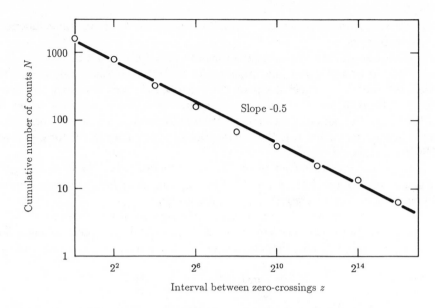

Figure 4   Distribution of time intervals without ruin.

Figure 5   Ruin-free time intervals (shown by thin lines) occur on all scales.

(i.e., lines at zero elevation) with $D_H = 1.2$, similar to the Hausdorff dimension of the west coast of Britain.

Levy Flights through the Universe

Another way to generate such point processes as the zero set of the gambler and to generalize them to more dimensions is the so-called *Levy flight* [Man 83]. In a Levy flight, named after the French mathematician Paul Lévy (1886–1971), one strings together independent increments ("flight paths") whose lengths z are (cumulatively) distributed according to a homogeneous power law:

$$P(z) = \text{const} \cdot z^{-D} \tag{12}$$

where D turns out to be the Hausdorff dimension of the resulting "dust." For $D = 0.5$ and one spatial dimension, equation 12 generates the voids in the gambler's ruin.

Figure 6A shows a *two*-dimensional isotropic Levy flight with exponent $D = 1.26$, making larger voids more probable than for $D = 0.5$. The turning points, that is, the "galaxies," generated by this process are shown in Figure 6B. For $D = 1.26$, the resemblance with the distribution of galaxies in the universe as seen from earth is astounding [Haw 88]. The implication, of course, is that the universe is a Cantor dust that has no natural length scales other than its own size. But the best-matching exponent, $D = 1.26$, for the galaxies and their soap-bubble-like aggregation still cries out for a proper explanation. One of the persistent puzzles in understanding the evolution—past, present, and future—of our universe is the mysterious role of dark matter, including black holes and, at the other end of the mass scale, the long-elusive but ubiquitous neutrino.[9]

Paradoxes from Probabilistic Power Laws

Probability distributions that follow a self-similar power law can have some rather paradoxical consequences. Consider a random variable $1 \leq x < \infty$ with a probability of exceeding a given value x given by x^{-D}. The *conditional* probability that the random variable exceeds the value x, given that $x > x_0$, equals $(x_0/x)^D$.

---

9. Enduring enigmas in the distribution of quasi-stellar objects ("quasars") at the far end of the universe have recently been interpreted as resulting from *gravitational lens* effects of dark matter surrounding "foreground" galaxies [BPS 87]. This optical effect, predicted by Einstein but believed by him to be unobservable, may yet turn out to be the most legible "fingerprint" of the dark universe. (The imaging properties of a gravitational lens are akin to those of the foot of a wine glass, producing multiple images.)

(A)

(B)

Figure 6   (A) Two-dimensional Levy Flight. (B) Resulting cluster of "galaxies."

For $D > 1$, the mean exists and equals $D/(D - 1)$. The conditional mean, given that $x > x_0$, equals, innocently enough,

$$\bar{x}_{x_0} = \frac{x_0 D}{D - 1} \tag{13}$$

As expected for a self-similar distribution, the conditional expectation depends linearly on x_0.

Now suppose that the completion times of a given human endeavor— writing a difficult report, for instance—are distributed according to a power law with an exponent $D = 1.5$, say. The expected completion time of the job is then $D/(D - 1) = 3$ hours or days or whatever unit of time one chooses. Intuitively, one would expect, 5 days after starting the work and not having completed it, that the expected completion time would be considerably less than 3 days— after all, it was only 3 days at the *start* of the project. However, equation 13 tells us that the expected completion time is now 15 days. And after another 60 days without finishing the job, the expected completion time has moved on to 180 days! In other words, the longer one works on such a project without actually concluding it, the more remote the expected completion date becomes.

Is this really such a perplexing paradox? No, on the contrary: human experience, all-too-familiar human experience, suggests that in fact many tasks suffer from similar runaway completion times. In short, such jobs either get done soon or they never get done. It is surprising, though, that this common conundrum can be modeled so simply by a self-similar power law. In fact, it might not be a bad idea to classify such dragging jobs by their characteristic exponents D— and pay the laggard contractor accordingly. For $D = 2$, $\bar{x}_{x_0} - x_0$ equals x_0, which is characteristic of the proverbial undertaking whose completion is always promised for *mañana*.

Invariant Distributions: Gauss, Cauchy, and Beyond

The sum of two Gaussian random variables is another Gaussian variable. The Gaussian distribution is therefore said to be *invariant under addition*. The variance σ^2 of the sum variable equals the sum of the individual variances σ_1^2 and σ_2^2:

$$\sigma^2 = \sigma_1^2 + \sigma_2^2 \tag{14}$$

The invariance of the Gaussian distribution is intimately connected with the *central limit theorem* of probability theory, which states that the suitably normalized sum of many independent random variables with *finite* variances converges on a Gaussian distribution.

For the summation rule in equation 14, the Gaussian distribution is the only one that is invariant under addition. However, if we introduce a *general* exponent D and replace the (possibly diverging) standard deviations σ by some other measure s of the width of a distribution that is guaranteed to exist (such as the interquartile range), we have instead of equation 14 the more general rule

$$s^D = s_1^D + s_2^D \tag{15}$$

Are there distributions that are invariant under addition with exponents D other than 2? There are indeed, and the solutions are related to the self-similar power-law distributions that we encountered in the preceding sections.

For the exponent D equal to 1, the invariant distribution is the bell-shaped *Cauchy* density

$$p(x) = \frac{1}{\pi(1 + x^2)} \tag{16}$$

which has the same functional form as the intensity resonance (the "resonance line") of a linear oscillator as a function of frequency.

The Cauchy distribution, named after the French mathematician Augustin Louis Cauchy (1789–1857), is the source of several noteworthy paradoxes. Its mean and variance do not exist, because the corresponding integrals diverge. One therefore characterizes the Cauchy distribution by its *median* and its *interquartile range*. The median is the value for which the distribution integrated from x to ∞:

$$P(x) := \int_x^\infty \frac{dy}{\pi(1 + y^2)} = \frac{1}{2} - \frac{1}{\pi} \arctan x \tag{17}$$

equals $\frac{1}{2}$. Here the median equals 0. The interquartile range is the difference of the two x-values for which $P(x)$ equals $\frac{3}{4}$ and $\frac{1}{4}$, respectively. For equation 16 it equals 2.

The Fourier transform ("characteristic function") of the Cauchy distribution (equation 16) is the symmetric exponential function $\exp(-|t|)$. (Once we recognize the Cauchy distribution as having the shape of a resonance line, the exponential is, of course, to be expected, because the amplitude of the oscillation of a linear resonator decays exponentially.)

Since adding two random variables means convolving their probability distributions or multiplying their Fourier transforms, we see that the sum of two variables distributed according to equation 16 has the Fourier transform $\exp(-2|t|)$. The corresponding probability density is thus the same as equation 16, but with the abscissa x stretched by a factor 2. More generally, the width s of the distribution of the sum of two Cauchy variables is the sum of their individual widths:

$$s = s_1 + s_2$$

Thus, for the Cauchy distribution, the exponent D in equation 15 is indeed equal to 1. The underlying reason for this linear scaling behavior is that the Cauchy distribution has an exponentially decreasing Fourier transform and that, upon multiplying two exponentials, their arguments *add linearly*.

As a result of this linear scaling, the distribution of the *average* of N identically distributed Cauchy variables is the same as the original distribution. Thus, averaging Cauchy variables does not improve the estimate; averaging begets no benefit. This is in stark contrast to all probability distributions with a *finite variance* σ^2, for which averaging over N variables reduces the uncertainties by a factor $1/\sqrt{N}$. This nonstandard behavior of the Cauchy distribution is a consequence of its weakly decaying "tails" that produce too many "outliers" to lead to stable averages.

The reader is encouraged to sample this paradoxical behavior by simulating the averaging of Cauchy variables on a pocket calculator or home computer. In general, a random variable with an integrated distribution $P(x)$ is obtained from a *uniform* variable u in the interval [0, 1], which is available on many calculators, by inverting the equation $u = P(x)$. With equation 17, this gives the "recipe"

$$x = \tan\left[\pi(0.5 - u)\right]$$

to produce a Cauchy variable x from a uniform variable u.

The Cauchy distribution occurs in numerous practical applications. For example, the *ratio* of two independent identically distributed zero-mean random variables is Cauchy-distributed. In other words, any two-dimensional isotropic distribution $p(x, y)$ centered on the origin has a Cauchy-distributed ratio x/y or y/x. This statement also implies that if a variable is Cauchy-distributed, so is its reciprocal. Thus, the logarithm z of a Cauchy variable also has a symmetric distribution: namely, $1/(\pi \cosh z)$, an important function in several branches of physics. (Light beams with a $1/\cosh z$ profile in time or space lead to *solitons* in optical fibers, for example. Sound velocity profiles in the ocean that vary according to $1/\cosh z$ with depth result in self-focusing and thereby engender low-loss transmission of acoustic energy over intercontinental distances.)

As we have seen, the exponent $D = 1$ in equation 15 for the Cauchy distribution follows directly from the fact that its Fourier transform is an *exponential* function, $\exp(-|t|)$, with a *linear* dependence of its argument on the Fourier variable t. Similarly, the exponent $D = 2$ for the Gaussian distribution results from the fact that its Fourier transform is an exponential function, $\exp(-t^2)$, with a *quadratic* dependence on the Fourier variable t. Following Cauchy, we therefore surmise that the inverse Fourier transform of $\exp(-|t|^D)$ will lead to a random variable that has an invariant distribution under addition with an exponent D. This is indeed the case in the range $0 < D \leq 2$. (For $D > 2$, Cauchy's prescription gives *verboten* negative probability values.)

Thus, the cherished Gaussian distribution with $D = 2$ stands revealed as but an extreme, albeit ubiquitous, member of an entire clan of distributions. And instead of just *one* central limit theorem, there are many. Depending on the scaling exponent D in equation 15, a properly normalized sum of random variables will converge on a specific limiting distribution that is invariant under summation of random variables.

For example, for $D = \frac{1}{2}$, the invariant distribution is

$$p(x) = \frac{1}{\sqrt{2\pi}} \; x^{-3/2} \exp\left(-\frac{1}{2x}\right) \tag{18}$$

which is the probability density that a Brownian noise function $B(t)$ starting out at the value 0, that is, $B(0) = 0$, equals 0 again in the interval $x \leq t \leq x + dx$.

The fact that a variable distributed according to equation 18 scales with an exponent $D = \frac{1}{2}$ is immediately obvious from the distribution integrated from x to ∞, which yields $P(x) = \text{erf}\,(1/\sqrt{2x})$. The interquartile range $s = 9.12$ of this distribution can be obtained from tables of the error integral [JE 51]. Because the Fourier transform of $p(x)$, equation 18, is of the form $\exp\left(-|t|^{1/2}\right)$, adding two independent random variables changes the Fourier variable t to $4t$ and therefore the random variable x to $x/4$. The interquartile range for the sum of two variables is therefore equal to $4s = 36.5$. Thus the *average* of two independent random variables of this sort has an interquartile range that is *twice as large* as that of an individual variable. In general, the average of N independent variables distributed according to equation 18 has a width that is *increased* by a factor $N-$ instead of being reduced by a factor $1/\sqrt{N}$ as in standard cases. No wonder that this kind of statistics is sometime called *non*standard. Yet the world harbors a lot more nonstandard statistics than many experts innocently expected.

For large x, the distribution $p(x)$ according to equation 18 is proportional to $x^{-3/2}$ and the integrated distribution is proportional to $x^{-1/2}$, that is, x^{-D}. More generally, for all invariant distributions other than the Gaussian, the integrated distribution is asymptotically proportional to x^{-D}, where D is the exponent in equation 15. Conversely, since we already know (see pages 155–156) that Levy flights, with integrated distributions following a power law x^{-D}, generate fractal sets with Hausdorff dimension D, we recognize the exponent in the equation $s^D = s_1^D + s_2^D$ as a bona fide fractal dimension for this geometric realization of the distribution. However, such exponents do not generally have the significance of a fractal dimension.

7

C antor Sets: Self-Similarity and Arithmetic Dust

*Eine Menge stelle ich mir
vor wie einen Abgrund.
(I imagine a set to be an abyss)*
—GEORG CANTOR, ca. 1888,
as related by Emmy Noether

In this chapter we shall further pursue one of the most important sources of self-similarity: Cantor sets. Originally constructed for purely abstract purposes, Cantor sets have of late turned into near perfect models for a host of phenomena in the real world—from strange attractors of nonlinear dynamic systems to the distribution of galaxies in the universe.

A Corner of Cantor's Paradise

*Most numbers in the continuum cannot be
defined by a finite set of words.*
—MARK KAC

In the midst of the animated debates during the nineteenth century on the foundations of mathematics—and the very meaning of the concept of *number*—Georg Cantor (1845–1918) wanted to present his colleagues with a set of numbers between 0 and 1 that has measure zero (i.e., a randomly thrown "dart" would be very unlikely to hit a member in the set) and, at the same time, has *so* many members that the set is in fact uncountable, just like *all* the real numbers between 0 and 1.

Many mathematicians, and even Cantor himself for a while, doubted that such a "crazy" set existed[1]—but it does exist, and its construction is in fact quite straightforward. Imagine the real line between 0 and 1 (drawn with chalk on a blackboard, if you will) and wipe out the open middle third, that is, the interval from $\frac{1}{3}$ to $\frac{2}{3}$, excluding the endpoints $\frac{1}{3}$ and $\frac{2}{3}$. Next erase the open middle third of each remaining third, and so forth *ad infinitum*. The result of the first five erasures was illustrated in Figure 10 in Chapter 1 (page 16), but there is no way to draw the final result, aptly called Cantor *dust* by Mandelbrot [Man 83]. In fact, the Cantor dust has holes *on all scales*[2]: No matter how powerful the "microscope" that we use to inspect the set, we always see holes; there is not a single continuous interval in the entire unit interval, but only isolated points. No Cantor number has another Cantor number as an immediate neighbor. The Cantor dust is totally discontinuous—yet infinitely divisible, just like a continuum. As much as it would have amazed the ancient Greeks, there is no fundamental antinomy or philosophical contradiction between the discontinuous (like matter composed of atoms) and the infinitely divisible; the Cantor dust is both.

Formally, a Cantor set is defined as a set that is *totally disconnected, closed, and perfect*. A totally disconnected set is a set that contains no intervals and therefore has no interior points. A closed set is one that contains all its boundary elements. (A boundary element is an element that contains elements both inside and outside the set in arbitrarily small neighborhoods.) A perfect set is a nonempty set that is equal to the set of its accumulation points. All three conditions are met by our middle-third–erasing construction, the original Cantor set.

In spite of its counterintuitive nature, there is a neat number-theoretic way to represent the Cantor dust, namely, by *ternary* fractions employing the digits 0, 1, and 2. For example, $0.5 = \frac{1}{3} + \frac{1}{9} + \frac{1}{27} + \cdots$ corresponds to $0.111\ldots$ in the ternary notation. Representations in the ternary system, like those in the decimal system, are not unique. For example, $\frac{1}{3}$ can be written either as 0.1 or $0.0\overline{2}$, where the bar over the 2 stands for an infinite sequence of 2s. One way to make such number representations unique is to outlaw terminating fractions. Thus, writing 0.1 for $\frac{1}{3}$ is illegal—one *has* to write $0.0\overline{2}$ for $\frac{1}{3}$.

Here we shall adopt the following convention for making ternary fractions unique: we forbid a 1 followed by all 0s or all 2s. Thus, $\frac{1}{3} = 0.1$ is represented by $0.0\overline{2}$, and $\frac{2}{3}$ becomes 0.2 (*not* $0.1\overline{2}$).

---

1. See the excellent new biography *Georg Cantor* [PI 87] for Cantor's arduous career and the genesis and eventual acceptance of set theory.

2. The expression *on all scales*, as we said before, reflects one of the central facts of self-similar structures. For example, in percolation at the critical point (see Chapter 15) there are clusters on all size scales; in spin glasses there are magnetic domains on all length scales; Of course, for real physical systems—as opposed to simple mathematical models—the "all" in the expression *on all scales* has to be taken with a grain of salt. For example, in percolation, cluster sizes lie, of course, somewhere between the size of an individual "atom" and the size of the entire sample.

With this convention, the numbers in the open interval $(\frac{1}{3}, \frac{2}{3})$ are precisely all those numbers that in the ternary system have the digit 1 in the first position to the right of the ternary point. Wiping these numbers away on our route to the Cantor dust, we are left with numbers that begin with 0.0 or 0.2.

Similarly, the second wiping (third line in Figure 10 of chapter 1) eliminates all numbers with a 1 in the *second* place to the right of the ternary point. In the end, having arrived at the Cantor dust, we are left with all those proper ternary fractions that have no 1s in *any* place, such as $0.0\overline{2}$, 0.2, 0.2002, and $0.\overline{2002}$.

The members of the Cantor set form a *self-similar* set in the following sense: take any line in Figure 10 of Chapter 1, leave out the right half, and magnify the remainder threefold. This results in the line immediately above it. More precisely, the Cantor set is invariant, modulo 1, to scaling by a factor of 3. In fact, in the ternary notation, this scaling is nothing but shifting all digits one place to the left and dropping any 2s that protrude to the left of the ternary point. For example, the Cantor number 0.202202 maps into 0.02202, another Cantor number.

With the ternary notation, it is easy to see why the Cantor set has measure zero: the probability that a random digit in the interval [0, 1] has not a single 1 in its ternary expansion is, of course, zero. More precisely, the expression for the probability that there is no 1 in n ternary places equals $(\frac{2}{3})^n$, which goes to zero when n goes to infinity.

The ternary number system is also useful to show that no two Cantor numbers could be adjacent to each other. For example, a Cantor number close to the Cantor number 0.2 would be 0.20 . . . 002. By making the sequence of 0s in this "neighbor" longer and longer, we can approach 0.2 arbitrarily closely, but there are always non-Cantor numbers between 0.2 and, say 0.20002—for example, 0.200012.

But how can we prove that the members of this extremely sparse set are so numerous that they are not even countable? (Note that the integers and the rational numbers—and even the algebraic irrational numbers—are countable.) The reason is that we can bring the members of the Cantor set, although it is very spotty, into a one-to-one correspondence with *all* the real numbers in the interval between 0 and 1. To accomplish this feat, we simply identify with each Cantor number the *binary* number obtained by changing all 2s to 1s. Thus, for example, 0.020222 corresponds to 0.010111 ($= \frac{23}{64}$). In this manner, each member of the Cantor set can be mapped into a real number and, conversely, *all* real numbers between 0 and 1 can be mapped into Cantor numbers, which have thus the same cardinality as the real numbers.

The well-known fact that the real numbers form an *uncountable* set was proved by Cantor using the "diagonal method" already known to Galileo. The diagonal method is used in an indirect proof that proceeds as follows.

Assume that all the real numbers between 0 and 1 form a countable set; they could then be written down, one after another, in a counting sequence. In decimal notation, for example, the list might look as follows:

$$0.91971 \ldots$$
$$0.29216 \ldots$$
$$0.36638 \ldots$$
$$0.55389 \ldots$$
$$\vdots$$

To make the notation unique, we use only nonterminating fractions. Thus, 0.5 is written as 0.49999 . . . , for example.

Now write down a number whose first digit to the right of the decimal point is different from the corresponding digit in the first number in the preceding list and whose second digit is different from the second digit in the second number, and so forth. To avoid ambiguities, do not use the digits 0 and 9 in the replacements. The resulting number, say,

$$0.88578 \ldots$$

cannot be found anywhere in the list because it differs from each number in at least one place. Thus, the list cannot be complete and our assumption that the real numbers form a countable set was false.

There are several ways to prove that *rational* numbers *are* countable. I find the following proof particularly appealing [Sag 89]. Write the rational number as m/n, where m and n are integers that are relatively prime. Let $m = p_1^{e_1} \cdot p_2^{e_2} \cdots p_k^{e_k}$ and $n = q_1^{f_1} \cdot q_2^{f_2} \cdots q_t^{f_t}$ be the prime-number decompositions of m and n. Then the desired counting function for the rational numbers is $f(1) = 1$ and

$$f\left(\frac{m}{n}\right) = p_1^{2e_1} \cdot p_2^{2e_2} \cdots p_k^{2e_k} \cdot q_1^{2f_1 - 1} \cdot q_2^{2f_2 - 1} \cdots q_t^{2f_t - 1}$$

This function is uniquely invertible. For example, the rational number $\frac{2}{3}$ is the twelfth number on the list; and the eighteenth number on the list is $\frac{3}{2}$.

Another, completely counterintuitive consequence of Cantor's set theory is the *equivalence* of two-dimensional areas and one-dimensional lines. Two sets are said to be equivalent if there is a one-to-one mapping between them. Thus, the unit square (an area) and the unit interval (a line) are equivalent: each point in the unit square corresponds uniquely to a point on the unit interval, and vice versa. In communicating this discovery to his friend Dedekind in Brunswick (on June 20, 1877), Cantor wrote, "I see it, but I don't believe it."

Actually, Cantor's mapping from the unit square to the unit interval is almost trivial. For example, the point with the rectangular coordinates $x = 0.123$ and $y = 0.456$ is mapped into the point 0.142536 on the unit interval. Can you see what is going on and do you believe it?

Perhaps equally astounding is the fact that *any* number in the interval [0, 2] can be represented by the sum of two Cantor numbers, their extreme sparseness in the unit interval not withstanding. The reader may find it instructive (and amusing) to prove this counterintuitive property either arithmetically or geometrically.

Cantor Sets as Invariant Sets

One of the arenas in which Cantor sets cavort is that of the so-called *invariant sets* of iterated mappings. (A member of an invariant set maps into a member of the set. An invariant set is the set of all such elements.) Let us consider the simple "tent map" (so called because of its tentlike shape)

$$x_{n+1} = 1.5 - 3|x_n - 0.5|$$

(see Figure 1). Because the slope of this mapping exceeds unity everywhere, there are no attractors, except $x = -\infty$. The two fixed points $x = 0$ and $x = 0.75$ are *repellors*: points near them diverge to infinity.

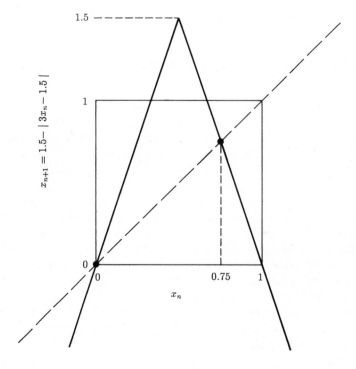

Figure 1   A tent map whose invariant set is the Cantor set of Figure 10 in Chapter 1.

Are there any points that do not diverge, that is, points whose iterates stay forever in the unit interval? To find the answer we write x_n in ternary notation. Then the mapping in Figure 1 says that, for $x_n \leq 0.5$, the next iterate, x_{n+1}, is given by a simple left shift of the digits of x_n. For $x_n > 0.5$, before left-shifting, we must first complement each digit, that is, replace 0 by 2 and 2 by 0, while digit 1 remains 1.

Now, if the initial value x_0 has a 1 anywhere in its expansion, this 1 will eventually be shifted to the left of the ternary point so that thereafter the iterates x_n will exceed 1 and diverge to infinity (unless the fraction terminates with a 1).

But suppose there is not a single 1 in the ternary expansion of x_0, that is, that x_0 is a Cantor number; then all the iterates will also be Cantor numbers (remember $0 \rightarrow 0$ or 2, and $2 \rightarrow 2$ or 0). In fact, the iterates will remain forever bounded in the unit interval. To see this, suppose that $x_n = 0.2022\ldots$, which is greater than 0.5. Thus, x_n has first to be complemented $(0.0200\ldots)$ and then left-shifted to yield $0.200\ldots$, which is smaller than 1. Of course, if x_n has 0 as a first digit, x_{n+1} is likewise smaller than 1. Thus, Cantor numbers—and *only* these—remain forever bounded, and they remain invariably Cantor numbers, which is why they are called the *invariant set* of the tent map $x_{n+1} = 1.5 - 3|x_n - 0.5|$.

A mapping similar to the one discussed here, but restricted to the unit interval, was described by Bau-Sen Du [Du 84].

Symbolic Dynamics and Deterministic Chaos

There is another important point that we can illustrate with the mapping described in the preceding section. Instead of characterizing a succession of iterates (called an "orbit") by the precise values of the x_n, they are often quantized to just two values. For example, if x_n is smaller than 0.5 that is, if it lies to the *left* of 0.5, we represent the values of x_n by a single symbol: L (for left). If x_n exceeds 0.5, then we write R.

Now, curiously, given any initial value x_0 from the invariant set, we can immediately predict the succession of L's and R's, called the *symbolic dynamics* of its iterates. For example, $x_0 = 0.022020002\ldots$ has the symbolic dynamics *LRLRRRLLR*\ldots, which is obtained by writing L if, in x_0, a digit and its left neighbor are the same and R otherwise. The symbolic dynamics describes the evolution in time of a *dynamic system* such as a playground swing—or the voting pattern of a political population.

Even curiouser, from the (discrete!) symbolic dynamics we can *uniquely* determine the exact value of x_0 (if it belongs to the invariant set). For example, the orbit with the symbolic dynamics *RLRLLLRRL*\ldots has started at $x_0 = 0.220000200\ldots$. Can the reader see why?

Understandably, invariant sets (and their complements) play a crucial role in dynamic systems in general because they tell the most important fact about any initial condition, namely, its eventual fate: will the iterates be bounded, or will they be unstable and diverge? Or will the orbit be periodic or aperiodic?

As we can see from these examples, invariant sets can be (and often are) self-similar Cantor sets, that is, uncountable sets with measure 0 and a scaling property. For the Cantor set, the similarity factor is 3, corresponding to a left shift by one digit in the ternary number system.

Representing the evolution in time of a dynamic system by left shifts of the digits in a suitable number system brings out another important property of systems for which such a representation is possible. No matter how accurately the initial condition x_0 of a coordinate is known, the accuracy will always be finite, that is, the digits of x_0 to the right of the last known digit will be unknown. As the dynamic system evolves in time, these unknown digits will be shifted to the left; that is, they will grow in significance, and will sooner or later arrive at the "decimal" point and thus *dominate* the behavior of the system. And because the digits are unknown, this behavior will be completely unpredictable. The resulting motion is called *chaotic*. To emphasize the fact that this chaos is caused by strictly causal, deterministic rules, it is called *deterministic chaos*. Thus we see that there is no contradiction between complete determinism and chaos. In fact, deterministic chaos can be found almost anywhere in nature from turbulence to population dynamics.

This analysis also tells us why the weather is so unpredictable. The reason is that the mathematical equations governing it are those of a chaotic system. To add a single day to reliable weather forecasting, the initial conditions of temperature, air pressure, wind velocity, and other variables at a large number of points on the earth would have to be known with a much greater accuracy than is presently feasible—not to mention the logistic and computational effort to deal with this mass of data. However, supercomputers now on the horizon, employing large-scale parallel processing, promise more reliable forecasting over a slightly extended time span.

Devil's Staircases and a Pinball Machine

One of the more interesting constructions based on Cantor sets is the devil's staircase. Take the original Cantor set, the middle-third–erasing set, and plot, as a function of x in the unit interval, the relative weight y of the set that lies to the left of x. In the first stage of construction, y rises from 0 to $\frac{1}{2}$ as x goes from 0 to $\frac{1}{3}$. Then y stays constant up to $x = \frac{2}{3}$. Beyond this plateau at $y = \frac{1}{2}$, y rises again from $\frac{1}{2}$ to 1 as x goes from $\frac{2}{3}$ to 1. The second stage of construction has two more plateaus, at $y = \frac{1}{4}$ and $y = \frac{3}{4}$ (see Figure 2A).

(A)

(B)

Figure 2 (A) Second stage and (B) a more advanced stage in construction of a devil's staircase. Plateaus from all earlier stages remain visible.

In the limit, the staircase function $y(x)$ has plateaus almost everywhere, yet it manages to rise from 0 to 1 at uncountably many values x; see Figure 2B, which shows an advanced stage of the resulting devil's staircase.

In order to know how high the staircase is for any given value of x, we have to write x as a ternary number and convert it into a binary fraction, replacing any digits 2 up to the first 1 (reading from left to right) by 1s. Keep the first 1 (if any) and write 0s for all following digits to the right. For example,

$$x = 0.20210012\ldots = \frac{1652}{2187}$$

is mapped into

$$y = 0.1011000\ldots = \frac{11}{16}$$

Thus, for every value of x, there is a unique y value.

To go from a given y value to the corresponding value(s) of x, we have to write y as a binary fraction and convert all 1s to 2s, except the last (if any) 1. After that, we replace each 0 by all combinations of 0, 1, and 2, thereby creating an interval. For example,

$$y = 0.1011 = \frac{11}{16}$$

goes into

$$x = 0.2021\overset{000}{\underset{222}{111}}\ldots$$

which represents all the numbers in the open interval $(0.2021\overline{0}, 0.2021\overline{2})$ or, in decimal notation, the interval $(\frac{61}{81}, \frac{62}{81})$. In other words, $y = \frac{11}{16}$ corresponds to one of the plateaus of the devil's staircase with a step width of $\frac{1}{81}$.

In general, any y value whose denominator (in lowest terms) is an nth power of 2 lies on a plateau with width 3^{-n}. All other y values (namely, all nonterminating binary fractions) have unique x coordinates. This is a devilish function indeed: it is constant (doesn't rise) almost everywhere, but it has uncountably many infinitely small discontinuities that allow it to "sneak up" all the way from 0 to 1.

Devil's staircases are excellent models for numerous complex situations in the real world—and the not so real world of mathematical physics. Michel Hénon once invented a kind of pinball machine (see Figure 3), whose symbolic dynamics for a given initial position x_0, $0 < x_0 < 1$, were obtained by computing the devil's function $y(x_0)$ as just described [Hen 88]. If $y(x_0)$ lies on a plateau of the devil's staircase, the ball will escape to plus or minus infinity. The devil's plateaus are, in effect, locked-in intervals for the attractor at infinity. However, if x_0 is a Cantor number, then $y(x_0)$ will not lie on a plateau and the orbit of the ball will forever remain confined to the interval $[0, 1]$. Its symbolic dynamics are then given by

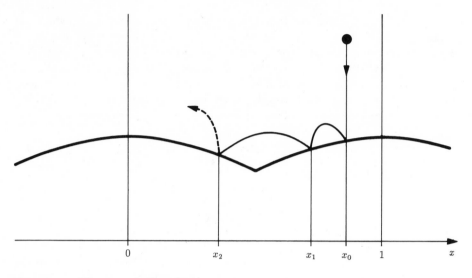

Figure 3 Hénon's pinball machine.

$y(x_0)$ written as a binary fraction with 0 interpreted as L (for left, i.e., $x_n < \frac{1}{2}$) and 1 interpreted as R (for right, i.e., $x_n > \frac{1}{2}$). This is much like the tent map illustrated in Figure 1. (However, Hénon assures me that there are also substantial differences between the tent map and his "inclined billiard," as he called his pinball machine.)

If Hénon's pinball machine were a game played on a computer and the aim were to keep the ball in constant confined motion, then the winning strategy would be to pick a Cantor number for the ball's initial position. This is reminiscent of *Sir Pinski's* game, discussed in Chapter 1 on pages 20–25, for which the Sierpinski gasket (a two-dimensional Cantor set) contains the winning points.

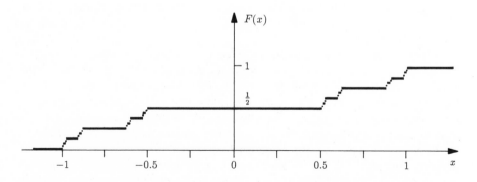

Figure 4 Distribution function for a random variable with a Cantor–set density.

Devil's staircases also appear in probability theory. Figure 4 shows the cumulative distribution of a random variable x given by

$$x = 3 \sum_{j=1}^{\infty} \sigma_j \beta^j \qquad \beta = \frac{1}{4}$$

where the σ_j are independent equiprobable $+1$s or -1s. It is easy to see that x can never fall inside the ranges $(-\frac{1}{2}, \frac{1}{2})$, $(-\frac{7}{8}, -\frac{5}{8})$, $(\frac{5}{8}, \frac{7}{8})$, and so forth. This entails the large plateaus in the cumulative distribution seen in Figure 4. Similar staircase distributions that are constant almost everywhere and rise only at a set of x values forming a Cantor set are obtained for any positive $\beta < \frac{1}{2}$.

Mode Locking in Swings and Clocks

One of the more pervasive oscillatory phenomena in nature goes by the names of *mode locking, frequency pulling, phase locking,* or simply *synchronization of two oscillators.* Here, too, devil's staircases play an illuminating role. In mode-locking applications, the height of the devil's staircase corresponds to the frequency ratio of the two oscillators and the plateaus represent locked-in frequency ratios, locked in at any rational number (not just terminating binary fractions, as in the staircase derived from the original Cantor set). Also, the self-similarity may be only asymptotic and not exact.

Think of a playground swing. It has a natural frequency with which it oscillates when driven by a child moving his or her center of gravity up or down at *twice* the natural frequency, as the child soon discovers—long before ever having heard of parametric amplifiers.

But a swing can also be driven externally by a patient parent who pushes and pulls with a frequency not necessarily equal to the swing's natural frequency. Provided that the external force is coupled strongly enough to the swing, the latter will follow the external force; that is, it will be synchronized with the external frequency over a certain range of driving frequencies.

The first scientist to describe such a synchronizing phenomenon was the Dutch mathematical physicist and astronomer Christian Huygens (1629–1695), the discoverer of the Huygens principle of wave propagation. In a letter from Paris to his father in Holland he described how two pendulum clocks hanging back to back on the same wall separating two rooms would synchronize their motions and tick away in perfect lockstep (see Huygens's book *Horologium Oscillatorium* [Huy 1673]). As this example shows, even the tiniest coupling force can "enslave" one oscillator to another if the ratio of their natural frequencies is close to a small-integer rational fraction, such as $\frac{1}{1}$.

Another early observation of synchronization, this time in outer space, came in 1812 when Gauss discovered that the orbit of the asteroid Pallas was locked to the orbital period of Jupiter in the precise integer ratio 7 to 18, two Lucas numbers L_n, which obey the same recursion as the Fibonacci numbers, namely

$L_n = L_{n-1} + L_{n-2}$, but with the initial condition $L_1 = 1$ and $L_2 = 3$. Gauss never published this epochal finding, except for a brief note in the *Gelehrte Anzeigen*, the rapid-communications bulletin of the Göttingen Academy—in encrypted form! But he did communicate his arresting result on May 5, 1812, to his close friend the astronomer Friedrich Wilhelm Bessel (1784–1846), imploring him to keep it completely secret "for the duration." It seems that Bessel functioned exactly as requested, because Gauss never got his due credit. In fact, the "prince of mathematicians" was later not a little miffed by this turn of events, apparently having forgotten his own secretiveness. (Why was Gauss so loath to let the world know about Pallas and Jupiter? Was he afraid to have uncovered some divine interference in the planetary clockwork? No, Gauss knew full well that it was pure and simple nonlinear mechanics. Perhaps he felt that the news would be too upsetting, as in the case of his non-Euclidean geometry, which he kept encased in his desk for decades.)

A similar synchronization can be observed in some radio (and television) receivers with automatic frequency control (AFC): the dial (if the set still has one) can be detuned manually over a certain frequency range, yet the chosen channel will remain locked in. Still another example is the synchronization by an external signal of the horizontal deflection (the "time base") of an oscilloscope or television set. The internal time-base generator will lock in to an external frequency over a certain frequency range and then jump discontinuously to another *rational* frequency ratio, preferably a ratio involving small integers in the numerator and denominator, such as $\frac{1}{2}$ or $\frac{3}{4}$. In fact, the frequency ranges over which the two frequencies are locked into a rational ratio depend, in many applications, on the magnitude of these integers and particularly the denominator. Thus, in a given mode-locking situation, the locked frequency range for the frequency ratio $\frac{1}{1}$ will be larger than that for frequency ratios such as $\frac{1}{2}$, $\frac{1}{3}$, or $\frac{2}{3}$.

Interestingly enough, these locked frequency ranges show a high degree of universality covering innumerable, seemingly unrelated phenomena such as the oscillations of superionic conductors [MM 86] or the heartbeat of periodically stimulated chicken embryos. In fact, these phenomena can be modeled by asymptotically self-similar fractals, which in many cases have identical Hausdorff dimensions: $D = 0.86 \ldots$. We will resume the topic of periodic, aperiodic, and chaotic oscillations in Chapter 14.

The Frustrated Manhattan Pedestrian

One of the more "charming" mode-locking situations is regularly suffered by this devoted pedestrian on his occasional forays to the Big Apple (also known as New York City). Walking along one of Manhattan's avenues, he is invariably caught, at every intersection, by a traffic light changing to red as he approaches.

Suppose the pedestrian's speed is just under two-thirds of the "speed" of the traffic lights (i.e., the distance between cross streets divided by the period of one complete green-yellow-red cycle). The red lights will force him to wait at every intersection and slow him down to a speed of one-half.

Assuming for simplicity that the green cycle, during which the walker can safely traverse the cross street, lasts exactly half a period and that all lights are perfectly synchronized (as they certainly would *not* be on a one-way avenue), then, in the walker's speed range $\frac{1}{2} \leq s < \frac{2}{3}$, he is locked into an effective velocity $v = \frac{1}{2}$. In general, in the speed range

$$\frac{1}{n+1} \leq s < \frac{2}{2n+1} \tag{1}$$

where $n = 1, 2, 3, \ldots$, he is locked into an effective velocity $v = 1/(n+1)$. But the walker can be locked into many more rational speeds (although they may not appear "rational" to him). In fact, for

$$\frac{2(k-1)}{2(k-1)n+1} \leq s < \frac{2k}{2kn+1} \tag{2}$$

$k = 2, 3, 4, \ldots$, the walker's effective velocity is locked into the lower limit of s.

The staircase function corresponding to these locked intervals is illustrated in Figure 5. Although the graph of v versus s is not exactly self-similar, the locking pattern in the interval $\frac{1}{2} \leq s < 1$ is approximately rescaled and repeated

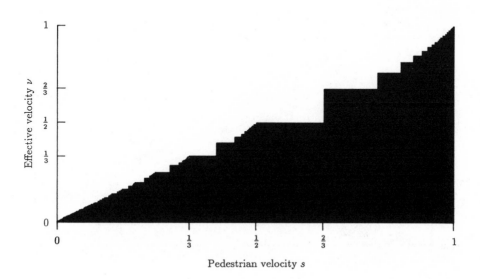

Figure 5   Progress of the frustrated Manhattan pedestrian as an illustration of mode locking.

in the intervals $1/(n + 1) \leq s < 1/n$. One also notices that the locked-in plateaus become smaller and smaller for increasingly larger denominators in inequalities 1 and 2 in the preceding paragraph. In fact, the locked-in speed intervals equal 2 divided by the product of the two denominators.

This scenario of locked intervals, being reciprocally related to the denominators of certain reduced fractions, goes far beyond the hapless Manhattan walker. In fact, we shall encounter staircase functions that, in contrast to Figure 5, have an *uncountable* number of steps.

Arnold Tongues

The plateaus of the devil's staircase encountered previously occur at all heights $y = (2k - 1)/2^n$, with $k, n = 1, 2, 3, \ldots$. However, there are even more satanic staircases that have plateaus at *every* rational number in the interval [0, 1]. While the staircase based on the Cantor set is exactly self-affine (with scaling factors of 3 in the x direction and 2 in the y direction), this is no longer true for satanic staircases, such as the one shown in Figure 6, obtained from the so-called *circle map*:

$$\theta_{n+1} = \theta_n + \Omega - \frac{K}{2\pi} \sin (2\pi\theta_n)$$

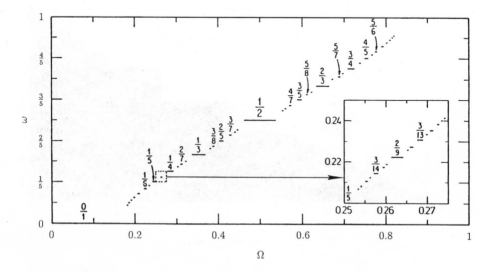

Figure 6   Satanic staircase with plateaus at every rational number.

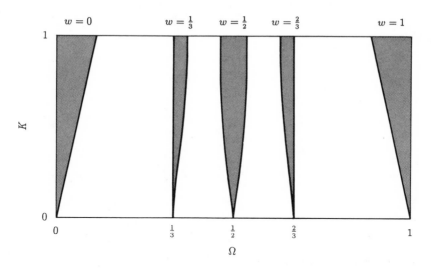

Figure 7   Arnold tongues: locked frequency intervals.

which models many mode-locking phenomena (see Chapter 14). Here K is a coupling strength parameter that controls the degree of nonlinearity and Ω is a frequency ratio, called the *bare winding number*. This frequency ratio may represent the ratio of a driving-force frequency and the resonance frequency of an oscillator (think of the swing, including the one "executed" on the dance floor; or contemplate the frequency ratios of planetary or lunar orbits and spins).

Without coupling ($K = 0$), the so-called *dressed* winding number w, defined as the limit as $n \to \infty$ of $(\theta_n - \theta_0)/n$, equals the bare winding number Ω. But for $K > 0$, w "locks" into rational (frequency) ratios, preferably ratios with small denominators.

Figure 7 shows some of the frequency-locked regions in the Ω-K plane. The shaded regions are called *Arnold tongues*, after their discoverer, the Russian mathematician V. I. Arnol'd. (There never seems to be a lack of suggestive terms in fractal heaven or hell.)

In other applications, the dressed winding number w may represent, for example, the relative number of up spins in an Ising model of an antiferromagnetic material or the relative abundance of a given element (or molecular structure) in a crystal or quasicrystal (see Chapter 13).

For the critical value $K = 1$ of the coupling parameter, the infinitely many locked frequency intervals corresponding to *all* the rational dressed winding numbers w between 0 and 1 actually cover the entire Ω range of bare winding numbers. Irrational values w correspond to an uncountably infinite set of zero measure of Ω values—in other words, a Cantor dust.

8
C H A P T E R

Fractals in Higher Dimensions and a Digital Sundial

Jetzt Mengenlehre (Now set theory)

—KURT GÖDEL, in 1937,
when he decided to turn his mental scalpel loose
on set theory, after completing his proof of
undecidability in number theory

Self-similar or self-affine sets in higher dimensions are models of strange attractors and their basins of attraction; of porous materials, dendritic crystal growth, and quasicrystals; of mountainscapes, Brownian motion, and related stochastic processes that describe an assortment of catastrophes (plus a few happier happenings). Some of these "practical fractals" will be visited in Chapter 10. Here we examine some of their foundations and design a *digital* sundial based on a Cantor set.

Cartesian Products of Cantor Sets

The original one-dimensional Cantor set can be generalized to dusty sets in two or more dimensions in several different ways. Consider the set of all points in the unit square for which both the abscissa x and the ordinate y belong to the Cantor set C. The resulting *Cartesian product* of the Cantor set with itself, usually written $C \times C$, is a *Cantor* dust embedded in two dimensions (see Figure 1).

What is the Hausdorff dimension D of this dust? The set $C \times C$ can obviously be covered by $N(r) = 4^n$ squares of side length $r = 1/3^n$. Thus

$$D(C \times C) = \lim_{n \to \infty} \frac{\log (4^n)}{\log (3^n)} = \frac{\log 4}{\log 3} \approx 1.26 \ldots$$

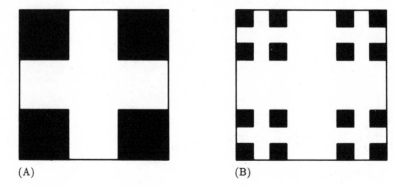

(A)   (B)

Figure 1   Toward a two-dimensional Cantor dust: the Cartesian product of the middle-third Cantor set with itself. (A) The first step and (B) the second step of construction. Fractal dimension = 1.26. . . .

or twice the value of the original one-dimensional Cantor set. In fact, it is easy to guess (and not difficult to prove) that for the three-dimensional Cartesian product $C \times C \times C$, the Hausdorff dimension equals

$$D(C \times C \times C) = \frac{\log 8}{\log 3} = 3\,\frac{\log 2}{\log 3} \approx 1.89 \ldots$$

and so forth: forming a k-fold Cartesian product multiplies the Hausdorff dimension by the factor k, just as for ordinary Euclidean dimensions. (Note that $C \times C \times C$, a dust floating around in three-dimensional space, is so thin that its Hausdorff dimension doesn't even reach the value 2.)

A Leaky Gasket, Soft Sponges, and Swiss Cheeses

Consider another Cartesian product, that of the *complement* C' of the Cantor set C with itself: $C' \times C'$. The complement of $C' \times C'$, that is, $(C' \times C')'$, can be constructed recursively in the following way. The initiator is the unit square, and the generator is the unit square with the central square of side length $\frac{1}{3}$ deleted. In the next iteration, the central squares of side length $\frac{1}{9}$ are removed from the eight remaining squares of side length $\frac{1}{3}$ (see Figure 2A–C). Infinite iteration produces the *Cantor gasket*, approximated by the dark area in Figure 2D.

What is *its* Hausdorff dimension? Figure 2 suggests that $(C' \times C')'$ has a *larger* dimension than $C \times C$. In fact, since the Cantor gasket is strictly self-similar, we need to consider only the generator. To cover it, we need eight

(A)   (B)   (C)   (D)

Figure 2   The first three steps in the construction of the Cantor gasket. Fractal dimension = 1.89 . . . .

squares of side length $\frac{1}{3}$. Thus,

$$D((C' \times C')') = \frac{\log 8}{\log 3} \approx 1.89$$

which is the same value we found for the three-dimensional dust $C \times C \times C$.

What about $(C' \times C' \times C')'$, nicknamed *Cantor cheese*? The generator can be covered by $27 - 1 = 26$ cubes of side length $\frac{1}{3}$. Thus,

$$D = \frac{\log 26}{\log 3} \approx 2.97$$

a value close to 3 because the Cantor cheese is quite solid and has only isolated holes.

The generalization of the Cantor cheese to k Euclidean dimensions, the set $(C' \times C' \times \cdots \times C')'$, has Hausdorff dimension $D = \log (3^k - 1)/\log 3 \approx k - 1/(3^k \log_e 3)$, a value just below the embedding dimension k.

There exists still another *symmetric* fractal set in three dimensions based on C (or any other fractal set in one dimension). It is called the *Menger sponge*, after its architect, Karl Menger, and is depicted in Figure 3 [Men 79]. It has no two-dimensional analogue. Its "holes" are open channels that penetrate the unit cube. Applying the inclusion-exclusion principle, one sees that the generator leaves $27 - 9 + 3 - 1 = 20$ cubes of side length $\frac{1}{3}$, giving a Hausdorff dimension $D = \log 20/\log 3 \approx 2.73$. Thus, the Menger sponge is intermediate between Cantor dust and cheese, but closer to the latter, as one might expect. Mandelbrot suggested the Menger sponge independently as a model of turbulent intermittency [Man 74].

A good set-theoretic description, which brings out the symmetry of the Menger sponge in the three Cartesian coordinates x, y, and z, can be constructed as follows. Call the set C in the x direction X, the set in the y direction Y, and

Figure 3   Menger sponge. Fractal dimension = 2.73 . . . .

that in the z direction Z. Then, just considering generators, $X' \cap Y'$ is the central square "hole" in the x-y plane (see Figure 2) and

$$(X' \cap Y') \cup (Y' \cap Z') \cup (Z' \cap X')$$

represents all the holes of the generator. The complement of this set is the generator of the Menger sponge. By De Morgan's rule [Hal 74], it is

$$(X \cup Y) \cap (Y \cup Z) \cap (Z \cup X) \tag{1}$$

Is the set whose *generator* is the complement of the sponge generator also fractal, with a Hausdorff dimension D between 0 and 3? (The complement of the sponge itself has, of course, a finite Lebesgue measure and $D = 3$.) The complementary generator has $27 - 20 = 7$ cubes of side length $\frac{1}{3}$: a cube bordered by 6 cubes. Hence, $D = \log 7/\log 3 = 1.77$ . . . . It is a rather spidery set.

The original Cantor set generalized to higher embedding dimensions and similar sets produce an increasingly varied zoo of dusty sets.

A Cantor-Set Sundial

By applying the Cantor construction in two dimensions, we obtain, as we have seen, a Cantor "dust" dispersed in the plane. It can be generated from the unit square by eliminating center thirds in both the x and y directions and repeating the process *ad infinitum* (see Figure 1). Mathematicians call the resulting set also the *direct product* of the Cantor set C with itself and denote it by $C \times C$. As we have seen, its Hausdorff dimension D follows directly from the generator, which consists of 4 remaining squares out of 9. Since $C \times C$ can be covered by 4^n (but no fewer) squares of side length 3^{-n}, we obtain

$$D = \frac{\log 4}{\log 3} = \frac{2 \log 2}{\log 3} = 1.26 \ldots$$

which equals twice the value for C itself. In general, we will find that the Hausdorff dimensions of Cartesian product sets are the *sums* of the dimensions of the individual sets. Thus, for example, the Cantor dust $C \times C \times C$ floating around in three-space has $D = 3 \log 2/\log 3 = 1.89 \ldots$.

Arithmetically, such n-dimensional Cantor sets are described by n-tuples of Cantor numbers $(x_1, x_2, \ldots, x_n)$, where each x_k is a Cantor number, that is, a ternary fraction using only 0s and 2s an no 1s, as described in pages 162–163 in Chapter 7.

It is interesting to note that fractal sets embedded in higher-dimensional Euclidean spaces, when projected into spaces with fewer Euclidean dimensions, generate fractal sets whose Hausdorff dimensions depend on the direction of the projection. Consider, for example, the Cantor-like one-dimensional set C_4, defined by eliminating central *quarters* of the unit interval. The dust in three-space constructed from the triple Cartesian product of this set has Hausdorff dimension $D = 3 \log 2/\log \frac{8}{3} = 2.12 \ldots$, which exceeds 2. This set, $C_4 \times C_4 \times C_4$, when projected along one of its three coordinate axes, generates the set $C_4 \times C_4$ with Hausdorff dimension $2 \log 2/\log \frac{8}{3} = 1.41 \ldots$, which is smaller than 2. But other projection directions can produce sets with dimensions equal to 2 having *connected* pieces. In other words, the Cantor-like dust $C_4 \times C_4 \times C_4$ will cast "weightless" shadows in some directions and shadows with visible patterns in other directions.

On the basis of this observations, K. J. Falconer [Fal 87] has proposed one of the most paradoxical sets ever conceived: a *digital* sundial (see Figure 4). Depending on the position of the sun in the sky, the set will cast a shadow that changes every minute to correspond to the local time. If desired, the set can

Figure 4   A digital sundial based on a Cantor set [Fal 87].

even be "enlarged" to show the correct date between winter and summer solstices. Here we have the ultimate timepiece driven by sun power.

Of course, the shadow-casting set is likely to be rather complicated, and the inventor understandably refrains from detailed instructions for its construction (presumably while patents are pending and diffraction limits are being circumvented). However, an inkling of how to set out constructing sets with projections of varying sizes is illustrated in Figure 5.

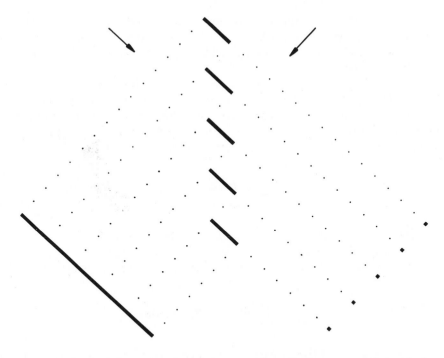

Figure 5   Idea underlying digital sundial: a set of bars that casts very different shadows, depending on the direction of projection.

Fat Fractals

The sundial fractal discussed in the previous section is an example of a fractal set that for certain projection direction has nonzero measure. There are numerous serious applications that are well characterized by fractal sets of nonzero measure. This is particularly true for nonlinear dynamic systems and their basins of attraction. For example, the parameter values for which the prototype of such systems, the logistic parabola (see Chapter 12), shows aperiodic behavior is such a set [Jak 81].

Corresponding to each periodic orbit is a finite interval of parameter values, called a periodic window. The union of periodic windows does not exhaust all parameter values. Thus, the parameter values for aperiodic orbits have a nonzero Lebesgue measure. On the other hand, the distribution of such parameter values has a fractal structure: it has holes (periodic windows) on all scales. Such fractal sets, with nonzero measure, have been called *fat fractals* [EU 86]. Another example of a fat fractal is the set of parameter values for which the subcritical circle maps show aperiodic behavior, that is, parameter values that do not lead to mode locking.

Clearly, such sets cannot be usefully characterized by their Hausdorff dimension, which would simply be equal to the embedding Euclidean dimension and would therefore not provide any additional information. Rather, fat fractals are distinguished by *scaling exponents*.

A simple example of a fat fractal is obtained by starting with the unit interval and removing the central $\frac{1}{3}$ in the first generation, the central $\frac{1}{9}$ of the two remaining thirds in the second generation, the central $\frac{1}{81}$ of the four resulting pieces, and so on, always cutting out central pieces of *relative* length $1/3^{2^k}$ (see Figure 6). After n iterations, we obtain 2^n pieces with a total length

$$\mu_n = \prod_{k=0}^{n-1} (1 - 3^{-2^k}) \tag{2}$$

which for $n \to \infty$ converges on a nonzero value, $\mu_\infty = 0.5851874\ldots$.

A somewhat leaner fat fractal is obtained by excising central pieces of relative length 3^{-k} at each iteration, resulting in a remaining length

$$\mu_\infty = \prod_{k=1}^{\infty} (1 - 3^{-k}) = 0.560\ldots \tag{3}$$

Fat fractals are distinguished by one or another of several scaling exponents. The most useful scaling exponent is defined as follows: Fill all holes of length up to ε and approximate the measure $\mu(\varepsilon)$ of the resulting set for $\varepsilon \to 0$ by the power law

$$\mu(\varepsilon) = \mu(0) + c\varepsilon^\beta \tag{4}$$

where c is a constant and β is the scaling exponent; β lies in the range $0 \leq \beta \leq \infty$.

Figure 6   Construction of a fat fractal. It has inifinitely many holes, but the remainders keep a total length that is greater than zero.

In the present example, with the excising rule 3^{-k}, setting $\varepsilon = 3^{-n}$ gives

$$\mu(\varepsilon) = \prod_{k=1}^{n-1} (1 - 3^{-k}) = \frac{\mu(0)}{\prod_{k=n}^{\infty} (1 - 3^{-k})}$$

or

$$\mu(\varepsilon) = \mu(0)(1 + 3^{-n} + 3^{-n-1} \cdots)$$

and, asymptotically for $n \to \infty$,

$$\mu(\varepsilon) - \mu(0) = \mu(0)(3^{-n}) = \mu(0)\varepsilon$$

Thus, for the fat fractal defined by equation 3, the scaling exponent β equals 1.

Denoting the measure of the holes smaller than ε by $F(\varepsilon)$, the scaling exponent β is also given by

$$\beta = \lim_{\varepsilon \to 0} \frac{\log F(\varepsilon)}{\log \varepsilon} \tag{5}$$

The exponent β is determined by the rate with which the measure of the small holes vanishes.

For the quadratic map, J. Doyne Farmer determined $\mu(0)$ numerically as 0.89795 ± 0.00005. Both for the quadratic map and for a trigonometric map with a quadratic maximum, $x_{n+1} = \gamma \sin(\pi x_n)$, Farmer found $\beta = 0.45 \pm 0.04$. This is a hint that the exponent β is universal, that is, the same for all maps with a quadratic maximum [Far 85].

Another scaling exponent, α, is obtained by fattening all holes by ε. This not only fills in the small holes but also reduces the size of the large holes. Let $G(\varepsilon)$ be the additional contribution to the measure. The exponent α is then defined by

$$\alpha = \lim_{\varepsilon \to 0} \frac{\log |F(\varepsilon) + G(\varepsilon)|}{\log \varepsilon} \tag{6}$$

It can be shown that $\alpha \leq \beta$. If $\alpha < \beta$, then α is determined by the large holes. Since, in most applications, the fine-grain fractal structure is more important than the coarse-grain structure, β is the more useful exponent. However, the exponent α can still contain useful information in the case of fat fractals, describing, for example, parameter values for which a nonlinear system shows chaotic motion. For $\alpha = \beta$, such systems exhibit sensitive dependence on the *parameter*; that is, arbitrarily close to some parameter value resulting in chaotic motion, there are other values for which the motion is periodic, while for $\alpha < \beta$ this is not the case. Thus, the equality $\alpha = \beta$ signals an important property of nonlinear dynamic systems called *parameter sensitivity*.

9

M ultifractals: Intimately Intertwined Fractals

Cantorism is a disease from which mathematics would have to recover.
—HENRI POINCARÉ

Fractals have immeasurably enlarged our ability to describe nature. The abstract constructions going back to Bernard Bolzano (1781–1848), Cantor, and Giuseppe Peano (1858–1932) have furnished us with models of reality much more realistic than the Euclidean empire of integer exponents and smooth shapes. Yet there are many phenomena in physics, chemistry, geology, and crystal growth, in particular, that require a generalization of the fractal concept to include intricate structures with more than one scaling exponent. Many of these matters are in fact characterized by an entire *spectrum* of exponents, of which the Hausdorff dimension is only one. The generalized fractals fashioned to cope with these cases are called *multifractals*. Their applications range from the distributions of people or minerals on the earth to energy dissipation in turbulence or fractal resistor networks. Diffusion-limited aggregation, viscous fingering, and the distribution of faults in computer networks, or that of impurities in semiconductors, are likewise well modeled by multifractals, as are certain games of chance. And many strange attractors of nonlinear dynamic systems are also clearly multifractals. In fact, it was mainly with *multi*fractals that the "fractal geometry of nature" overtook pure geometry to conquer the natural sciences.

The Distributions of People and Ore

Gold is everywhere on earth. Not only can it be found in a few rich veins; there are thousands of lesser deposits where gold can be profitably produced. And there are millions of sites on earth where gold is known to exist but not worth mining. In fact, gold is all around us and even *in* us. The total amount of gold in the oceans is estimated at billions of tons, but its concentration is less than 6 parts in 1 trillion parts of seawater. Thus, there seem to be no total voids in the distribution of gold on earth.

And what is true of gold is manifest for many minerals too. As the Dutch geologist H. J. de Wijs once wisely observed, a mineral is typically not concentrated exclusively in ore veins, but can be found between the veins, too, albeit in lesser concentrations. And the veins themselves show characteristic variations of the concentration [deW 51]. In fact, every time a volume of ore is bisected along the vein, the relative amounts of the mineral in the two half volumes are p and $1 - p$, respectively. Interestingly, the maximum value of the parameter p, which measures the variability of the mineral's concentration, stays roughly constant from bisection to bisection. But does this law live for ore alone?

How are *people* distributed over a large connected landmass, say, Eurasia or the Americas? If we cut the total land area into two equal-area pieces, we may find that perhaps 70 percent of the people live on one side of the dividing line, while only 30 percent live on the other side. In general, the proportion of people on one side may be p, with $p > 0.5$, whereas it is only $1 - p$ on the other side. In the present example, $p = 0.7$. The excess proportion p depends, of course on the direction of the cut. Let us therefore assume that the cut direction is chosen to maximize p.

If we now proceed to cut the denser half area into quarter areas, again positioning the cut to maximize the proportion in one of the quarters, we may find that the population percentages are approximately p^2 and $p(1 - p)$, respectively. Similarly, cutting the sparser half into two equal areas will result in quarter areas with densities near $(1 - p)p$ and $(1 - p)^2$.

Iterating this bisecting process results in an asymptotically self-affine distribution (see Figure 1). We begin with a uniform probability distribution over the unit interval (Figure 1A). After bisecting the interval once, we find two probabilities, $1 - p$ and p, for the two halves of the unit interval. With $1 - p < p$, we obtain the single-step distribution shown in Figure 1B for $p = \frac{2}{3}$. Bisecting once more results in four intervals and the probability distribution shown in Figure 1C. A third bisecting results in distributions shown in Figure 1D. In the limit of infinitely many bisectings, we obtain a self-affine probability distribution: the left half of the distribution stretched by a factor of 2 in the horizontal direction and by a factor of $1/(1 - p)$ in the vertical direction reproduces the entire distribution [PS 86].

If we consider $p = 0.7$ as characteristic for the distribution of people on the earth as a whole, then 18 bisections of the earth's land area will leave just two hermits living on an area of $576 \text{ km}^2 = 24 \text{ km} \times 24 \text{ km}$ in the sparsest region, say, in central Siberia, while 8 million people will share the same area in a dense megalopolis. According to this simple bisecting model, most people (3.5 billion) live in 60,000 communities of 20,000 to 300,000 people each.

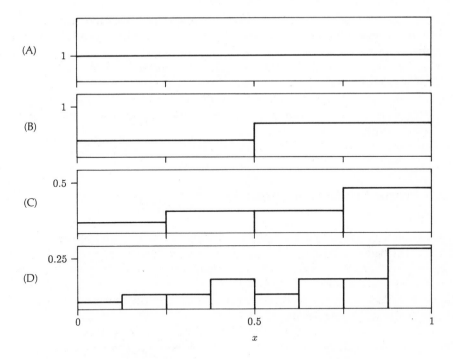

Figure 1   Constructing a self-similar distribution by iterated bisecting [PS 86].

And what is true of people and ore obtains for photons too. Take a beam of light from an old-fashioned incandescent lamp and cut it into two equal portions. The number of photons in the two halves will not be the same, nor will the numbers be equal in the four quarter beams, and so on, for each bisection. Or take electromagnetic cavity radiation, produced in a hollow space whose walls are heated to a given temperature. (A tiny hole in the cavity's wall will emit the justly famous blackbody radiation, as described by Planck's law.) The number of photons in one of the cavity's phase cells or modes of oscillation is distributed according to a geometric distribution: add one photon, and the probability decreases by a constant factor $m/(m + 1)$, where m is the expected number of photons (given by the "Bose-Einstein" distribution). The variance σ^2 of this distribution equals $m^2 + m$, where the first term (m^2) stems from the heat-induced random fluctuations of the classical electromagnetic field. The second term (m) reflects the "granularity" of the energy due to Einstein's photons, the particles of light whose existence he deduced from the added m in $\sigma^2 = m^2 + m$. (This granularity was originally introduced by Planck to match theory to experiment.)

For large m, $\sigma^2 \approx m^2$; and bisecting the cavity's volume will result in an expected proportion p of photons in one half and $1 - p$ in the other half, with $p \approx 0.6$ for a cutting direction that maximizes p—independent of the *number* of photons. For large m the geometric distribution is in fact scale-invariant or self-similar, because $\sigma \sim m$. Thus, repeated bisections will continue to apportion the available photons in a self-similar branching ratio $p/(1-p) \approx 1.5$ until their number becomes so small that σ/m is no longer constant and the existence of individual photons destroys strict self-similarity.

In laser light, by contrast, photons obey a Poisson distribution with $\sigma^2 = m$. Thus, the scale of the distribution, $\sigma \sim \sqrt{m}$, is not proportional to its mean m, and self-similarity does not obtain.

Self-Affine Fractals without Holes

The repeated bisecting and multiplying of proportions in each half by $(1-p)$ for each left half and by p for the right half intervals, which we exercised in the preceding section, is a special case of a multiplicative random process. Infinite iteration of such a process results in a *self-affine* distribution of densities (see Figure 1, which shows the initial uniform density over the unit interval and the results of the first three iterations).

After n iterations, the probability in the interval $m \cdot 2^{-n} < x < (m+1)(2^{-n})$ is given by $p^k(1-p)^{n-k}$, where k is the number of 1s in the first n binary places of x. For example, for $n=6$ and $x = \frac{1}{5} = 0.00110011 \ldots$, $k=2$. Hence the proportion equals $p^2(1-p)^4$, as it does for the entire interval $0.001100 = \frac{12}{64} < x < 0.001101 = \frac{13}{64}$ of length 2^{-6} that contains $x = \frac{1}{5}$. In the sixth iteration, there are $\binom{n}{k} = \binom{6}{2} = 15$ intervals with this density, namely, those 15 intervals whose x values have precisely two 1s in their first six binary places. The most frequent density for $n=6$ corresponds to $k=3$. It occurs $\binom{6}{3} = 20$ times, the leftmost interval beginning at $x = 0.000111 = \frac{7}{64}$. The reason the binary notation for x describes this kind of distribution is that each 0 in the binary expansion of x corresponds to a left half interval and each 1 to a right half interval.

Note the incipient self-affinity in this recursive construction: the right half of each distribution (see Figure 1) equals the left half times $p/(1-p)$, and the entire distribution tends to be invariant as the left half is stretched by a factor of 2 in the horizontal direction and a factor of $1/(1-p)$ in the vertical direction

In the limit as $n \to \infty$, the distribution $P(x)$ over the entire unit interval equals the one over the left half interval stretched horizontally by a factor of 2 and vertically by the factor $1/(1-p)$:

$$P(x) = \frac{1}{1-p} P\left(\frac{x}{2}\right) \tag{1}$$

This is a functional equation that we shall encounter again in an intriguing gambling strategy (see pages 207–210). The factors 2 and $1/p$ are the two scaling factors of this self-affine fractal.

How else can we characterize this fractal function? The usual Hausdorff dimension D, based on the limit as $r \to 0$ of log N/log $(1/r)$, is of little help here. After n iterations the number of pieces N equals 2^n and the length r of each piece equals 2^{-n}. Thus $D = 1$, reflecting the fact that the fractal shown in Figure 1 has no holes.

On the other hand, if we focus on the percentages and their distribution, we find that after $n = 2m$ iterations a large number of the segments, namely, $\binom{n}{m}$, have a probability of $(1 - p)^m p^m$. Their locations within the unit interval are precisely all those half-open intervals of length 2^{-n} whose abscissa values x have an equal number of 0s and 1s in the first n binary places of x. For example, for $n = 4$, these $\binom{4}{2} = 6$ special intervals are given by $x = 0.0011 \ldots \triangleq (\frac{3}{16}, \frac{4}{16})$, $x = 0.0101 \ldots$, $x = 0.0110 \ldots$, $x = 0.1001 \ldots$, and $x = 0.1010 \ldots$, and $x = 0.1100 \ldots$. (Here the triple dots indicate all possible combinations of 0s and 1s, thereby defining an *interval* and not just a single value.)

In general, there are $\binom{n}{k}$ segments of length 2^{-n} with density $(1 - p)^k p^{n-k}$, representing a total probability

$$P_{k/n} = \binom{n}{k}(1 - p)^k p^{n-k}(2^{-n})$$

Note that

$$\sum_{k=0}^{n} P_{k/n} = 1$$

Using Stirling's formula for factorials and ignoring an immaterial factor $(n/2\pi k(n - k))^{1/2}$, we can write

$$\binom{n}{k} \approx \left(\frac{k}{n}\right)^{-k}\left(1 - \frac{k}{n}\right)^{k-n} = 2^{nH(k/n)} \tag{2}$$

where H is the entropy function, well known from thermodynamics and information theory:

$$H(\xi) := -\xi \log_2 \xi - (1 - \xi) \log_2 (1 - \xi)$$

shown in Figure 2.

The fractal dimension of the set of all those subintervals having the common probability $(1 - p)^k (p)^{n-k}$ is given by

$$\tilde{f}(\xi) = \lim_{n \to \infty} \frac{\log \binom{n}{k}}{n \log (1/r)} \tag{3}$$

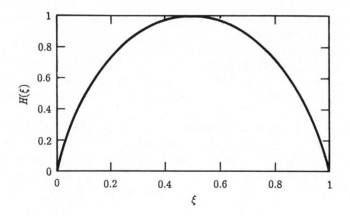

Figure 2   The entropy function.

where $\xi = k/n$. With $r = \frac{1}{2}$ and the approximation in equation 2, we have

$$\tilde{f}(\xi) = H(\xi) \tag{4}$$

Thus, depending on the value of ξ, representing a given probability, we get different fractal dimensions for the support of that probability. In fact, $\tilde{f}(\xi)$ ranges from the smallest value $\tilde{f}(\xi) = 0$ for the lowest and the highest densities ($\xi = 0$ and $\xi = 1$, respectively) to a maximum value $\tilde{f}(x) = 1$ for $\xi = 0.5$. This is one reason why such fractals are called *multi*fractals. Note that the subsets of the unit interval that correspond to a given fractal dimension $\tilde{f}(\xi)$ are scattered all over the unit interval and are intimately intertwined with subsets of other dimensions. This is another characteristic feature of multifractals. Thus, for $\xi = 0.25$, for example, the fractal dimension $\tilde{f}(\xi)$ equals approximately 0.811 and the corresponding subsets are all those *points* $0 \le x \le 1$ for which the binary expansion of x has 25 percent 0s and 75 percent 1s. For example, $x = 0.\overline{0111} = \frac{7}{15}$ is one of uncountably many such points. Again, the fractal dimension of this Cantor-like dust equals $0.811 \ldots$.

After having considered the fractal dimension $f(\xi)$ of the *support* of a multifractal, we ask how the *probabilities* $(1 - p)^k p^{n-k}(2^{-n})$ of equation 1 scale as we let $n \to \infty$. For this purpose we introduce the *Lipshitz-Hölder exponent* $\alpha(\xi)$, which is defined to ensure that the product $p^k(1 - p)^{n-k} r^{-n\alpha(\xi)}$ does not diverge to zero or infinity as $n \to \infty$. Thus, the Lipshitz-Hölder exponent, which characterizes the singularities of the probabilities, is given by

$$\alpha(\xi) = \frac{\xi \log p + (1 - \xi) \log (1 - p)}{\log r} \tag{5}$$

where $\xi = k/n$, as before, and $r = \frac{1}{2}$ for our particular bisecting process.

As equation 5 shows, $\alpha(\xi)$ is a *linear* function of ξ that is monotonically increasing for $p < 0.5$. For $\xi = 0$, $\alpha = \alpha_{min} = -\log_2(1 - p)$; and for $\xi = 1$, $\alpha = \alpha_{max} = -\log_2 p$. Thus, for $p = 0.3$, for example, the Lipshitz-Hölder exponent α ranges from $\alpha_{min} = 0.51$ to $\alpha_{max} = 1.74$. The value α_{min} represents the least probable part of the multifractal and α_{max} the most probable.

Although the fractal subsets of a multifractal are perfectly deterministic, as opposed to random fractals, they exhibit much less geometric regularity than the original Cantor set. For example, at the twelfth stage of construction, the triadic Cantor set consists of 2^{12} pieces of length $3^{-12} \approx 2^{-19}$, which form a regular geometric pattern. Its Hausdorff dimension, we recall, is $D \approx 0.631$.

A multifractal without holes having about the same Hausdorff dimension ($D \approx 0.629$) is characterized by binary fractions with a proportion of 1s equal to 3/19. Its nineteenth stage of construction consists of $\binom{19}{3} = 969$ pieces, which like the triadic Cantor set, have length 2^{-19}. But these intervals form a rather irregular pattern, as dictated by the binary fractions of length 19 containing three 1s. By contrast, *ternary* fractions with missing 1s, which describe the triadic Cantor set, result in a well-ordered pattern.

Another irregularity of our multifractal is betrayed by the *number* of pieces. If one estimated its Hausdorff dimension by the number of pieces at the nineteenth stage of construction, the result would be $\tilde{D} = \log 969/\log 2^{19} \approx 0.522$, which is considerably less than the asymptotic value $D \approx 0.629$. Even at the 190th stage of construction, generating $7.74 \cdot 10^{34}$ pieces, the estimated value of $D = 0.610$ still falls short of the final value by 3 percent. Such subsets of multifractals thus demonstrate how slowly estimates of fractal dimensions can converge if a fractal is not self-similar—even if the structure from which the fractal is derived is perfectly self-affine (Figure 1).

To illustrate the irregularity of such multifractals, Figure 3 shows successive stages of construction of the subset with $k = \lfloor n/2 \rfloor$ of the multifractal illustrated in Figure 1.

Figure 3 Successive stages of construction of the subset with the highest probability of occurence of the multifractal illustrated in Figure 1.

The Multifractal Spectrum: Turbulence and Diffusion-Limited Aggregation

In most applications, one has no direct access to the variable ξ [Fed 88]. In fact, ξ, being related to a very specific bisecting process, is often irrelevant. The important parameters for describing a multiplicative random process, like the one considered in the preceding sections, are the fractal dimension $\tilde{f}$ of the support, the Lipshitz-Hölder exponent α of the density distributions, and their relation $f(\alpha) := \tilde{f}(\xi(\alpha))$, called the "strength of the singularity" α, that is, the Hausdorff dimension of its support, or simply the *multifractal spectrum*. In our example, in which α is a linear function of ξ, the spectrum $f(\alpha)$ is simply a stretched and transposed version of $\tilde{f}(\xi)$ (see Figure 4).

The maximum of $f(\alpha)$ occurs for $\xi = 0.5$. According to equation 5, the corresponding value of α is $\alpha_0 = -\frac{1}{2} \log_2 p(1 - p) = (\alpha_{min} + \alpha_{max})/2$. The value of the multifractal spectrum equals $f(\alpha_0) = 1$, that is, the Hausdorff dimension of the unit interval (or any interval). For multiplicative random processes on a fractal (as opposed to an interval) with Hausdorff dimension D, the highest value of $f(\alpha)$ equals D. In other words, the maximum of the multifractal spectrum $f(\alpha)$ equals the Hausdorff dimension of the support of the process.

Another special point of $f(\alpha)$ is the value $f(\alpha_1)$ at which its slope, $df/d\alpha$, equals 1. With

$$\frac{df}{d\alpha} = \frac{d\tilde{f}}{d\xi} \cdot \frac{d\xi}{d\alpha} = \frac{\log \xi - \log (1 - \xi)}{\log p - \log (1 - p)} \tag{6}$$

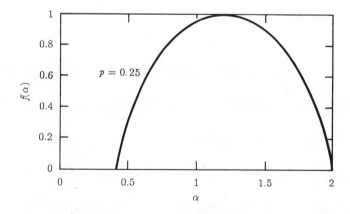

Figure 4   Multifractal spectrum of multiplicative random process.

we see that, for $df/d\alpha = 1$, the corresponding value of ξ is given by $\xi_1 = p$. With equation 5 we then obtain $\alpha(\xi_1) = \alpha_1 = H(p)$, which, according to equation 4, equals $\tilde{f}(p) = f(\alpha_1)$. Hence, $f(\alpha_1) = \alpha_1$, and $f(\alpha_1)$ lies on the tangent of the $f(\alpha)$ curve with slope 1 through the origin. The value $f(\alpha_1)$ equals the information dimension D_1 (see pages 203–207).

Figure 5 shows the multifractal spectrum for the energy dissipation in fully developed turbulence along a one-dimensional straight-line path through the turbulent flow. The turbulent regions form the *support* of the multifractal. The experimental points are from different physical realizations of turbulence (such as atmospheric turbulence, boundary-layer turbulence, and turbulence in the wake behind a circular cylinder or wire grid). Note that these measurements are well matched by a single $f(\alpha)$ curve, the best match being obtained for $p = 0.3$. Thus, it seems that turbulence is indeed well modeled by multifractals as originally suggested by Mandelbrot [Man 74].

Another beautiful example of a multifractal phenomenon is diffusion-limited aggregation (DLA) as analyzed by Meakin and his coworkers [MSCW 85, MCSW

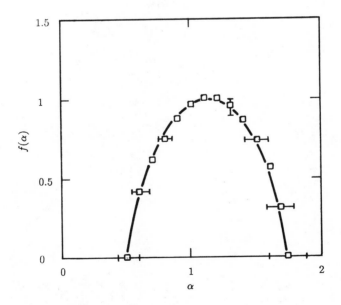

Figure 5   Multifractal spectrum in turbulence [MS 87].

86]. In DLA, single molecules perform a random walk until they become "stuck" on the aggregate, producing attractive random fractals (see Figure 6), reminiscent of certain biological growth patterns and "Lichtenberg" figures of electrical breakdown on insulating surfaces (see Figure 7 [NPW 84]). These patterns are characterized by a dendritic design with "fjords" on many size scales. The reason for this structure in DLA is that a wandering molecule will settle preferentially near one of the tips of the fractal, rather than inside a deep fjord; the probability

Figure 6   Crystal growth by diffusion-limited aggregation (DLA) [HFJ 87].

of penetrating a deep fjord without having become stuck earlier is simply too low. Thus, different sites have different growth probabilities, which are high near the tips and decrease with increasing depth inside a fjord. This is the precise paradigm for which multifractals are tailor-made.

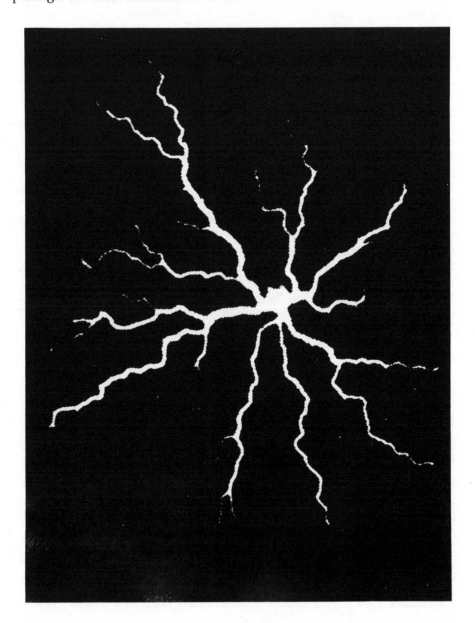

Figure 7   Lichtenberg figure of electrical discharge. Fractal dimension $D \approx 1.7$ [NPW 84].

Figure 8A shows a two-dimensional DLA cluster and Figure 8B shows the sites with a relatively high growth probability. As expected, these high-probability growth sites favor the tips of the cluster. The experimental multifractal spectrum $f(\alpha)$, which extends from $\alpha_{min} \approx 0.5$ to $\alpha_{max} > 5$, models the theoretical one quite well; see the review by Stanley and Meakin and the references cited there [SM 88].

Figure 9 shows the result of an early computer simulation of DLA by Witten and Sander. The fractal dimension for two-dimensional DLA is found to lie near 1.7. This means that the mass of the aggregate increases with its linear dimension L as $L^{1.7}$ and the average density goes as $L^{1.7}/L^2 = L^{-0.3}$—that is, it decreases, in accordance with the visual appearance of such growth patterns. In three-dimensional DLA the fractal dimension is typically near 2.5 [WS 81].

The visual similarity between Lichtenberg figures and DLA patterns is not accidental. Both processes are governed by the Laplace equation of potential theory, the gradient of the potential corresponding to the diffusion field in DLA. The surface of the DLA cluster is an equipotential surface. In this approach to DLA, particles will attach themselves preferentially at those sites of the cluster for which the potential gradient is high, which is near the tips.

In lightning and Lichtenberg figures, and similar electrical breakdowns, the potential is, of course, the *electrical* potential. The growth of a lightning stroke or a discharge occurs preferentially in the direction of the highest gradient of the potential. The deep "fjords" of the pattern, by contrast, are well shielded electrically and therefore experience little or no growth. This correspondence between potential theory and fractal growth has been fully confirmed in careful measurements and numerical solutions of the potential equation [NPW 84].

(A)   (B)

Figure 8   (A) DLA cluster. (B) Sites with high probability of growth [MCSW 86].

Figure 9 Fractal cluster grown by computer simulation of diffusion-limited aggregation
[WS 83].

Viscous Fingering

A related problem is *viscous fingering*, observed most conveniently at the interface
of two liquids between two glass plates. For two miscible liquids, like gelatin
and water, a DLA-like structure appears when one liquid invades the other see
(Figure 10A). This growth process is the result of a hydrodynamic instability
between the liquids. As in DLA, any small "bump" on the interface will tend to
grow predominantly at the tips because the pressure gradient, which drives the
growth, is largest at the tips.

(A)   (B)

Figure 10   Viscous fingering at the interface of two liquids. The shape of the "fingers" depends on the mutual miscibility of the liquids. (A) A dark-colored water was injected through a central cannula (visible as a black bar) into fluid gelatin, which is highly miscible with water. (B) A less miscible concentrated sugar solution was injected into the gelatin.

For two *immiscible* liquids, like glycerine and oil, the fingers are much wider because the surface tension between the two liquids prevents the formation of thin dendrites—that is, tips with high curvature (see Figure 10B). Because of this kind of fingering, much of the oil stays in the ground when water is injected with high pressure into oil-containing shale. By increasing the surface tension of the water through additives, the tips can be made even rounder, thereby lowering the fractal dimension and increasing the amount of oil that can be extracted before the first water arrives at the point where the oil is extracted.

Multifractals on Fractals

Considering the awe that multifractals have inspired in some quarters, the generalization that leads from fractals to multifractals is surprisingly simple. We will next consider multifractals on a *fractal* support. Take the unit interval from which a number of open intervals have been removed, leaving individual line segments of lengths r_i, separated by empty "holes." We associate with each line segment r_i a weight or *probability* p_i. Iterating this generalized process of removal and assignment of probability, we arrive at a generalization of the Cantor set with probabilities associated with each speck of "dust." This is the prototypical multifractal on a fractal support that we want to study in this chapter.

Associating a probability p_i with each segment r_i will allow us to model fractal growth processes in which the different segments correspond to the different sites at which growth takes place. The probabilities p_i represent the different growth rates at these sites as in diffusion-limited aggregation (see pages 193–199). In the application to strange attractors, the segments r_i converge on the different values a dynamic variable can assume, called the support of the attractor, while the probabilities p_i model the frequencies with which the segments are "visited."

While a self-similar Cantor set, generated from segments of equal length, is characterized by a single scaling exponent, the Hausdorff dimension D, multifractals are described by *two* scaling exponents, one for the supporting fractal and one for the probabilities.

To properly introduce these two scaling exponents, we first recall the definition of the Hausdorff dimension. The Hausdorff dimension D of a set is given by the limit as $r \to 0$ of the expression $\log N / \log (1/r)$, where N is the smallest number of pieces of diameter r to completely cover the set:

$$D := \lim_{r \to 0} \frac{\log N}{\log (1/r)} \qquad (7)$$

This definition of D can also be rendered in the following implicit form:

$$\lim_{r \to 0} N r^D = c \qquad 0 < c < \infty \qquad (8)$$

where c is a constant. Equation 8 brings out an important property of the Hausdorff dimension: it is *the* exponent that keeps the product $N r^D$ finite and nonzero as $r \to 0$. If D is altered even by an infinitesimal amount, this product will diverge either to 0 or to ∞.

In the recursive construction of a self-similar set, the number of pieces N after n iterations is N_G^n, where N_G is the number of pieces of the generator. Similarly, r equals r_G^n, *where* r_G is the length of the segments of the generator. (These segments are assumed to have equal lengths at this point.) Thus, instead of equation 8, we may write

$$\lim_{n \to \infty} (N_G r_G^D)^n = c \qquad (9)$$

which of course requires

$$N_G r_G^D = 1$$

or

$$D = \frac{\log N_G}{\log (1/r_G)} \tag{10}$$

Hence, for a strictly self-similar set it is not necessary to take the limit as $r \to 0$ as in equation 7. It suffices to use the parameters of the generator N_G and r_G.

For generator segments of different length r_i, equation 9 becomes

$$\lim_{n \to \infty} \left(\sum_{i=1}^{N} r_i^D \right)^n = c \tag{11}$$

which implies

$$\sum_{i=1}^{N} r_i^D = 1 \tag{12}$$

For the generalized generator with line segments r_i and probability p_i, we introduce *two* exponents, the exponent τ for the support intervals r_i, and the exponent q for the probabilities p_i. Thus, instead of the limit in equation 11, we consider

$$\lim_{n \to \infty} \left(\sum_{i=1}^{N} p_i^q r_i^\tau \right)^n \tag{13}$$

and we ask for the values of q and τ for which expression 13 stays finite—in other words, which q and τ satisfy

$$\sum_{i=1}^{N} p_i^q r_i^\tau = 1 \tag{14}$$

It is obvious from equation 14 that there are no unique values of q and τ. Rather, there is a continuous range of exponents $\tau = \tau(q)$ corresponding, as we shall see, to a continuum of fractal dimensions.

In the case of the original Cantor set ($N = 2$, $r_i = \frac{1}{3}$), equation 14 reads, with $p_i = \frac{1}{2}$,

$$\left(\frac{1}{2} \right)^q \left(\frac{1}{3} \right)^\tau + \left(\frac{1}{2} \right)^q \left(\frac{1}{3} \right)^\tau = 1$$

which has the solution $\tau = (1 - q) \log 2/\log 3$. Injecting the Hausdorff dimension $D = \log 2/\log 3$ for the original Cantor set, we can also write $\tau = (1 - q)D$.

This relation between τ and D is also borne out by comparing expression 13 with equation 11: for $q = 0$, τ corresponds to D, in agreement with the relation $\tau = (1 - q)D$. This comparison suggests that $\tau/(1 - q)$ may play the role of a *generalized dimension* D_q that agrees with the Hausdorff dimension for $q = 0$ but may be different for other values of q.

Another method of creating new dimensions is discussed in the following section.

Fractal Dimensions from Generalized Entropies

In his attempt to generalize the concept of entropy of a probability distribution, the Hungarian mathematician A. Rényi introduced the following expression based on the moments of order q of the probabilities p_i:

$$S_q := \frac{1}{q - 1} \log \sum_{i=1}^{N} p_i^q \tag{15}$$

where q is not necessarily an integer [Ren 55]. For $q \to 1$, the definition in equation 15 yields the well-known entropy

$$S_1 = -\sum_{i=1}^{N} p_i \log p_i \tag{16}$$

of a discrete probability distribution. The definition in equation 15 can therefore be considered, as was Rényi's intent, a *generalized entropy*.

Taking a cue from Rényi, we define the generalized dimensions

$$D_q := \lim_{r \to 0} \frac{1}{q - 1} \frac{\log \sum_{i=1}^{N} p_i^q}{\log r} \tag{17}$$

where p_i is the probability that the random variable falls into the ith "bin" of size r. The parameter q ranges from $-\infty$ to $+\infty$. Note that, for a self-similar fractal with equal probabilities $p_i = 1/N$, the definition in equation 17 gives $D_q = D_0$ for all values of q. For such a fractal, going to the limit as $r \to 0$ is not necessary. Thus

$$D_q = \frac{1}{q - 1} \frac{\log N(1/N)^q}{\log r} = \frac{\log N}{\log (1/r)}$$

which is independent of q.

It is clear that for $q = 0$, the definition in equation 17 agrees with that for the Hausdorff dimension D. For this reason we call the D_q *generalized dimensions*, hoping that they will prove another potent tool in describing multifractals. This is indeed the case. In fact, the D_q are uniquely related to the two exponents q and τ for the general multifractal.

This relation can easily be deduced from the limit in expression 13 by introducing N *constant* bin sizes $r_i = r$, which, for $n \to \infty$, does not affect the values of τ and q for which the limit in expression 13 converges. Thus,

$$\tau = \tau(q) = -\lim_{r \to 0} \frac{\log \sum_{i=1}^{N} p_i^q}{\log r}$$

and, with the definition in equation 17.

$$\tau(q) = (1 - q)D_q \tag{18}$$

For a self-similar fractal, the dimensions D_q can be obtained directly form the p_i and r_i of the generator using equation 14 and the identity in equation 18:

$$\sum_{i=1}^{N} p_i^q r_i^{(1-q)D_q} = 1 \tag{19}$$

For $q = 1$, $\tau(q) = 0$ and D_q is given by

$$D_1 = -\frac{d\tau}{dq}\bigg|_{q=1}$$

which, with equation 14, becomes

$$D_1 = \frac{\sum_{i=1}^{N} p_i \log p_i}{\sum_{i=1}^{N} p_i \log r_i} \tag{20}$$

or, for N equal probabilities $p_i = 1/N$,

$$D_1 = \frac{N \log N}{\sum_{i=1}^{N} \log (1/r_i)} \tag{21}$$

For $q \to 1$, the definition in equation 17 yields

$$D_1 = \lim_{r \to 0} \frac{S_1}{\log r}$$

where S_1 is the entropy of the probabilities p_i given by equation 16.

This entropy and D_1, also called the *information dimension,* play an important role in the analysis of nonlinear dynamic systems, especially in describing the loss of information as a chaotic system evolves in time. In this context, the entropy S_1 is called the *Kolmogorov entropy.*

For $q = 2$, equation 17 yields the so-called *correlation dimension*

$$D_2 = \lim_{r \to 0} \frac{\log \sum_{i=1}^{N} p_i^2}{\log r}$$

which, in addition to D_0 and D_1, is another important fractal dimension. Its main practical advantage is the relative ease with which it can be determined for "practical" fractals (see Chapter 10). The theoretical importance of D_2 lies in its close relation with the fundamental concept of correlation. In fact, we will show in Chapter 10 that D_2 is determined by the "correlation function" of the fractal set, that is, the probability of finding, within a distance r of a given member of the set, another member. Thus, measuring D_2 comes down to a simple counting process.

In principle D_q can be determined for all q in accordance with its definition (equation 17). In practical applications, however, one sometimes encounters difficulties for $q > 0$ because positive q diminish the terms with small p_i (corresponding to the "rarely visited" parts of the fractal). As a result the limit as $r \to 0$ converges very slowly. This drawback can be overcome by calculating the numerator in equation 17 for both r and $r/2$ and requiring that their *ratio* equal 1 as $r \to 0$; see Halsey et al. [HJKPS 86].

The Relation between the Multifractal
Spectrum $f(\alpha)$ and the Mass Exponents $\tau(q)$

In the preceding sections, we have introduced two different functions for describing a multifractal:

- the multifractal spectrum $f(\alpha)$ that describes the fractal dimension f of a subset with a given Lipshitz-Hölder mass exponent α, and

- the generalized fractal dimensions D_q or, equivalently, the exponents $\tau(q) = (1 - q)D_q$

Since both functions, $f(\alpha)$ and $\tau(q)$, describe the same aspects of a multifractal, they must be related to each other. In fact, the relationships are

$$\tau(q) = f(\alpha) - q\alpha \qquad (22)$$

where α is given as a function of q by the solution of the equation

$$\frac{d}{d\alpha}(q\alpha - f(\alpha)) = 0 \tag{23}$$

Conversely, if the fractal dimension D_q or the exponents $\tau(q)$ are known, the multifractal spectrum is given by

$$f(\alpha(q)) = \tau(q) + q\alpha(q) \tag{24}$$

where $\alpha(q)$ is given by

$$\alpha(q) = -\frac{d}{dq}\tau(q) \tag{25}$$

which, with equation 24, implies

$$\frac{df}{d\alpha} = q \tag{26}$$

Equations 24 and 25 give a parametric description of the $f(\alpha)$ curve in terms of q. These two equations represent a *Legendre transform* from the variables q and τ to the variables α and f. Such transformations play an important role in thermodynamics in converting, for example, energy as function of volume and entropy into free energy as a function of volume and temperature.

In fact, the analogies between multifractals and statistical mechanics go much further than a change of variables mediated by the Legendre transform. As the physicist will appreciate, equation 14 is patterned after one of the most powerful analytical tools in statistical mechanics, the *partition function* invented by the noted theoretical physicist and chemist J. Willard Gibbs (1839–1903). Because of this mathematical analogy, several of our parameters are formally equivalent to such thermodynamic concepts as energy (α), free energy (τ/q), entropy (f), and temperature ($1/q$).

Strange Attractors as Multifractals

Strange attractors are among the most important realizations of multifractals. An attractor of an iteration $x_{n+1} = f(x_n)$ is a single point or an indecomposable bounded set of points to which starting values x_0 from the attractor's "basin of attraction" converge as $n \to \infty$. A *strange* attractor is an attractor for which the iterates x_n depend sensitively on the initial x_0; that is, arbitrarily close initial values will become macroscopically separated for a sufficiently large n. Strange attractors are fractal dusts with a fractal dimension smaller than the Euclidean dimension of the embedding space.

2 Self-Similar Leaf

(B)

3 Peaceful Border between Three Countries

4 "Rainbow to Infinity"

(A)

(B)

(A)

(B)

(C)

(D)

(A)

(B)

(C)

(D)

(A)

(B)

(C)

(D)

A prototypical strange attractor is furnished by the *logistic parabola* (see Chapter 12) defined by the "quadratic map" $f(x) = rx(1 - x)$. For $r < r_\infty = 3.5699456\ldots$, the iterates of $f(x)$, $f^{(n)}(x)$, with $0 \le x \le 1$, are periodic with period length 2^m. But for $r = r_\infty$, the iterates are aperiodic and converge on a strange attractor. This strange attractor is a Cantor set that is asymptotically well modeled by a generator with two intervals with lengths $r_1 = 0.408$ and $r_2 = r_1^2$, and with equal weight $p_1 = p_2 = 0.5$.

The Hausdorff dimension $D = D_0$ of this model attractor is given by equation 19 for $q = 0$:

$$r_1^D + r_2^D = 1$$

which, for $r_2 = r_1^2$, has the solution

$$r_1^D = \frac{\sqrt{5} - 1}{2} = 0.618\ldots$$

Thus, $D = D_0 \approx \log 0.618/\log 0.408 \approx 0.537$.

The information dimension is given by equation 21. With $N = 2$ and the values for r_i just used one obtains $D_1 = 0.515$. For all other q and $p_1 = p_2 = 0.5$, equation 14 gives

$$r_1^\tau + r_1^{2\tau} = 2^q$$

which yields

$$D_q = \frac{\log\left[\frac{1}{2}(\sqrt{(1 + 2^{q+2})} - 1)\right]}{(1 - q)\log r_1} \qquad q \ne 1$$

With $r_1 = 0.408$, this gives $D_{-\infty} = -1/\log_2 r_1 = 0.773,\ldots$, $D_{-1} = 0.561$, $D_0 = 0.537$, $D_2 = 0.497$, $D_3 = 0.482,\ldots$, $D_\infty = -1/\log_2 r_1^2 = 0.387$.

The entire multifractal spectrum $f(\alpha)$, calculated from D_q, is shown in Figure 11. Because $df(\alpha)/d\alpha$ equals q, the maximum of $f(\alpha)$ equals $D_0 = 0.537$, while $f(0)$ corresponds to $df(\alpha)/d\alpha = \pm\infty$, yielding the values $\alpha_{min} = D_\infty = 0.387$ and $\alpha_{max} = D_{-\infty} = 0.774$, which are within 2.5 percent of the best numerical values [HJKPS 86]. This discrepancy, albeit small, reflects the fact that, contrary to our assumption, the period-doubling strange attractor is not exactly self-similar.

A Greedy Algorithm for Unfavorable Odds

In Chapter 6 (pages 150–152), we outlined a strategy for optimally playing a game of chance when the odds are favorable. And we advised against gambling when the odds are unfavorable. Nevertheless, for those who persist in the face

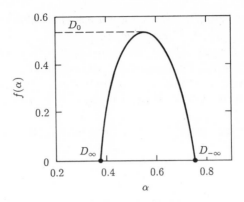

Figure 11   Multifractal spectrum of iterates of quadratic map at period-doubling accumulation point [HJKPS 86].

of adversity and are willing to risk everything, there is a small chance of "making a killing" if they follow a simple rule based on maximum avarice: a *greedy algorithm*.

Suppose a player has an initial capital of K dollars and the bank has B dollars, and the probability of winning a single play is p, with $p < 0.5$. Of course, a reasonable person, intent on a positive expected gain, *should not* play against odds that are less than even ($p = 0.5$). But suppose our player is willing to risk his entire capital for a small chance of winning all the money in the bank. From equation 8 in Chapter 6 we see that the probability of achieving this goal, while betting always exactly one dollar, is given by

$$p_K = 1 - q_K = \frac{r^K - 1}{r^B - 1} \qquad (27)$$

where $r := (1 - p)/p > 1$. An unbiased roulette wheel with 36 positive numbers and only one 0 gives $p = \frac{18}{37}$, or $r = \frac{19}{18}$, for an "even" chance. More generally, K in equation 27 is the player's initial capital divided by the size of his bet and B is the bank's capital divided by the bet.

Say that the player has an initial capital of $1000 ($K = 1000$) and the bank has $10,000 ($B = 10,000$); then, according to equation 27, the probability p_K of breaking the bank is approximately 10^{-211}. In other words, our player looses all his money with probability indistinguishable from certainty.

A less timid player might wager $10 on every play, thereby changing K to 100 and B to 1000 in equation 27. This "improves" his chances of winning $10,000 to $p_K \approx 10^{-21}$. And an even bolder player, always wagering $100, corresponding to $K = 10$ and $B = 100$, is rewarded with $p_K \approx 0.0025$. This suggests that the higher the wager, the better the chances of winning the pot.

Why should this be so? Obviously, a player who wagers very little, like $1 or $10, has to play very often reaching $10,000 (*if ever*), and every time one plays, one gives the unfavorable odds another chance to come into play. This suggests that always betting all one's money would be an optimum strategy in the face of unfavorable odds. More specifically, as long as $x = K/B$ is smaller than $\frac{1}{2}$, one would bet all and, upon winning, would double one's money. Thus, if $P(x)$ is the probability under this greedy algorithm of ultimately winning all, then we have the following functional recursion:

$$P(x) = pP(2x) + (1 - p) \cdot 0 \qquad 0 \leq x \leq \frac{1}{2} \tag{28}$$

Of course, if $x > \frac{1}{2}$, one only has to wager the difference $B - xB$ to reach B. In this case, even if one loses, one still has $K - (B - K) = 2K - B$ dollars to continue playing [Bil 83]. Hence,

$$P(x) = p \cdot 1 + (1 - p)P(2x - 1) \qquad \frac{1}{2} \leq x \leq 1 \tag{29}$$

Now the only remaining question is how to calculate $P(x)$ from the equations 28 and 29. Of course, for $0 < p < \frac{1}{2}$, $P(0) = 0$ by equation 28 and $P(1) = 1$ by equation 29. For other values of x, we note that $P(x)$ must be a *self-affine* function. Indeed, according to equation 28, compressing x by a factor of 2 and multiplying P by p reproduces the left half of $P(x)$.

In fact, $P(x)$ is precisely the function that we encountered earlier in this chapter (with p replaced by $1 - p$, see Figure 1 and equation 1). However, like many other self-similar or self-affine functions, $P(x)$ is not well behaved: its derivative vanishes almost everywhere except at values of x which form a Cantor set. In other words, $P(x)$ is a devil's staircase—see Figure 12, where $P(x)$ is plotted for $p = 0.25$.

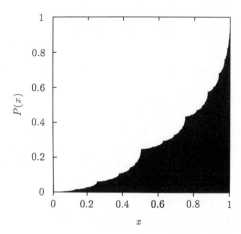

Figure 12   The cumulative probability of winning the pot when playing according to a "greedy" algorithm: a self-affine devil's staircase.

To calculate $P(x)$, we express x as a binary fraction. If this fraction does not terminate, we truncate it after n places. In our example, $x = \frac{1}{10}$ has the periodic binary expansion $x = 0.0\overline{0011}$, which we can approximate by $\tilde{x} = 0.00011 = \frac{3}{32}$. Now $P(\tilde{x})$ is given by the following expression:

$$P(\tilde{x}) = p_0 p_0 p_0 p_0 p_0 + p_0 p_0 p_0 p_0 p_1 + p_0 p_0 p_0 p_1 p_0 \tag{30}$$

where $p_0 = p$ and $p_1 = (1 - p)$. The three sequences of indices correspond to all binary fractions with $n = 5$ places which are smaller than $\tilde{x} = 0.00011$; that is, 0.00000, 0.00001, and 0.00010. Thus, for $p_0 = \frac{18}{37}$ and $p_1 = \frac{19}{37}$, $P(\tilde{x}) = p_0^4(p_0 + 2p_1) \approx 0.084$. Since $\tilde{x} < x$ and $P(x)$ is a nondecreasing function, $P(x)$ is actually somewhat larger than 0.084. This is quite an improvement over the winning probability $p_K = 0.0025$ for constant \$100 bets. And even constant \$1000 bets give only $p_K = 0.077$.

The three terms in equation 30 reflect the three different routes to success under our greedy algorithm if we stop after five plays. For example, the third term, $p_0 p_0 p_0 p_1 p_0$, represents three wins, followed by a loss and a final win. The corresponding capital sequence is, in thousands of dollars: 1, 2, 4, 8, 6, 10. The capital sequence corresponding to the second term in equation 30, $p_0 p_0 p_0 p_0 p_1$, is 1, 2, 4, 8, 10, 10. The first term, $p_0 p_0 p_0 p_0 p_0$, has the same capital sequence because once the goal (\$10,000) is reached, the player stops playing.

Of course, if the player isn't ruined after five plays, he can continue to play and improve his chances of breaking the bank. This corresponds to truncating x in $P(x)$ after more than five places. For example, $\frac{51}{512}$ is an excellent approximation to $\frac{1}{10}$, giving $P(\frac{51}{512}) > 0.09$. For irrational x, which have aperiodic binary fractions, the calculation of $P(x)$ is of course an infinite process. But for rational x, with periodic binary fractions, there is a closed formula, which we leave the reader to discover. So what is $P(\frac{1}{10})$?

And what is the expected duration of the game until either the player or the bank is broke under the greedy algorithm? Like $P(x)$, it should be a self-affine function, and it should be less than the expected duration of the game with any constant wager [Fel 68]:

$$D_K = \begin{cases} \dfrac{1}{p(r-1)}\left[K - B\,\dfrac{r^K - 1}{r^B - 1} \right] & r \neq 1 \\[2ex] \dfrac{1}{p}\,K(B-K) & r = 1 \end{cases} \tag{31}$$

For $p = \frac{18}{37}$, $r = \frac{19}{18}$, and \$1 bets, $K = 1000$ and $B = 10,000$, giving an expected duration of $D_K = 37,000$ plays. For \$100 wagers, $K = 10$ and $B = 100$, giving a duration of 358 plays. And for \$1000 wagers, $K = 1$ and $B = 10$, resulting in a duration of only 25 plays. Thus, the greedy algorithm should have an expected duration of less than 25 plays.

10

C   H   A   P   T   E   R

Some Practical Fractals and Their Measurement

The equality that we demand is the most endurable degree of inequality.
—GEORG CHRISTOPH LICHTENBERG

In this chapter we shall pay a belated visit to the world of practical fractals. As Mandelbrot and others have so aptly observed, nature loves fractals at least as much as regular shapes. For every smooth curve or surface seen around us, there are as many, if not many more, that are highly irregular and often in fact fractal, with detailed structure on many size scales.

Why are fractals so prevalent in nature? The overriding reason is that a smoothly curved surface embodies an inherent length scale: the radius of curvature. And for such an inherent length there must of course be a reason. For example, the earth, seen from afar, is a ball with a curved surface with a radius of curvature, R. What is this R? It can be expressed in terms of the total mass M of the earth and its mean density ρ:

$$R = \left(\frac{3}{4\pi} \frac{M}{\rho} \right)^{1/3}$$

With $M \approx 6 \cdot 10^{24}$ kg and $\rho \approx 6 \cdot 10^{3}$ kg/m^3, the radius becomes about 6000 km. Thus, the earth ball's natural length scale is ultimately determined by the amount of aggregating dust, and its mean density, that formed the primordial earth 4.7 billion years ago.

But many objects and laws of nature, by luck, lack such a natural scale within the range of observation. Thus, whatever is true at one magnification must be true over a whole range of magnifications. In other words, the object must exhibit self-similarity—statistical, asymptotic, or even strict self-similarity. And if the object has any structure, then a similar structure must appear on many

Figure 1   Determining the fractal dimension of the coast of Norway by counting how many boxes the outline of the coast penetrates [Fed 88].

size scales. In other words, lack of scales breeds self-similarity, and a self-similar object, exempting a few featureless entities such as the infinite straight line, must be fractal.

Let us look at Mother Earth again, not from outer space, but from distance at which her gross curvature becomes irrelevant. For many coastlines (as for some man-drawn boundaries) there is no natural length scale. The processes that shape many an interface between water and land are similar over a wide range of scales. Figure 1 shows a piece of the coast of Norway, the homeland of Jens Feder, from whose book *Fractals* [Fed 88] this illustration is drawn. There are large fjords and smaller fjords and ever littler inlets. And if we consult maps showing more and more detail and finally walk (or row) along the beach, we see that the little inlets harbor still littler inlets and so forth, down to the level of the seawater penetrating the spaces between individual pebbles.

How do we measure the length of such a fractal coast? Obviously, the more detail we "consult," the longer the apparent length L. In fact, if the coast is self-similar, we shall find a self-similar *power law* connecting L and the scale unit r employed in its measurement:

$$L(r) \sim r^{\varepsilon} \tag{1}$$

where the exponent ε is negative if L increases as r decreases. By contrast, for a smooth curve, the measured length approaches an asymptotic value as $r \to 0$—that is, the exponent ε is zero.

As we have seen in earlier chapters, fractal objects are characterized by a fractal dimension, such as the Hausdorff dimension

$$D = \lim_{r \to 0} \frac{\log N(r)}{\log (1/r)} \tag{2}$$

where $N(r)$ is the minimum number of disks of diameter r needed to cover the fractal. If we measure L in this manner, $N(r)$ equals $L(r)/r$ and the exponent ε in equation 1 is seen to equal $1 - D$. Thus, the Hausdorff dimension is given by

$$D = 1 - \varepsilon \tag{3}$$

which, for $\varepsilon < 0$, will exceed 1 (the value for a smooth curve).

Dimensions from Box Counting

The Hausdorff prescription of covering the fractal with disks is not always the most convenient way to measure a fractal dimension, nor is the Minkowski sausage recipe that we divulged in pages 41–43 in Chapter 1. Rather, for

Figure 2   Measured length of Norwegian coast against grid size r. The slope of the straight line gives the fractal dimension of the coast $D \approx 1.52$ [Fed 88].

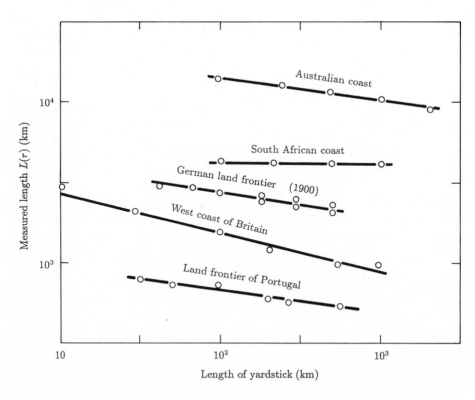

Figure 3   Apparent lengths and fractal dimension of different coasts and land frontiers [Man 83].

many practical fractals, *box counting* is the method of choice. In box counting, we superimpose a square grid over the fractal (see Figure 1) and count the number of boxes $N(r)$ that are penetrated by the fractal. Again, as r, the spacing between grid lines, becomes very small, we find that $\log N(r)/\log (1/r)$ converges to a finite value, the Hausdorff dimension D.

Figure 2 shows the result of Feder's counting the coast of Norway with grid sizes ranging from $r = 0.6$ km to 80 km. The measured length $L(r)$ decreases with increasing r by a factor of 12—from about 30,000 km to 2500 km. On a double logarithmic plot, all measurement points fall near a straight line with a slope of $\varepsilon = -0.52$. Thus, the Hausdorff dimension, according to equation 3, is $D = 1 - \varepsilon = 1.52$—halfway between the Euclidean dimension of a smooth curve and that of a smooth surface.

Figure 3 shows data publicized by Mandelbrot [Man 83] for the apparent lengths $L(r)$ of several other coasts and land frontiers, whose Hausdorff dimensions range from a smooth $D \approx 1$ for the coast of South Africa to a ragged $D \approx 1.3$ for the west coast of Britain. But no coast or country can match Norway's $D \approx 1.52$.

The Mass Dimension

For many purposes, the box-counting dimension is still not the most appropriate or convenient fractal measure. Look at a "Lichtenberg figure" (see Figure 4), one of the first physical fractals made by man. It is the electrical discharge pattern from a metallic tip placed on an insulator, first made visible in 1777 (the year of Gauss's birth in nearby Brunswick) by the Göttingen physicist and aphorist Georg Christoph Lichtenberg (1742–1799).[1] If we measure the bright area M of Lichtenberg's figure, we find that it increases with the characteristic radius R according to a simple, homogeneous power law:

$$M \sim R^D \tag{4}$$

But the exponent D does not equal 2, as for a homogeneous figure in the plane (e.g., a circular disk, whose area M equals πR^2). Rather, the exponent D for the Lichtenberg figure lies between 1.7 and 1.9.

The exponent D in equation 4 can serve as another fractal dimension, more conveniently measured and easier to grasp than the dimensions introduced so

---

1. Lichtenberg traveled twice to England, once as a guest of George III. There he encountered and came to admire British science and urbanity. His fame rests on the plethora of penetrating aphorisms with which he needled fellow scientists and citizens. Lichtenberg's major literary oeuvres are his illuminating comments on William Hogarth's engravings (*including Marriage à la Mode and*, *The Rake's Progress*).

Figure 4 Another Lichtenberg figure: an electrical discharge pattern on the surface of a glass plate. Mass dimension $D \approx 1.9$ [NPW 86].

far. It applies to many fractals, from the man-made Cantor dust (see Figure 5) to the natural soft down (see Figure 6) that still fills a few (all too few!) pillows. In each case, the "mass" inside a radius R does not increase with the Euclidean dimension as an exponent but with some lesser power, such as about 1.6 for the down (depending on its price—smaller D costs *more!*).

In the Sierpinski gasket, too (see pages 17–19 in Chapter 1), D is smaller than 2 because the area M enclosed by a circle of radius R increases by a factor of 3 (not 4) every time the radius is doubled. Thus $M \sim R^D$ with $D = \log 3 / \log 2 = 1.58 \ldots .$

For strictly self-similar mathematical fractals, such as the Sierpinski gasket and the Cantor dust, the mass dimension is the same as the Hausdorff dimension (and any other fractal dimension considered here). They are all given by the similarity dimension of the scaling law, defined by the initiator and the generator that generates the fractal. But for practical fractals, there are significant differences, as we saw when we introduced the Minkowski dimension in pages 41–43 in Chapter 1 to account for the number of vibrating modes of a drum with a fractal perimeter.

Figure 5   Cantor dust from the output of a digital-to-analog converter. Mass dimension $D \approx 1.26$.

The mass dimension is particularly appropriate to parameterize the packing of powders. Primary powder particles are apt to form clusters with a packing density of, say, p. Assume that these clusters have radii that are r times larger than those of the primary particles. Now very often these clusters will again cluster with the same or similar values of p and r, and so forth, for several generations; see Figure 7 [OT 86].

After n such generations, the density of the powder P equals p^n and the cluster radius R is equal to r^n. The total mass M is of course proportional to PR^d, where d is the Euclidean dimension in which the powder resides ($d = 3$ for most powders in a three-dimensional world). Thus,

$$M \sim PR^d = R^D$$

with the mass dimension $D = d + \log p / \log r$, which, since $p < 1$ and $r > 1$, is smaller than d. In Figure 7, d equals 2, r is about 7, and p is roughly 0.7, giving a mass dimension $D \approx 1.82$. For the Sierpinski gasket with $d = 2$, $r = 2$, and $p = \frac{3}{4}$, for comparison, we get a smaller mass dimension, namely, $2 + \log \frac{3}{4} / \log 2 \approx 1.58$, the same value as its Hausdorff dimension.

Figure 6   Natural down. Its softness results from its low mass dimension $D \approx 1.6$.

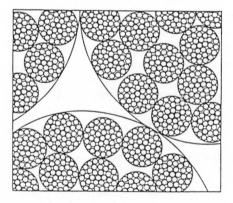

Figure 7   Three generations of a self-similar agglomerate in a powder [OT 86].

The mass dimension, as the name suggest, can be applied in higher Euclidean dimension, too, especially to spongy substances, such as the Menger sponge shown in Figure 3 of Chapter 8, with $D = \log 20/\log 3 \approx 2.73$, and crystals grown by *diffusion-limited aggregation* (DLA), for which $D = 2.4$ is a typical value [Mea 87]. In DLA, (see Figure 9 in Chapter 9 for the result of two-dimensional computer simulation), one "atom" at a time is released at a large distance from the growing aggregate and allowed to diffuse until it attaches itself to the aggregate once it comes under the short-range attractive forces of the atoms already captured. Simple probability tells us that new atoms will attach themselves preferentially near the tips of the outreaching "dendrites" rather than deep inside the crystal's "fjords."

Measurement of the mass dimension gives values of $D \approx 2.4$ for three-dimensional DLA and $D \approx 1.7$ in two-dimensional DLA. However, the exact values of D do depend on the physical and chemical parameters of the process and contain important clues for the manufacture of new materials and for practical applications in which fractal processes dominate.

Take viscous fingering (see Figure 10 in Chapter 9), produced by the surface instability as one liquid or gas invades the "territory" of another, more viscous liquid. (The reader is encouraged to produce his own viscous fingers with the help of water and glycerol, squeezed between two glass plates.)

Viscous fingering has also been observed as one liquid (say, water) replaces another (oil) in a porous medium (shale), a standard method of squeezing oil form bituminous rock (see Figure 8). The fractal mass dimension of the watery fingers depends sensitively on the liquid's viscosity, the porosity of the rock,

Figure 8   Squeezing oil from bituminous rock.

and wetting properties of the liquid and the rock surface [SWGDRCL 86]. If the water's edge were smooth ($D = 2$), most of the oil could be extracted from a distant well before the water reached the extracting pump. But, unfortunately for an oil-hungry world, D is much larger than 2, and the first water reaches the pump long before all the oil has been pumped out. However, the addition to the water of viscosity-increasing polymers reduces the fractal dimension and the oil industry hopes thereby to double the amount of oil that can be recovered by injecting water into shale. (The additives, the industry also hopes, will cost less than the value of the extra crude oil extracted.)

The Correlation Dimension

One of the most widely used fractal dimensions is the *correlation dimension*, because it is experimentally the most accessible, especially if the fractal comes as a "dust"—isolated points, thinly sprinkled over some region of space.

To determine the correlation dimension, one first counts how many points have a smaller (Euclidean) distance than some given distance r. As r varies, so does the relative count $C(r)$, defined as the total count divided by the squared number of points. The quantity $C(r)$ is also called the *correlation sum* (or *correlation integral*) [GP 83, Gra 83].

The correlation dimension is then defined by

$$D_2 := \lim_{r \to 0} \frac{\log C(r)}{\log r} \tag{5}$$

Figure 9 shows the experimental determination of D_2 for the strange attractor of the iterated *Hénon map*, which yields a straight-line dependence of $\log C(r)$ on $\log r$ over six orders of magnitude, with a slope $D_2 = 1.21$ [GP 83].

Infinitely Many Dimensions

The correlation dimension D_2 belongs to an infinite family of dimensions D_q defined by

$$D_q := \lim_{r \to 0} \frac{1}{q-1} \frac{\log \sum_k p_k^q}{\log r} \qquad -\infty \leq q \leq \infty \tag{6}$$

where the sum is over all the cells of linear size r into which the space has been subdivided and p_k is the relative frequency or probability with which points of the dusty fractal fall inside cell k [HP 83a, Gra 83] (see also pages 202–205 in Chapter 9).

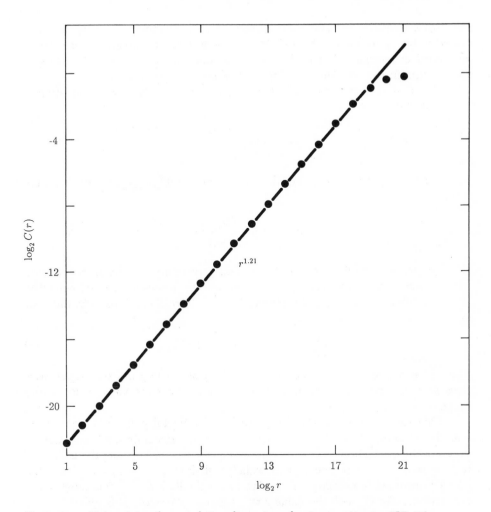

Figure 9   Determining the correlation dimension of a strange attractor [GP 83].

For $q = 0$, we recover our old friend, the box-counting or Hausdorff dimension D_H, because the sum $\sum_k p_k^0 = \sum_k 1$ simply counts how many boxes or cells are "invaded" by the fractal. Thus, $D_H = D_0$.

For $q = 2$, it is easy to show that, in the limit as $r \to 0$, the sum $\sum_k p_k^2$ equals the relative count $C(r)$, which yields the correlation dimension introduced in the preceding section [Schu 84].

For $q \to 1$, equation 6 yields the so-called *information dimension*

$$D_1 = \lim_{r \to 0} \frac{-\sum_k p_k \log p_k}{\log (1/r)} \qquad (7)$$

—so called because the numerator in equation 7 is Shannon's entropy, as introduced in his "information theory" (following Boltzmann's invention of entropy in statistical mechanics). The dimension D_1 does in fact measure the loss of information in the dynamic development of chaotic systems.

For $q \to \infty$, only the highest probability p_{max}, in the sum in equation 6, counts. Hence

$$D_\infty = \lim_{r \to 0} \frac{\log p_{max}}{\log r}$$

Conversely, for $q \to -\infty$, the smallest probability p_{min}, controls the sum. Thus,

$$D_{-\infty} = \lim_{r \to 0} \frac{\log p_{min}}{\log r}$$

$D_{-\infty}$, depending, as it does, on the smallest probability, is difficult to measure for practical fractals. The sites with low probability are visited too infrequently.

Note that $D_{-\infty} \geq D_\infty$. In general, for any two dimensions with different q,

$$D_q \geq D_{q'} \qquad \text{for} \qquad q < q' \tag{8}$$

Thus, D_q is a monotone nonincreasing function of q. Only in exceptional cases does D_q not depend on q at all and does it have the same value in the entire range $-\infty \leq q \leq \infty$.

One such exception is a strictly self-similar fractal generated from "non-erased" segments of equal length, such as those that generate the triadic Cantor dust. For the calculation of D_q according to equation 6, we consider the generator to consist of $N = 2$ segments with equal probabilities $p_1 = p_2 = \frac{1}{2}$. Since the triadic Cantor set is a strictly self-similar fractal, the limit as $r \to 0$ is superfluous; we can calculate D_q with the value $r = \frac{1}{3}$ for the generator. This gives

$$D_q = \frac{1}{q-1} \frac{\log [2(1/2)^q]}{\log (1/3)} = \frac{\log 2}{\log 3}$$

which is a value that is indeed independent of q.

If the generator consists of segments of *different* lengths r_k, then, for $D_q = D$ to hold for all q, the probabilities p_k must be proportional to $r_i^{D_q}$. To show this, we use equation 14 from Chapter 9, the formula for self-similar fractals:

$$\sum_i p_i^q r_i^\tau = 1 \tag{9}$$

where $\tau = (1 - q)D_q$.

With $p_i \sim r_i^{D_q}$ in equation 9, we get

$$\sum_i r_i^{Dq} = 1$$

which means, of course, that D_q is independent of q.

For example, for $r_1 = \frac{1}{2}$ and $r_2 = \frac{1}{4}$, $D = \log \gamma/\log (\frac{1}{2}) \approx 0.694$, where $\gamma = (\sqrt{5} - 1)/2$ is the golden mean. Thus, $p_1 = (\frac{1}{2})^D = \gamma$ and $p_2 = (\frac{1}{4})^D = \gamma^2$. Probabilities $p_i = r_i^D$, except for $r_i = r$, are, of course, rather artificial, and D_q in general *does* depend on q. For example, for $r_2 = r_1^2$ and $p_1 = p_2 = \frac{1}{2}$, equation 9 gives

$$D_q = \frac{1}{1-q} \frac{\log [\frac{1}{2}\sqrt{1 + 2^{q+2}} - 1/2]}{\log r_1} \qquad q \neq 1$$

which ranges from $D_{-\infty} = \log 2/\log (1/r_1)$ down to $D_\infty = \log 2/\log (1/r_2)$.

Note that, for $r_1 > \frac{1}{2}$, $D_{-\infty}$, which measures the densest part of the fractal, exceeds 1. In general, for constant p_k, $D_{-\infty}$ is determined by the longest segment, r_{max}, and D_∞ by the shortest segment, r_{min} (see Chapter 9).

The Determination of Fractal Dimensions from Time Series

A large quantity of the data that humans take in comes in the form of a "time series," that is, a temporal sequence of measured values, such as electroencephalographic (EEG) potentials (see Figure 10). Is this just random noise, or is it deterministic chaos, generated by some underlying deterministic, albeit chaotic, process?

In order to decide this often crucial question, one constructs d-dimensional data vectors from d measurements spaced equidistant in time and determines the correlation dimension D_2 of this d-dimensional point set. If the data were truly

Figure 10   An electroencephalogram [Rös 86].

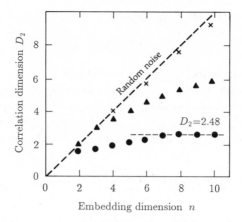

Figure 11   Correlation dimension as a function of embedding dimension [LH 86].

random, then, as d is increased, the calculated D_2 would go up accordingly. But for a deterministic system, no matter how chaotic it appears to the "naked eye," the calculated correlation dimension will not increase further, once the so-called *embedding dimension d* exceeds the correlation dimension D_2 of the data (see Figure 11). In this manner the author's student J. Röschke found that the seemingly noiselike EEGs recorded from the acoustic cortex of cats are in fact not noise at all, but deterministic chaos whose correlation dimension depends on the state of the cat's wakefulness [Rös 86].

This "embedding" method of determining the fractal dimension of experimental data, and thereby distinguishing deterministic chaos from random noise, has been applied successfully to a wide variety of physical, meteorological, biological, and physiological observation [HP 83b].

Abstract Concrete

Self-similarity is not only amenable to *measurement*; self-similarity can also be employed profitably in the *design* of fractal structures and materials with increased durability or lower cost (or both). A case in point is the construction of field walls on many New England farms; see Figure 12, a snapshot taken by the author during a biking tour through Connecticut. There are large stones whose gaps are filled by smaller stones, whose interstices, in turn, are filled by still smaller stones. As a result of this roughly self-similar composition, the wall keeps standing upright without the customary edifying intervention of expensive cement to fill and fixate the cracks. If the number of stones as a function of their size is selected according to a power law, what would be a good exponent?

Figure 12   The hierarchical construction of a New England field wall.

In a column in *Nature* entitled "Abstract Concrete," David Jones argues that by employing ever finer particles, from the coarsest gravel to the finest dust, the volume to be filled by expensive binder can be made arbitrarily small, thereby cutting cost—or allowing more expensive, high-tenacity binding materials like epoxy or even polyimide to be used [Jon 88]. Likewise, many other composite materials, such as fiberglass, could probably be improved by a self-similar composition.

Fractal Interfaces Enforce Fractional Frequency Exponents

Finite electrical circuits constructed of passive "lumped elements," like resistors and capacitors, have input impedances that are rational functions of frequency. For example, a capacitor, with capacitance C, has an impedance $Z = (i\omega C)^{-1}$, where $i = \sqrt{(-1)}$ and ω is the radian frequency (2π times the frequency). Thus, for a capacitor, $Z \sim \omega^{-1}$, which is a rational function of ω.

Electrical engineers found out a long time ago that a finite electrical circuit, constructed in any way from a finite number of lumped elements, always has an impedance that is a rational function of frequency. This is a consequence of the fact that rational functions form a group under composition—that is, if $R(x)$ and $S(x)$ are rational functions, so is $S(R(x))$.

This rationality is unfortunate in a way, because the *characteristic impedance* Z_0 of a transmission line (such as a TV cable), which is needed for a "matching" echo-free connection, is *not* a rational function of frequency. Instead, Z_0 typically involves square roots like $\omega^{-1/2}$. Thus, no finite network can match such a characteristic impedance exactly; matching networks are always approximations (and the standards of approximation, alas, vary from country to country).

However, *infinite* networks can produce an irrational frequency dependence. For example, the infinite ladder network shown in Figure 13 has an input impedance Z_0 that is best written as a continued fraction:

$$Z_0 = R + \cfrac{1}{G +} \cfrac{1}{R +} \cfrac{1}{G +} \cdots$$

A closed form for the value of Z_0 can be obtained by exploiting the periodicity of this continued fraction, which permits us to write

$$Z_0 = R + \frac{1}{G + (1/Z_0)}$$

This is a quadratic equation for Z_0. The physically meaningful solution (the one with positive real part) is

$$Z_0 = \frac{1}{2}(R + \sqrt{(R^2 + 4R/G)}) \tag{10}$$

For $R = 1/G = 1$ ohm (Ω) and a finite number of elements, Z_0 equals a rational number, namely, the ratio of two successive Fibonacci numbers: $1/1$, $2/1$, $3/2$, $5/3$, ... But for an infinite number of elements, as equation 10 shows, Z_0 is no longer rational. In fact, it equals the reciprocal golden mean $(\sqrt{5} + 1)/2 = 1.618\ldots$, an irrational number.

Figure 13   Electrical ladder network composed of resistances R and conductances G.

In order to realize an irrational *frequency* dependence, we can make G a capacitor, with reciprocal impedance $G = i\omega C$. For small frequencies, $\omega \ll 1/RC$, Z_0 then depends on ω in simple power-law manner $\omega^{-\beta}$, with the fractional exponent $\beta = \frac{1}{2}$:

$$Z_0 = R\left(i\frac{\omega}{\omega_c}\right)^{-1/2} \tag{11}$$

where $\omega_c = 1/RC$ is the upper "cutoff frequency" of this approximation.

However, the real world of electrical conduction goes beyond simple half-integer exponents like $\beta = \frac{1}{2}$. This is noteworthy because *uniform* networks are described by *periodic* continued fractions, which lead invariably to square roots and half-integer exponents. Hence, the observation in a physical system of a nonstandard frequency dependence with a fractional value of 2β implies some kind of *non*uniform structure.

In fact, such nonstandard frequency behavior is frequently observed in electrical conduction across *rough surfaces*, such as between an electrode and the electrolyte of a car battery. Indeed, simple models of the current-carrying interface as *fractals* have established a unique relationship between the fractal geometry of the interface and the frequency exponent η.

Figure 14A shows a highly schematized, two-dimensional model of an interface between an electrode (white) in contact with an electrolyte (black) [Liu 85]. The model is based on a Cantor set whose generator consists of two segments of equal length r. In Figure 14A, r equals 0.3. To model the roughness of the interface, the "grooves" in the electrode have increasing depth, becoming deeper by a constant amount with each generation of constructing the fractal, as shown in Figure 14A.

Figure 14B, shows the treelike electrical circuit representing current conduction through this interface. Note that the resistances increase by a factor $1/r > 2$ with each generation, reflecting the fact that the grooves filled by the electrolyte become narrower and narrower. This network is in fact a *self-similar* Cayley tree, that is, a tree with a constant branching ratio (here equal to 2) and a scale factor for the resistances equal to $1/r$, representing the resistance increases in the progressively narrower grooves.

The input impedance of this tree is given by the continued fraction

$$Z(\omega) = R + \cfrac{1}{i\omega C +} \cfrac{2}{R/r +} \cfrac{1}{i\omega C +} \cfrac{2}{R/r^2 +} \cdots \tag{12}$$

Here the 2s in the numerators stem from the 2:1 branching ratio of the tree.

This continued fraction, too, would be periodic if it were not for the fact that the scaling factor r is different from $\frac{1}{2}$. It can be written in the following closed form:

$$Z(\omega) = R + \cfrac{1}{i\omega C +} \cfrac{2r}{Z(\omega/r)} \tag{13}$$

(A)

(B)

Figure 14   (A) Middle-third-erasing model of interface between electrode (white) and electrolyte (black) in a car battery. (B) Treelike electrical circuit representing current flow through the fractal interface shown in part A [Liu 85].

Note that $Z(\infty) = R$, as one would expect from the circuit diagram in Figure 14B. For finite frequencies, $|Z(\omega)| > R$.

For

$$R \ll \frac{|Z(\omega)|}{2r} \ll \frac{1}{\omega C}$$

which implies $\omega \ll 1/RC$, equation 13 simplifies to the scaling law

$$Z(\omega) = \frac{1}{2r} Z\left(\frac{\omega}{r}\right) \tag{14}$$

which is solved by the power law

$$Z(\omega) \sim \omega^{-\beta}$$

with the exponent given by

$$\beta = 1 - \frac{\log 2}{\log (1/r)} = 1 - D \tag{15}$$

Here D is the Hausdorff dimension of the Cantor set used in the model (Figure 14A). Because $0 \leq D \leq 1$, the exponent β too lies between 0 and 1; it is not necessarily restricted to $\beta = \frac{1}{2}$.

The Hausdorff dimension of the one-dimensional electrode-electrolyte interface in two dimensions equals $D + 1 = 2 - \beta$. For a two-dimensional interface with isotropic roughness, the relations for β and D are unchanged. The corresponding tree model would employ a branching ratio of 4 instead of 2, and a resistance scale factor equal to $1/r^2$ instead of $1/r$. However, the Hausdorff dimension of the interface becomes $2D + 1 = 3 - 2\beta$. These, then, are the anticipated relations between conductor geometry and frequency exponent. Note that the rougher the interface (large D), the smaller the frequency exponent β.

Intuitively, the reason for the increase in impedance as the frequency is lowered is that the current reaches deeper and deeper into narrower and narrower crevices of the interface before being shunted away by the capacitances. For the treelike fractal interface, the penetration depths scale with an exponent β which can assume a range of values depending on the roughness of the interface.

If it were not for the exponential growth in the number of needed components, treelike networks, like the one shown in Figure 14B, would also be useful for generating noises or filtering signals with a fractional-exponent frequency dependence. However, the $1/f$ noises observed in many electronic materials may in some cases be generated by a fractal composition that is amenable to the modeling discussed here.

The Fractal Dimensions of Fracture Surfaces

God made the bulk;
surfaces were invented
by the devil.
—WOLFGANG PAULI

Of course, fractal surfaces are not limited to rough electrodes. *Fracture* is another omnipresent source of two-dimensional fractals. In a classic study, Mandelbrot, Passoja, and Paullay investigated the structure of fractured samples of a low-carbon steel which they found to be self-affine [MPP 84]. By plating the fracture surface with a thick layer of nickel and subsequent planing, they could create little islands of steel that grew in size as the planing parallel to the surface progressed. Figure 15 shows the areas A of these steel islands versus their perimeters L in a double logarithmic plot. The data are well fitted over four orders of magnitude by the power law $A \sim L^{1.56}$, implying that the perimeters of the steel islands are self-similar. The value of the exponent, 1.56, is quite distinct from the Euclidean law $A \sim L^2$. The exponent 1.56 also means that the Hausdorff dimension D of the perimeter equals $2/1.56 = 1.28$. This value is close to that of the coast of Britain ($D \approx 1.3$), but considerably greater than that of the "fractal hexagon" ($D = \log 9/\log 7 \approx 1.13$) that we encountered in Chapter 1, pages 13–15. For the fractal hexagon, too, the reader may recall, the perimeter had to be raised to the power $2/D \approx 1.77$ (not 2) to properly predict the "hexagon's" area from its perimeter.

The value $D = 1.28$ for the steel islands, incidentally, implies that the self-affine fracture surface itself has a fractal dimension of $D + 1 = 2.28$, a value typical for rough mountains. A vertical cut through such a fracture mountainscape would produce a profile with a fractal dimension $2.28 - 1 = D$. Mandelbrot and his collaborators confirmed this result by measuring the spatial frequency spectrum $P(f)$ of such fracture profiles. They found $P(f) \sim f^{-4.5}$. From the frequency ex-

Figure 15 Areas of steel "islands" plotted against their perimeters in fractured low-carbon steel [MPP 84].

ponent $\beta = 4.5$, the fractal dimension is obtained by the formula $D = 3 - (\beta - 1)/2 = 1.25$ (see page 137 in Chapter 5), in good agreement with their independent "island" measurements. Thus, metal fracture is free from an inherent scale over several orders of magnitude.

Interestingly, the fractal dimension and the impact energy required for fracture were found to be related to the temperature used in tempering the steel. The precise metallurgical basis of this dependence of fracture energy and topography on heat treatment remains to be rendered intelligible, though.

The Fractal Shapes of Clouds and Rain Areas

Heavy rains interfere with microwave transmission—the medium of choice, for the last half-century, of long-distance telephone transmission. A patch of dense rain between two microwave towers, ubiquitous in the United States and other countries, necessitates the rerouting of communications links. Small wonder, then, that telephone engineers have shown great interest in the temporal distributions of precipitation and the geometric shapes of rain areas. Rain comes, of course, from clouds (although there seem to be people, quite erudite by the way, who confessed surprise when first told so).

Thus, nothing seems more natural than to study the statistics of rain and clouds together, as Lovejoy among others has done [Lov 82]. Figure 16 shows

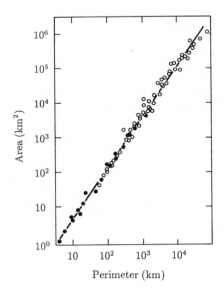

Figure 16 Areas of rain patches (filled circles) and clouds (open circles) plotted against their perimeters [Lov 82].

the area A of rain patches (filled dots) and clouds (open circles) versus the perimeter L. On a double logarithmic plot, like Figure 16, the data hew close to a straight line over a combined range of six orders of magnitude, from 1 km^2 to 1 million km^2! The slope is just under 1.5, implying a fractal dimension D of the perimeters equal to $2/1.5 \approx 1.35$. Again, as in fracture and so many other natural phenomena, there seems to be no natural length scale.

The numerical value of the exponent is in good agreement with a thermodynamic model by Hentschel and Procaccia [HP 84]. Lovejoy and Mandelbrot proposed a mathematical model in which rain areas are assumed to be generated by the superposition of individual rain patches with hyperbolic size distribution [LM 85]. Their model is capable of generating eminently realistic images of clouds. Color Plate 5 shows a photograph, taken by the author, of a natural cloud that rivals the realism of computer-generated nebulosities. Color Plate 6A and B shows two other fractals observed in nature—one generated by natural decay, the other by natural growth.

Cluster Agglomeration

In diffusion-limited aggregation (DLA), aggregates of molecules grow by adding one molecule at a time. Another important growth process that leads to fractal structures is the agglomeration of aerosols and colloids. Figure 17 shows an electron microscope image of a gold colloid grown by cluster aggregation by Weitz and Oliveria [WO 84]. The fractal structure of this colloid is so sparse that, even in the two-dimensional projection shown here, the cluster is transparent.

The agglomeration of clusters is illustrated in Figure 18. Initially, individual particles are distributed roughly uniformly within a finite volume (Figure 18A). They are then allowed to migrate randomly as in Brownian motion. When two particles touch each other, they stick together and from then on move together as a little "cluster." When these little clusters run into each other, they stick together and form larger clusters as observed by Meakin [Mea 83]. In such agglomeration processes, larger and larger clusters are formed as shown in Figure 18B and C. The large clusters are in fact statistically self-similar fractals. In two-dimensional computer simulations, Hausdorff dimensions D near 1.4 are found. In three dimensions, Kolb, Botet, and Jullien found $D \approx 1.8$ [KBJ 83].

Experimentally, the fractal dimension can be determined by scattering of light, x rays, or neutrons from the fractal. For spatial frequencies (reciprocal wavelengths) f in the range $1/R \ll f \ll 1/r$, where R is the size of the entire fractal and r that of the individual particle, one expects the scattered intensity $I(f)$ to follow a simple power law: $I(f) \sim f^{-D}$ (see the next section). The fractal dimension D determined in this manner by Schaefer and his collaborators for silica particles is about 2.1; see Figure 19 [SMWC 84]. This value, being considerably higher than the computer simulation results, points to a different mechanism for the agglomeration of silica. Simulations in which clusters form only *slowly*

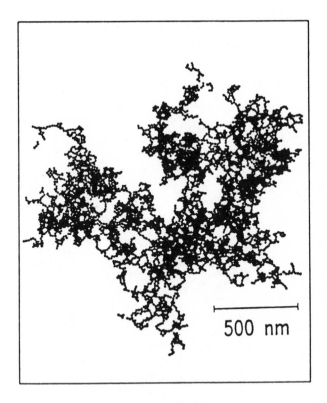

Figure 17   Gold colloid grown by cluster aggregation [WO 84].

after multiple collisions give $D \approx 2.0$, in better agreement with the results obtained by scattering [FL 84, SO 85].

Agglomeration plays a decisive role also in electrolytic deposition and catalytic reactions. Cluster formation is likewise prevalent in the spreading of epidemics, gossip, and opinions. Grassberger found that, in opinions surveys, fractal structures can lead to strongly biased and therefore false results [Gra 85].

Diffraction from Fractals

For incoherent diffraction from a fractal that consists of independent particles, such as a colloid, the scattered intensity $I(f)$ as a function of spatial frequency f is proportional to the total "mass" M contained in a volume of radius $\rho = 1/f$. With $M \sim \rho^D$, one obtains

$$I(f) \sim f^{-D} \tag{16}$$

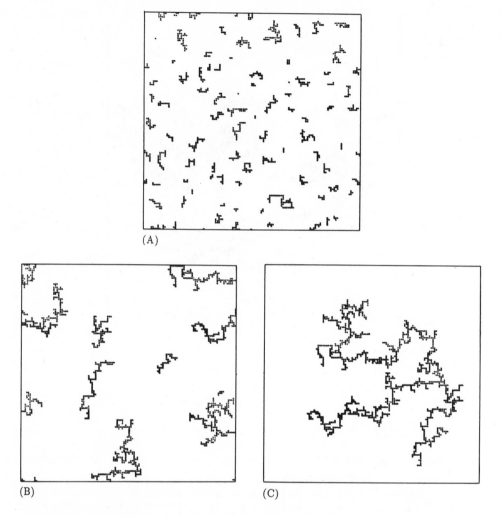

Figure 18   Agglomeration of clusters: (A) uniform random distribution; (B) formation of small clusters following random motion; (C) formation of larger clusters [Mea 83].

For *surface* fractals, $I(f) \sim S$, where S is the surface area. But $I(f)$ can also be written as $M^2 F(f\rho)$, where F is some universal function [MH 87]. With $S \sim \rho^{D_s}$ and $M \sim \rho^d$, this implies $F(f\rho) \sim (f\rho)^{D_s - 2d}$ and

$$I(f) \sim f^{D_s - 2d} \tag{17}$$

where D_s is the fractal dimension for the scattering surface and d is the Euclidean embedding dimension. For $d = 3$, we thus have

$$I(f) \sim f^{D_s - 6} \tag{18}$$

For a smooth surface, D_s equals 2 and $I(f)$ is proportional to f^{-4}, a well-known classical result for the scattering regime considered here. As already noted, equation 18 permits the determination of D_s of a fractal surface by wave diffraction.

Figure 19 shows that scattered intensity of x rays, scattered from the colloidal aggregate of silica that we mentioned before, as a function of spatial frequency in reciprocal angstroms (10^{-10} m). The measurements are restricted to small scattering angles, because at large angles the x rays would resolve the *molecular* structure, as opposed to the cluster structure that is of interest here. For $f < 1/r$, where $r = 27$ Å is the radius of the nonfractal monomers that form the fractal colloid, the scattered intensities fall on a straight line with a slope of -2.1 in a double logarithmic plot, which is thus the Hausdorff dimension of the fractal colloid.

For $f > 1/r$, there is another linear regime of the scattered intensities with a slope of -4. This slope is the one that is theoretically expected for the nonfractal monomers that form the fractal. The experimental result is therefore an indication that the monomers remain intact in the aggregate.

Here then we have a quintessential application of wave diffraction to the analysis of a fractal structure, that of colloids. Understanding the processes that govern colloidal aggregation has been a long-standing aim in several branches

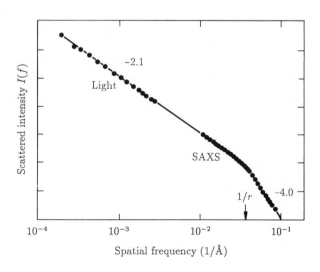

Figure 19   Determining the fractal dimension of silica particles. Intensity of x rays scattered from a colloidal aggregate of silica. SAXS = small-angle x-ray scattering [SMWC 84].

of physics and chemistry, and for a wide variety of commercial applications. Wave diffraction, always the tool of choice for structure analysis, is now being successfully extended from homogeneous bodies to fractal substance.

Another fractal structure that has been elucidated by small-angle x-ray and neutron scattering is lignite, or "brown coal." Lignite is pervaded by microscopic pores with a fractal inner surface. These pores and their surfaces are, of course, what makes "active" coal active and interesting for air filters and other purifying applications. Theory predicts that the scattering intensity $I(f)$ should be proportional to f^{-4} for smooth pores. For rough pores, with a fractal surface, $I(f)$ according to equation 18 should be proportional to f^{D-6}, where D is the fractal dimension of the inner surfaces of the pores. This law should hold for spatial frequencies corresponding to the reciprocal surface roughness of the pores.

Experimental results by Bale and Schmidt gave an exponent of -3.44 with an error smaller than 1 percent over an intensity range exceeding seven orders of magnitude [BS 84]. Thus, the pore surfaces are fractal, and their fractal dimension equals $6 - 3.44 = 2.56$.

Interestingly, the same power law, $I(f) \sim f^{D-6}$, would also obtain for *smooth* pores with a self-similar *size* distribution. Specifically, if the number of pores $N(r)$ with a radius larger than r is proportional to r^{-D}, then, according to Pfeifer and Avnir, the scattered intensities would follow the law $I(f) \sim f^{D-6}$ [PA 83]. However, the spatial frequency range would be different: it would be given by the pore *sizes*, not by their roughness. This is another example of how relatively simple things can become if scaling similarity prevails and if one properly exploits it.

Several years ago, the author proposed various number-theoretic concepts (quadratic residues, primitive polynomials and primitive roots in finite number fields, and Zech logarithms) as design principles for *reflection phase gratings* with very broad scattering of the incident energy over wide frequency bands [Schr 90]. The frequency bands efficiently scattered by such gratings can be further widened by recruiting self-similarity for their design, resulting in fractal diffraction gratings [D'An 90].

11

C H A P T E R

Iteration, Strange Mappings, and a Billion Digits for π

What good your beautiful proof on [the transcendence of] π: Why investigate such problems, given that irrational numbers do not even exist?

—LEOPOLD KRONECKER
to Ferdinand Lindemann

Apart from power laws, *iteration* is one of the prime sources of self-similiarity. Iteration here means the repeated application of some rule or operation—doing the same thing over and over again. (As the cartoon, Figure 1, illustrates, sometimes even a single repetition can lead to self-similarity.) In this chapter we continue to explore some of the strangely attractive consequences of iteration— one of our recurring themes.

A concept closely related to iteration is *recursion*. In an age of increasing automation and computation, many processes and calculations are recursive, and if a recursive algorithm is in fact repetitious, self-similarity is waiting in the wings. Think of the recursive calculation of the golden mean $(\sqrt{5} - 1)/2 = 0.618\ldots$, obtained by the rule "add 1 and take the reciprocal." Beginning with 1, we obtain 1, $\frac{1}{2}$, $\frac{2}{3}$, $\frac{3}{5}$, $\frac{5}{8}$, $\frac{8}{13}$, and so on, a sequence of fractions that converges to the golden mean. (These fractions, in fact, approach the golden mean with an error that decreases geometrically, forming an asymptotically self-similar error sequence with a similarity factor equal to the golden mean squared.) Recursion is also one of the main themes of Hofstadter's *Gödel, Escher, Bach* [Hof 80].

A very early example of an iterative algorithm is Euclid's method for determining the greatest common divider (GCD) of two natural numbers: divide the larger number by the smaller, take the reciprocal of the remainder, and *iterate*

Figure 1 Iteration leads to cartoon "self-similarity." (Drawing by Sempé; © 1985 The New Yorker Magazine, Inc.)

until the remainder is zero. The denominator of the last quotient is the GCD. For example, for 182 and 78, $\frac{182}{78} = 2 + \frac{26}{78}$ and $\frac{78}{26} = 3 + 0$. Thus, 26 is revealed as the GCD of 182 and 78. (The principle behind Euclid's method is the simple fact that the integers in the remainder fraction, (26 and 78), have the same GCD as the original integers, (182 and 78.)

Another ancient problem that is solved by a recursive algorithm is the "tower of Hanoi," in which a stack of disks of different sizes has to be transferred from one peg to another peg in such a manner that no larger disk ever comes to rest on a smaller one in the restacking process. A third peg is used as a "parking lot" for disks; otherwise the task would, of course, be impossible. For a large number of disks, trial and error would lead nowhere, but a simple recursive recipe gives the optimum solution without any guesswork.

We shall see how a simple iterative scheme, invented by Newton, for finding the zeros of a function produces basins of attraction of these zeros that are intimately intertwined self-similar multifractals. On a more mundane level, Newton's iteration is applied to calculate reciprocals and roots in high-precision arithmetic, using only multiplication. Thus $1/z$ is given by the recursion $x_{n+1} = 2x_n - zx_n^2$, and $z^{-1/2}$, for example, is calculated by iterating $x_{n+1} = x_n(3 - zx_n^2)/2$. (Initial values x_0 have to be chosen aptly, of course, to make the recursion converge to the desired result.)

The efficacy of iteration is also at the basis of the efficient algorithms for the fast Fourier transform (the famous FFT) and the less well-known *Hadamard transform* [HS 79]. The important point is that the matrices describing these transforms can be factored into the *direct* product of smaller matrices [Schr 90]. Specifically, a Fourier or Hadamard matrix with 2^n rows and 2^n columns can be iteratively factored into n 2×2 matrices, and it is precisely this iterative decomposition that results in the substantial computational savings by a factor $2n/2^n$. Not surprisingly, Hadamard matrices generated in this manner by iteration show self-similarity when represented as graphic images or optical masks.

The iterated *baker's transformation* is revealed as playing musical chairs with binary digits while simulating nonlinear transformations in two-dimensional spaces. Arnol'd's cat map leads to another chaos-producing recursion for modeling area-preserving transformations. (The *logistic parabola*, the paradigm of nonlinear mapping, and its two-dimensional generalization, the *Hénon map*, are analyzed in Chapter 12.)

But recursion, properly controlled, engenders beauty, too, be it in the form of handsome designs for needlework or shapely trees and flowers. And in mathematics proper, recursion weaves some veritable wonders—such as the computation of a billion digits of π in just 15 steps, a computational coup that is based on the work of Srinivasa Ramanujan (1887–1920), the great Indian mathematician whose awesome intuition remains beyond human comprehension.

Looking for Zeros and Encountering Chaos

Some 300 years ago, Isaac Newton (1642–1727) suggested finding the zeros of a function $f(z)$ by an *iteration*, based on drawing tangents. Given an approximate value of z_n to the solution of $f(z) = 0$, Newton finds the next approximation by calculating

$$z_{n+1} = z_n - \frac{f(z_n)}{f'(z_n)} \tag{1}$$

where $f'(z_n)$ is the derivative of $f'(z)$ at $z = z_n$. For equation 1 to be applicable, the slope $f'(z_n)$ of the tangent must be different from zero.

For $f(z) = z^2 - 1$, equation 1 reads

$$z_{n+1} = \frac{1}{2}\left(z_n + \frac{1}{z_n}\right) \tag{2}$$

Not surprisingly, for an initial value z_0 with a positive real part, z_n converges toward the positive root of $z^2 - 1 = 0$, namely, $z = 1$. Similarly, for a negative real part of z_0, the solution converges to the negative root, $z = -1$.

But what happens for purely imaginary $z_0 = ir_0$, where $r_0 \neq 0$ is real? Interestingly, it does not converge at all; the iteration in equation 2 cannot make up its mind, so to speak, and hops all over the imaginary axis, according to the mapping

$$r_{n+1} = \frac{1}{2}\left(r_n - \frac{1}{r_n}\right) \tag{3}$$

For example, the golden mean $0.618\ldots$ maps iteratively into -0.5, 0.75. $-0.291\overline{6}$, $1.56845\ldots$ and so on. But some r_0 behave quite differently, such as $r_0 = 1 + \sqrt{2}$, which maps into 1, 0, and ∞, a kind of fixed point, if rather distant.

How can we inject order into this chaotic mapping? A trigonometric substitution will do the trick:

$$r = -\cot(\pi\alpha) \tag{4}$$

which turns equation 3 into the exceedingly simple iteration

$$\alpha_{n+1} = 2\alpha_n \bmod 1 \tag{5}$$

where "mod 1" means subtracting the integer part and keeping only the fractional part, which lies in the half-open interval $[0, 1)$. For example, 2.618 mod 1 equals 0.618.

In terms of the new variable α, the chaotic mapping of r becomes totally transparent. If we express α_n as a binary fraction, then the digits of α_{n+1} will be simply the same as those of α_n shifted one place to the left. A 1 that moves to the left of the binary point is dropped. Thus, a periodic binary number α_0 will lead to periodic *orbits* (i.e., periodic sequences of iterates). For example, $\alpha_0 = \frac{1}{3} = 0.\overline{01}$ will map into $0.\overline{10} = \frac{2}{3}$, which will map right back into $\alpha_2 = 0.\overline{01} = \alpha_0$. (Indeed, $r_0 = -\cot(\pi/3) = -1/\sqrt{3}$ is mapped by equation 3 into $r_1 = 1/\sqrt{3}$, which maps back to $r_2 = -1/\sqrt{3} = r_0$.)

Similarly, *preperiodic* binary fractions, which begin aperiodically and end in a periodic tail, such as $\alpha_0 = \frac{5}{6} = 0.1\overline{10}$, lead to *preperiodic* orbits:

$$r_0 = \sqrt{3}, \quad r_1 = \frac{1}{\sqrt{3}}, \quad r_2 = -\frac{1}{\sqrt{3}}, \quad r_3 = r_1, \quad \text{etc.}$$

Terminating binary fractions are but a special kind of preperiodic binary fraction, the periodic tail being $\overline{0}$. Where will *they* end up? Continued left-shifting and taking fractional parts (remainders modulo 1) will, sooner or later, produce $\alpha = 0$, which corresponds to $r = \infty$. For example, $\alpha_0 = \frac{7}{8} = 0.111$ will map into $0.11 = \frac{3}{4}$ and, then $0.1 = \frac{1}{2}$, which maps into 0. In fact, the corresponding $r_0 = -\cot(7\pi/8) = 1 + \sqrt{2}$ maps into 1, 0, and ∞, as we already noted.

We also see by this analysis that any irrational α_0 will lead to an aperiodic "orbit" along the imaginary axis in the z plane. Thus, the simple mapping given by Newton's iteration for the function $z^2 - 1 = 0$ has rather strange consequences for initial values on the imaginary axis: in terms of the corresponding values of α, numbers are classified into three categories:

1 Periodic binary rational numbers

2 Preperiodic binary rational numbers

3 Irrational numbers

Periodic and preperiodic binary rational numbers α_0 converge on a fixed point or lead to periodic orbits. By contrast, irrational numbers α_0, an uncountable set, give aperiodic orbits: the same value never occurs twice, nor is α_n ever rational. Surprisingly, the simple map in equation 5 even makes a subtle distinction between different kinds of irrational numbers—not the usual number-theoretic distinction between algebraic (such as $\sqrt{2}$) and transcendental irrational numbers (for example, π), but between *normal* and *nonnormal* numbers, including "Liouville" numbers.

A normal number (in a given number system) is defined as a number in which every possible block of digits is equally likely to occur. For example, on the evidence of its first 100 million digits, π appears to be normal in the decimal system [Wag 85]. This means, for example, that somewhere in the decimal expansion of π a string of eight 7s will occur—in fact, there is a good probability that this will occur in the first 10^8 digits of π (or any other normal decimal number). For up-to-date evidence, see the book by Klee and Wagon [KW 89]. Iteration of normal numbers under the rule of equation 5 gives rise to chaotic orbits.

How awesome an object a normal number is can perhaps best be appreciated by the following reflection. The entire contents of the *Encyclopaedia Britannica* can be coded into a single decimal number (about 10^{10} digits long). And somewhere in the expansion of a normal number this block of digits will occur. In fact, the contents of the encyclopedia will occur infinitely often! (But don't ask where!)

Are there any *nonnormal* irrational numbers? In fact there are uncountably many in any number system! For example, the Cantor numbers (see Chapter 7) are nonnormal in the ternary system, because they eschew the digit 1. In subsequent chapters we shall get to know the Morse-Thue constant $= 0.01101001\ldots$ and the rabbit constant $= 0.10110101\ldots$ neither of which can boast a triple 1 in its binary expansion. Thus, they cannot be normal binary numbers.

But there are even stranger nonnormal numbers, such as the binary

$$L = \sum_{k=1}^{\infty} 2^{-k!} = 0.110001000000000000000001000\ldots$$

and other Liouville numbers[1] that are irrational yet very close to rational numbers.

In general, a Liouville number β is defined as an irrational number for which rational numbers p and q exist such that

$$\left| \beta - \frac{p}{q} \right| < \frac{1}{q^n} \tag{6}$$

for *any* n. In fact, to satisfy inequality 6 for the Liouville number L, one sets $q = 2^{n!}$. The resulting approximation error is then $2^{-(n+1)!} + 2^{-(n+2)!} + \cdots$, which (for $n > 1$) is smaller than $2^{-n!n} = q^{-n}$. (For $n = 1$, one sets $q = 4$ and notes that $\left| L - \frac{3}{4} \right| < \frac{1}{4}$.) The fact that for algebraic irrational numbers of degree n the absolute difference in inequality 6 *exeeds* $1/q^{n+1}$ demonstrates the existence of numbers, such as L, that "transcend" the algebraic numbers, namely, the *transcendental numbers*.

The iterates according to equation 5 of normal binary numbers α_0 will, by their definition, fill the unit interval densely and with equal probabilities for any subintervals of equal lengths. Thus their stationary distribution under the iteration, called the *invariant distribution*, is in fact flat. By contrast, *nonnormal* numbers will do nothing of the kind. For example, iterates of L will accumulate at 0 and all negative powers of 2 (i.e., $\frac{1}{2}$, $\frac{1}{4}$, $\frac{1}{8}$, and so on.)

Thus, we have to face the curious fact that the tiniest distinctions—the differences between rational and irrational numbers and, among the irrational numbers, between normal and nonnormal numbers—make a decisive difference in the final fate of a numerical iteration. Ordinarily, one should think that physics, and certainly the *tangible* world at large, would be untouched by the purely mathematical dichotomies between rational and irrational numbers or normal and nonnormal numbers. But in reality this is just not so. While, true enough, every-thing in the real world can be adequately described by rational numbers, it so happens that a mathematical *model* that distinguishes between different kinds of

---

1.   Named after Joseph Liouville (1809–1882), who exhibited the first transcendental number of which L is a prototype. The proofs that e and π are transcendental came later from Charles Hermite (1822–1901) and Ferdinand Lindemann (1852–1939), respectively.

Another (existence) proof of transcendental numbers was furnished by Cantor, who showed that algebraic numbers (i.e., the solutions of polynomial equations with rational coefficients) form a countable set. Since the continuum, by contrast, is an uncountable set, there must be numbers (in fact, uncountably many) that "transcend" the algebraic irrational numbers, namely, the so-called transcendental numbers.

numbers is not only extremely useful but catches the true, possibly hidden, spirit of the problem.

More specifically, two different initial conditions of a physical system that are completely indistinguishable by any finite measurement precision will sooner or later lead to a total divergence as the system evolves in time or space. The essential condition for this to happen is that the corresponding iteration, called *Poincaré mapping* (see chapter 14), be sufficiently *nonmonotonic*, such as equation 5, which has a sawtooth nonmonotonicity. The rate of divergence is measured by the so-called *Lyapunov exponent* $\lambda := \log(\alpha_{n+1}/\alpha_n)$ for $n \to \infty$. In our bare-bones example based on the iteration in equation 5, $\lambda = \log 2 \approx 0.693$ (if we take the natural logarithm).

Although the iteration in equation 5 may look unrealistically simple, it captures the essence of innumerable nonlinear problems which show period-doubling bifurcations and thus follow one of the two outstanding routes to chaos (see Chapter 12). (The other route is by quasi periodicity modeled by the so-called *circle map*, discussed in Chapter 14.)

The Strange Sets of Julia

Newton's method (equation 1) applied to the cubic equation $f(z) = z^3 - 1$ gives the iteration

$$z_{n+1} = \frac{2z_n^3 + 1}{3z_n^2} \tag{7}$$

which we already encountered in pages 38–40 in Chapter 1.

Now the mapping is even crazier. For one thing, the naive conjecture that all z_0 will converge toward the closest of the three roots is false. For example, $z_0 = -1$ will converge on 1, the root most distant from it.

Color Plate 3 shows the intimately intertwined basins of attraction of the three roots—a real, or rather imaginary, crazy quilt. In fact, it can be shown that for the mapping in equation 7 two basins (colors) can never meet if the third is not present also. This may sound impossible, and in fact it would be—if it were not for the fractal nature of the boundaries as intimated in Color Plate 3, which also shows the attractive self-similarity resulting from Newton's iteration.

Why are the three basins of attraction not simply shaped like three pieces of pie or sectors, each $120°$ wide? After all, Newton's iteration for $z^2 = 1$ leads to two basins that are half planes. That this cannot be so for $z^3 = 1$ becomes clear when one looks at the point $z = -2^{-1/3}$, which is mapped to the origin, $z = 0$, by equation 7. The neighborhood of the point $z = 0$ contains points from all three basins of attraction (because of the $120°$ rotational symmetry of the problem). Thus, because the inverse map of equation 7 is continuous, the point

$z = -2^{-1/3}$ must also be surrounded by points from all three basins of attraction. In fact, its infinitesimal neighborhood must be a scaled-down version of the "cloverleaf" at the origin. As a consequence, there are points below the negative real line that belong to the basin of attraction for the attractor $z_1 = \omega :=$ $\exp(i\pi/3)$, which lies *above* the real line. Similarly, there are points above the negative real line that will converge on $z_2 = \omega^2$, which lies *below* the real line. Thus, the basins cannot be simple sectors; the basins "nibble" at each other's pies.

In fact, quite generally for $z^n = 1$, Newton's method will produce at the origin basins that form a cloverleaf with n leaves. A preimage of the origin that falls on the boundary between two attractors will therefore intermingle *all* basins at that point, and for $n > 2$, this intermingling must result in a fractal boundary, because in two dimensions only two attractors can meet in a smooth boundary. Furthermore, since *all* boundary points are preimages of the origin, they are boundary points of *all* n basins. Such strange sets of boundary points are customarily called *Julia sets*. By definition, the Julia set J of a rational function $R(z)$ is the set of points z for which $R(z)$ is *not normal*. (Normal points are those for which $R(z)$ is equicontinuous in a neighborhood of z.)

The Julia set of a rational function has the following astounding property. If z_k is a periodic attractor and A_k its basin of attraction, then $J = \partial A_k$ for all k. Here ∂A_k is the *boundary* of the basin A_k, that is, all those points in whose neighborhood, no matter how small, one finds points both in A_k and outside A_k. Hence, if we have a point z that lies on the boundary of A_1, say, we know that it also lies on the boundaries of $A_2, A_3, \ldots$

Another important property of a Julia set J is that periodic repellors of $R(z)$ are dense in J. In fact, if z, is a periodic repellor, then J is the closure of the set of all the preimages of z,. As a result, the dynamics on a Julia set are chaotic, that is, sensitive to initial conditions, as we already saw in pages 239–241 for the Julia set of $R(z) = (z + 1/z)/2$. A Julia set always contains an uncountable number of points, but it is not necessarily fractal.

Julia sets, consisting of repellors, are difficult to plot because of the chaotic sensitivity, which requires unrealistic numerical precision. However, another property of J comes to the plotter's aid: for any z in J, the *inverse orbit* $R^{-n}(z)$, $n = 1$, $2, 3, \ldots$, is dense in J. For points in J, $R^{-n}(z)$ is attractive and so there are no problems with numerical divergence. However, $R^{-1}(z)$ is in general multivalued, so that clever algorithms are needed to cover the entire Julia set uniformly. Such algorithms are described by Peitgen and Saupe in *The Science of Fractal Images* [PS 88]. Figure 2 shows the dustlike Julia set for the mapping of equation 7 generated by such an algorithm.

The simple example of a rational mapping with three attractors and its Julia set discussed here is not devoid of physical implications. Consider, for example, a pendulum consisting of an iron bob at the end of a string. Below the pendulum are three magnets to which the iron bob is attracted. After some oscillations, the pendulum will come to rest with the bob directly above one of the three magnets. But will the bob always go the attractor nearest to its initial position?

Figure 2   Julia dust for Newton iteration [PS 88].

It will not. Try it! The iron will, for many initial conditions, follow a rather tortuous route, and the end position will seem totally unpredictable—and it often not only will *seem* unpredictable but will be unpredictable, unless the initial position can be given with a totally unrealistic precision. In other words, the basins of attraction of the three magnets are bounded by fractal sets—just as the Julia set in our Newton iteration.

A Multifractal Julia Set

As we have seen in the preceding section, Julia sets of rational functions with more than two attractors are fractals—in fact, they are *multi*fractals. Such sets have been traditionally analyzed by generating individual points of the set by numerical backward iteration. However, for some Julia sets, *analytical* methods, which offer much higher accuracy and require less computing, are feasible. A prime example is the recent analysis by Nauenberg and Schellnhuber of the multifractal properties of the Julia set associated with the Newtonian map (equation 7) for the solution of $z^3 - 1 = 0$ [NS 89].

   The first order of business is to construct the *support* of the fractal set. For the original Cantor set the support is the unit interval: all members of the set "live" on the straight unit interval. For our Newtonian Julia set, by contrast, the support is already a highly complicated manifold, an infinitely nested "cobweb." To construct this spidery support, Nauenberg and Schellnhuber note that one of the three preimages of the negative real line is the interval $-\infty < z \le -2^{-1/3}$, called M_0 in Figure 3A. The two other preimages are obtained from M_0 by rotations of $\pm 120°$.

(A)

(B)

(C)

Figure 3   (A) Constructing the Julia flower: the first-generation preimages. (B) Support of the Julia-flower set at fourth stage of construction. (C) Julia petal: the first three steps of construction [NS 89].

The two preimages M_+ and M_- of M_0 that do not fall on the negative real line are given by

$$M_\pm = \xi \pm i(\xi^{1/2} - \xi^{1/2}) \qquad 0 \le \xi \le 2^{-4/3} \qquad (8)$$

The other four arclike second-order preimages are obtained by $\pm 120°$ rotations. The six arcs together form the central "flower" shown in Figure 3A. The six arcs and the three straight-line spikes at the end of the petals together form the first generation of the support of the fractal. The second generation consists of $3 \cdot 9 = 27$ parts, and so on. Figure 3B shows the support at the fourth stage of construction, made up of the first four generations obtained by backward iteration from the central flower. The nested character of this Julia fractal is already clearly evident.

The higher-order preimages x_n of $z = 0$ falling on the negative real line are given by the recursion

$$x_{n+1} = \frac{x_n}{2} + x_n \cosh\left[\frac{1}{3}\cosh^{-1}(1 - 2x_n^{-3})\right] \qquad (9)$$

where $x_1 = -2^{-1/3}$. The asymptotic scaling factor of the x_n is $\frac{3}{2}$, as can be deduced from the derivative of the forward iteration, $dN/dz \to \frac{2}{3}$ for large z.

The preimages $\zeta_n = \xi_n + i\eta_n$ of the x_n on M_+ are given by

$$\xi_{n+1} = -x_n \sinh^2\left[\frac{1}{3}\sinh^{-1}((-x_n)^{-3/2})\right] \qquad (10)$$

and $\eta_{n+1} = \pm(\xi_{n+1}^{1/2} - \xi_{n+1}^2)$ according to equation 8. The arc lengths l_n on M_+ between ξ_n and ξ_{n+1} scale asymptotically as $l_n \sim (\frac{2}{3})^{n/2}$. The longest arc has length $l_1 = 0.3834$.

For the calculation of the fractal dimensions D_q, we focus on one of the three petals of the central flower (see Figure 3C). In the first generation the petal consists of just two large arcs. As the next step, we consider the infinite succession of smaller arcs whose vertices, lying on these arcs, are given by the first-order preimages of the points x_n on the negative real line. Note that each arc generates a succession of *double* arcs. The bottom of Figure 3C shows the result of the third step, in which each smaller arc has sprouted an infinite succession of even smaller double arcs.

The Julia set that we are interested in, namely, the common boundary points of three basins of attraction, consists of the *branch* points of this support. (This is analogous to the original Cantor set, which is given by the *end* points of the surviving intervals.) The fractal dimensions D_q of this Julia set are given by the expression

$$2 \sum_m p_m^q l_m^\tau = 1 \qquad \tau = (1 - q)D_q \qquad (11)$$

(see Chapter 9) where the summation is over successive generations of the support-generating process and the factor 2 reflects the fact that each mother arc gives birth to twin daughter arcs. If we choose the three preimages with equal weight, then $p_m = (\frac{1}{3})^m$. The l_m are the arc lengths computed with equations 8 to 10.

Of all the dimensions, D_∞ is the easiest to compute, because for $q \to \infty$, $\tau \to -\infty$, leaving only the largest l_m in equation 11 to make a contribution. The largest l_m is $l_1 \approx 0.3834$. Hence, with $p_1 = \frac{1}{3}$,

$$D_\infty = \frac{\log p}{\log l_1} = 1.146 \dots$$

For $q = \pm\infty$, the factor 2 in equation 11 is irrelevant. The other extremal dimension, $D_{-\infty}$, is also easy to calculate. Here the *smallest* l_m in equation 11 dominate the sum, that is, l_m for $m \to \infty$. With $l_m \sim (\frac{2}{3})^{m/2}$ for large m and $p_m = (\frac{1}{3})^m$, equation 11 yields

$$D_{-\infty} = \lim_{m \to \infty} \frac{\log 3}{\frac{1}{2}\log \frac{3}{2}} = 5.419 \dots$$

Note that $D_{-\infty}$, unlike the Hausdorff dimension D_0, has no simple geometric meaning. Its value, being larger than 2 (the Euclidean dimension of the space embedding the Julia fractal), is therefore no contradiction.

Interestingly, in the present approach, the numerical value of $D_{-\infty}$ is given by a simple analytical fact: namely, that the l_m are proportional to $(\frac{2}{3})^{m/2}$. By contrast, numerical methods are not even feasible for $q = -\infty$, because the computer would have to run forever to explore the sparsest regions of the Julia set that are characterized by $D_{-\infty}$. Thus, one is compelled to estimate $D_{-\infty}$ from D_q for large q. But this access is hampered by a frustratingly slow numerical convergence. The other dimensions, including the Hausdorff dimension D_0, are likewise obtained by elementary calculations. For good approximations, only a few of the l_m in equation 11 need to be calculated explicitly; for the remaining terms, the approximation $l_m = 0.1986(\frac{2}{3})^{m/2}$ suffices. The result for the Hausdorff dimension is $D_0 = 1.429 \dots$, a considerably more accurate result than the one obtained by number-crunching methods on the basis of 1 million points of the Julia set. Note that $D_0 < 2$, as behooves a two-dimensional "dust." For the important information dimension (see Chapter 9),

$$D_1 = \frac{\sum_m p_m \log p_m}{\sum_m p_m \log l_m}$$

just *two* rough values of l_m, $l_1 \approx 0.38$ and $l_2 \approx 0.18$, suffice for a good estimate, $D_1 \approx 1.2$.

The Beauty of Broken Linear Relationships

Figure 4 shows the result of applying the piecewise linear "sawtooth" map $\tilde{b} = 8b$ mod 1 to the local brightness $0 \leq b \leq 1$ of the photograph of a human face and quantizing the result $\tilde{b}$ to 0 if $\tilde{b} > b$ and to 1 for $\tilde{b} \leq b$ [Schr 69]. Figure 5 shows the beautiful pattern that results from the iteration of another simple piecewise linear function [PR 84].

Such maps are useful mathematical models of *deterministic* diffusion [Schu 84]. While traditional (thermodynamic) diffusion is a typical random process, piecewise linear maps, such as that shown in Figure 6, cause a particle to drift in a seemingly random manner, although the drifting is a strictly deterministic process. This drifting is another example of a chaotic motion which depends strongly on the precise (and not exactly knowable) initial coordinate value. As in traditional diffusion, the spatial correlation between two initially close particles decays to zero as time evolves [Gro 82]. As a consequence, the mean squared drift $\overline{x^2}$ increases linearly with time, as in ordinary diffusion:

$$\overline{x^2} = 2Dt \qquad \text{for } t \gg 1$$

Figure 4   Sawtooth map of "Karen," an early computer graphic by the author [Schr 69].

Figure 5   Embroidery by broken-linear relationship [PR 84].

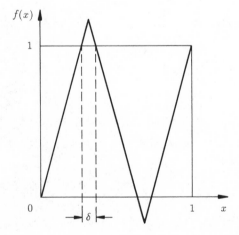

Figure 6   Broken-linear map for modeling deterministic diffusion [Gro 82].

with a diffusion constant D that is proportional to the width δ over which the iterated function "sticks out" from the unit square.

The Baker's Transformation and Digital Musical Chairs

Bakers mix their dough by rolling it out, folding it over, and rolling it out again, in a seemingly endless iteration—roll-fold-roll-fold . . .—until they are satisfied that they have achieved a sufficiently uniform mixture of the dough's ingredients.

A mathematically sanitized version of the dough rolling and folding, called the *baker's transformation,* is illustrated in Figure 7. It is a useful model of all kinds of mixing processes, including the chaotic mixing of fluids. Arithmetically, a point (x, y) in the unit square is transformed to a point $(2x, y/2)$ by rolling out and to $(\langle 2x \rangle_1, y/2 + \lfloor 2x \rfloor/2)$ by subsequently cutting the rolled-out dough in half and putting the right half on top of the left half. (This operation is mathematically simpler than the folding over.) Here, as before, the pointed brackets

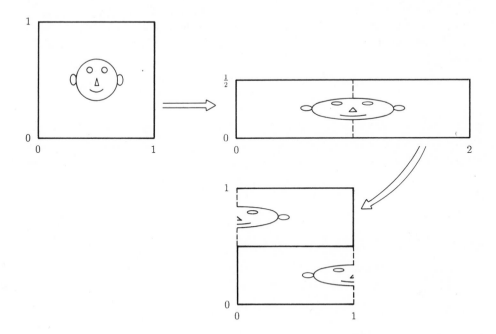

Figure 7   Baker's map, a recipe for chaotic mixing. The unit square is first "rolled out" to a rectangle. The right half is then cut off and stacked on top of the left half.

mean "take the fractional part" and the open-top brackets "round down to the nearest integer."

If we express x and y in binary notation, then the baker's transformation becomes particularly simple: the digits of x are left-shifted by one place, the digits of y are right-shifted by one place, and the leftmost digit of x becomes the leftmost digit of y. In fact, the binary digits of x and y play "musical chairs." For example, the coordinate pair

$$x_1 = 0.10110001\dots$$
$$y_1 = 0.01110100\dots$$

is mapped into

$$x_2 = 0.0110001\dots$$
$$y_2 = 0.101110100\dots$$

which goes into

$$x_3 = 0.110001\dots$$
$$y_3 = 0.0101110100\dots$$

and so on. Thus, any terminating fraction for x will asymptotically approach the origin $(0, 0)$, which is therefore an attractor for such x values.

Periodic binary x values will converge on periodic orbits. For example, $x = \frac{1}{3} = \overline{0.01}$, $y = \frac{2}{3} = \overline{0.10}$ will lead to a periodic orbit of period length 2. By contrast, normal binary numbers will follow chaotic orbits, in which even initially close points are soon separate and follow independent orbits; see Figure 8. Both chaotic and nonchaotic flows are discussed in Ottino's "The Mixing of Fluids" [Ott 89].

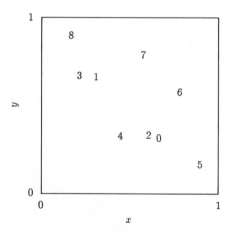

Figure 8   A chaotic orbit for the baker's map starting *near* the period-2 point $x_0 = \frac{2}{3}$, $y_0 = \frac{1}{3}$.

Figure 9   Smale's horseshoe map. A space is stretched in one direction, squeezed in another, and then folded. When the process is repeated several times, a pair of points that end up together may have begun far apart.

Sometimes the baker throws away some dough during every iteration, in which case the remaining dough turns into a Cantor dust in one direction. This generalized baker's transformation is a simple model of the phase spaces of nonlinear dynamic systems that contract in some directions and thus sport strange attractors, such as the Julia sets discussed in pages 243−248.

Related transformations are Smale's horseshoe map [Sma 67] (see Figure 9) and Hénon's map [Hen 76] (see Figure 10), which are characteristic of dissipative physical systems with strange attractors.

Where is the self-similarity in Hénon's map? Take another look at Figure 10D and the Hénon attractor after 10^4 iteractions shown in Figure 11. While strange attractors may be strange in many ways, they do maintain self-similar order in their chaotic orbits.

Arnol'd's Cat Map

Another important two-dimensional map, for the description of Hamiltonian nonlinear systems, is Arnol'd's area-preserving cat map,

$$x_{n+1} = x_n + y_n \bmod 1$$
$$y_{n+1} = x_n + 2y_n \bmod 1$$

(12)

illustrated in Figure 12A and B. The notation "mod 1" means, as before, that only fractions in the half-open unit interval [0, 1) are considered. The distortion engendered by this transformation is reminiscent of a maladjusted, out-of-sync

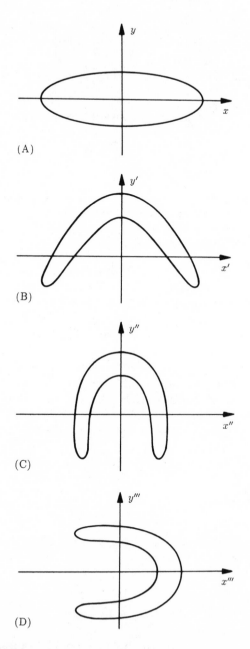

Figure 10   Hénon's map. (A) Initial ellipse. (B) Area-preserving bending: $x' = x$, $y' = 1 - ax^2 + y$. (C) Contraction in the x direction: $x'' = bx'$, $y'' = y'$. (D) Rotation by 90°: $x''' = y''$, $y''' = -x''$ [Schu 84].

Figure 11   Self-similarity of the Hénon attractor. (A) The entire attractor. (B) Enlargement of portion shown by small square in part A. (C) The result of another enlargement. Note the similarity of the streak patterns between parts B and C, attesting to the ultimate self-similarity of the Hénon attractor [Far 82].

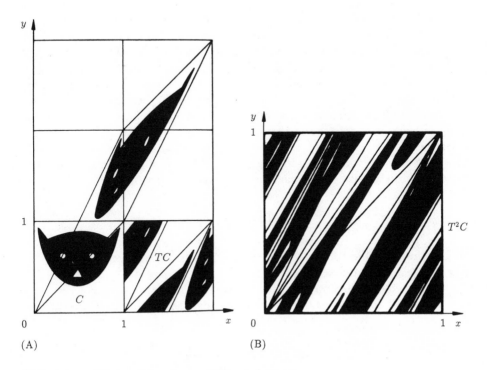

Figure 12   (A) Arnol'd's cat map; (B) iterated cat map.

television set. The cat map's two eigenvalues, λ_1 and λ_2, are the squared reciprocal of the golden mean γ and its inverse: $\lambda_1 = (3 + \sqrt{5})/2 > 1$ and $\lambda_2 = \lambda_1^{-1} = (3 - \sqrt{5})/2 < 1$. Because one eigenvalue is larger than 1 and the other is smaller than 1, its fixed points are all *hyperbolic*, meaning that the map is expanding in one direction (corresponding to the eigenvalue λ_1) and contracting in the orthogonal direction (corresponding to λ_2). Hyperbolic fixed points are therefore neither repellors nor attractors; they are both, depending on the direction of approach. In geometric representations, hyperbolic fixed points are *saddle* points: a ball rolls down the mountain *toward* a saddle point (mountain pass), but then it will move *away* from the saddle point toward the valley. The mountain pass first *attracts* and then *repels* water running down the mountain in its direction.

Maps with hyperbolic fixed points are the hallmark of chaotic motion in energy-conserving physical assemblies, called Hamiltonian systems [Arn 89]. For example, $(x, y) = (0.4, 0.2)$ is a fixed point of the once-iterated cat map; that is, it belongs to an orbit with period length 2. It is transformed into $(0.6, 0.8)$, which is mapped back into $(0.4, 0.2)$. A perturbation of the initial point $(0.4, 0.2)$ in the direction corresponding to the eigenvalue λ_2, $\Delta x/\Delta y = -(\sqrt{5} + 1)/2 = -1/\gamma$, is contractive. Indeed, the initial even-numbered iterates x_{2n} of $x_0 = 0.4 + 1/100\gamma$, $y_0 = 0.2 - 1/100$ are, beginning with x_2, approximately 0.402, 0.4003, 0.40005, 0.400007, 0.400001, 0.4000002. Thus, the point $(0.4, 0.2)$ acts like an attractor when approached from the direction that corresponds to λ_2. But following this convergence, using a 12-digit calculator, which introduces a small perturbation in the λ_1-direction, numerical inaccuracies cause eventual divergence, as evidenced by subsequent iterates: 0.4000005, 0.400003, 0.40002, 0.4002, 0.401, 0.407, 0.45, 0.75, and so on to chaos. The point is that, because of this attractive-repelling nature, any numerical calculation, or any physical system modeled by the cat map, will sooner or later show chaotic motion, except for rare initial conditions with measure zero. Such initial conditions, of course, cannot be physically realized; they correspond to unstable equilibria—such as a pencil standing on its tip.

One of the advantages of the cat map is that its iterates are easy to analyze. In matrix form, the cat map (equation 12) is given by the matrix

$$T \begin{pmatrix} 1 & 1 \\ 1 & 2 \end{pmatrix}$$

which transforms the column vector (x_n, y_n) into the column vector $(x_{n+1}\ y_{n+1})$, where all x and y are taken modulo 1.

Given the recursion for the Fibonacci numbers $F_n = F_{n-1} + F_{n-2}$, $F_1 = F_2 = 1$, it is easy to show by induction that the nth iterate of the cat map is

$$T^n = \begin{pmatrix} F_{2n-1} & F_{2n} \\ F_{2n} & F_{2n+1} \end{pmatrix}$$

which inherits its area preservation from that of T. (Indeed, the determinant of T^n, $F_{2n-1}F_{2n+1} - F_{2n}^2$, equals 1.) Since T^n is a symmetric matrix, it has only real eigenvalues, namely, $\lambda_1^{(n)} = \lambda_1^n > 1$ and $\lambda_2^{(n)} = \lambda_2^n < 1$.

Fixed points of T^n correspond to orbits of period length n and any divisors of n. For $n = 2$, for example,

$$T^2 = \begin{pmatrix} 2 & 3 \\ 3 & 5 \end{pmatrix}$$

To find the fixed points of period length 2, one has to solve the equations

$$2x + 3y = x \bmod 1$$

$$3x + 5y = y \bmod 1$$

Apart from the solution $x = y = 0$ (which has period length 1), eliminating y results in $5x = 0 \bmod 1$, that is, $x = k/5$, where k is an integer. The only allowed values for $0 < x < 1$ are $k = 1, 2, 3, 4$, each of which, in fact, yields a solution. The corresponding values of y are $y = k'/5$, with $k' = -2k \bmod 5$. These four periodic points of period length 2 form two orbits, namely, $(\frac{2}{5}, \frac{1}{5}) \to (\frac{3}{5}, \frac{4}{5})$, which we already encountered, and $(\frac{1}{5}, \frac{3}{5}) \to (\frac{4}{5}, \frac{2}{5})$.

We leave the highly instructive analysis of the complete orbit structure of the cat map to the interested reader.

A Billion Digits for π

Iteration is one of the most powerful mathematical tools. To calculate the value of π by means of the Gregory-Leibniz series

$$\frac{\pi}{4} = 1 - \frac{1}{3} + \frac{1}{5} - \frac{1}{7} + \cdots$$

to an accuracy of just three decimal places, one needs to sum 500 terms. By contrast, a recursion involving the arithmetic-geometric mean doubles the number of correct digits with every iteration [BB 87, KW 89]. And there are iterative algorithms, based on Ramanujan's work, that multiply the number of decimal digits for each iteration by 4 or even 5 [BBB 89]. Thirteen iterations of such an algorithm have yielded more than 134 million digits of π, and just two more iterations would give π to an accuracy exceeding two *billion* digits. This is a relative error of 10^{-10^9} (not just 10^{-9}), an awesome accuracy.

Of course, nobody needs π that accurately. Among the most accurate measurements in physics are those exploiting the Mössbauer effect, with a precision of, say, 14 decimal digits, corresponding to an error of 1 second in 3 million years. And 39 digits for π suffice to calculate the circumference of the known universe from its radius to within the diameter of the hydrogen atom. However, computations of the digits of π have become benchmark tests for supercomputers and superfast algorithms. Needless to say, these calculations do not run on pocket calculators; for these do not have enough memory and their displays are too limited for the purpose at hand.

One of Ramanujan's astonishing results is the following formula:

$$\frac{1}{\pi} = \frac{\sqrt{8}}{9801} \sum_{n=0}^{\infty} \frac{(4n)![1103 + 26390n]}{(n!)^4 (396^{4n})} \tag{13}$$

whose very first term gives π with a relative accuracy better than $3 \cdot 10^{-8}$. Each additional term adds about eight more decimal digits (i.e., multiplies the accuracy by 100 million).

Very rapid approximations to π are based on a seminal paper by Ramanujan which established a close connection with the transformation theory for elliptic integrals [Ram 14]. One of the recursive algorithms resulting from this connection is the following. Let $\alpha_0 = 6 - 4\sqrt{2}$ and

$$y_{n+1} = \frac{1 - (1 - y_n^4)^{1/4}}{1 + (1 - y_n^4)^{1/4}} \qquad y_0 = \sqrt{2} - 1$$

Then

$$\alpha_{n+1} = (1 + y_{n+1})^4 \alpha_n - 2^{2n+3} y_{n+1}(1 + y_{n+1} + y_{n+1}^2)$$

approaches $1/\pi$ with an error smaller than $16 \cdot 4^{n+1} \exp(-2\pi \cdot 4^{n+1})$. The first approximation, α_1, has already an accuracy of 9 digits, and α_2 has 40 correct digits. The number of correct digits of α_n is greater than $2 \cdot 4^n$; asymptotically it *quadruples* with each iteration and exceeds 1 billion after 15 steps.

There is even a *quintic* algorithm which multiplies the number of correct digits by a factor of 5 with each step [BBB 89].

How random are the first billion digits of π? Gregory and David Chudnovsky of Columbia University found that the digits of π are more random than the strings of digits produced by standard pseudorandom number generators, which, as finite-state algorithms, are, of course, ultimately periodic. Specifically, when used to generate Brownian motions, the digits of π generate random walks that seem to satisfy the iterated-logarithm law (in contrast to the usual pseudorandom number generators).

Even so, there is no mathematical proof that π is a normal number, in which all groups of digits occur with the same asymptotic probability. In fact, given that π is determined by such compact and rapidly converging fomulas as equation 13, experts suspect the π is *not* normal in some appropriate base.

Bushes and Flowers from Iterations

Iteration of simple rules is one of the more potent prescriptions for generating not only mathematical fractals but interesting biological shapes too [Pru 87]. A preferred paradigm for these applications is the so-called turtle algorithm for producing line drawings:

A *state* of the "painting turtle" is defined as a triplet (x, y, α), where the Cartesian coordinates (x, y) represent the turtle's *position* and the angle α, called the turtle's *heading*, is interpreted as the direction in which the turtle is facing.

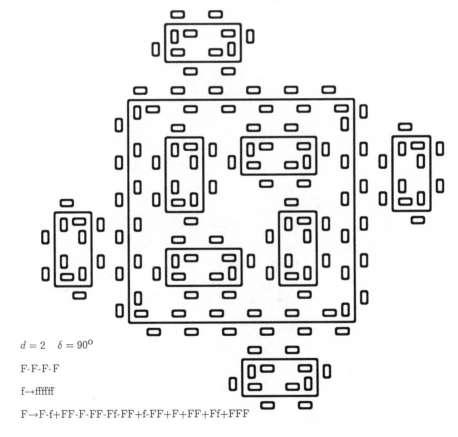

$d = 2$ $\delta = 90^\circ$

F-F-F-F

f→ffffff

F→F-f+FF-F-FF-Ff-FF+f-FF+F+FF+Ff+FFF

Figure 13   Nested "Islands and Lakes" drawn by the turtle algorithm [Man 83].

Given the *step size d* and the *angle increment* δ, the turtle can respond to commands represented by the following symbols:

F   Move forward a step of length d. This motion changes the state of the turtle to (x', y', α) where $x' = x + d \cos \alpha$ and $y' = y + d \sin \alpha$. The turtle also leaves a trail, drawing a line segment between points (x, y) and (x', y').

f   Move forward a step of length d without drawing a line.

+   Turn right (clockwise) by angle δ. The next state of the turtle is $(x, y, \alpha + \delta)$.

−   Turn left by angle δ. The next state of the turtle is $(x, y, \alpha - \delta)$.

Let v be a string of commands, (x_0, y_0, α_0) the initial state of the turtle, and d and δ fixed parameters. The picture (set of lines) drawn by the turtle responding to the string v is called the *turtle interpretation* of v. Figure 13 shows as an example

$$d = 4 \qquad \delta = 22.5°$$
$$F \rightarrow FF+[+F\text{-}F\text{-}F]\text{-}[\text{-}F+F+F]$$

Figure 14   "Breezy Bush" (watch its self-similarity) [Pru 87].

the drawing of the fractal "Islands and Lakes," starting with the unit square, given by the command string F — F — F — F. According to these rules, the turtle interprets a character string as a sequence of line segments, connected "head to tail."

In a further elaboration, Lindenmayer introduced a notation for representing graph-theoretic trees using strings with *brackets* [Lin 68]. The motivation was to formally describe branching structures found in many plants, from algae to trees. An extension of the turtle interpretations to bracketed strings uses two additional symbols interpreted by the turtle:

[Remember the current turtle state (x, y, α) for later retrieval.

] Recall the turtle state at the *corresponding* "open" bracket ([) and continue executing the instructions to the right of the "closed" bracket (]). Note that brackets can be nested.

An example of a bracketed string and its turtle interpretation of a bush in a breeze is shown in Figure 14.

Color Plate 7A through F shows six plantlike structures generated by Prusinkiewicz in this manner [Pru 87]. By adding color to the constructions, Prusinkiewicz has obtained breathtakingly beautiful and realistic-looking flowers and bushes, as evidenced in the color plates, which are reproduced here with his kind permission.

In another approach to image coding, called *iterated function systems*, Barnsley has created sunflowers, ferns, and forests from an astonishingly small set of parameters (see pages 28−30 in Chapter 1).

12
C H A P T E R

A Self-Similar Sequence, the Logistic Parabola, and Symbolic Dynamics

> *I work in statistical mechanics, but I am*
> *not interested in getting to the moon.*
> —MARSTON MORSE

In this chapter we shall delve deeper into the self-similarities engendered by iteration. We shall focus our attention on the so-called *logistic parabola*, a simple quadratic equation that models the waxing and waning of warring species and restrained growth processes—restrained by a lack of "logistical" resources and supplies. This simple law has found widespread application in many fields. Its iteration gives rise to numerous universal features and self-similarities. We shall attempt to illuminate such signal attributes of the logistic parabola as *stable* and *unstable orbits, deterministic chaos, tangent bifurcations, intermittency*, the *hierarchy of orbits*, the *bifurcation of chaos bands*, and *invariant distributions*. We shall also touch upon some of the mathematical tools that exploit the self-similarity inherent in the quadratic map and explore some noteworthy transformations that shed a lot of light on this and other iterations.

One of the predictions of the logistic parabola, borne out by observation of many natural phenomena, is the occurrence of periodic cycles, especially those of period lengths 2, 4, 8, 16, 32, and so on. This is the famous *period-doubling* "scenario," which is born of self-similarity and begets, in the end, total *chaos*, albeit *deterministic* chaos. We shall attempt to bridge the seemingly impossible gulf between the self-similarity of the binary integers and period doubling in the logistic parabola, forging a strong bond, dubbed *symbolic dynamics*, between these two fundamental phenomena. But, as we shall also see, deterministic chaos is intimately linked to a simple operation on the digits of (binary) numbers: an incessant left shift until their totally unpredictable tails are exposed to full view.

After having reveled long enough in the *real* world of the logistic parabola, we shall follow Mandelbrot in an imaginative leap in the *imaginary* direction into the complex plane, where much becomes plain that was obscure on the real line. In other words, we shall *complexify* to *simplify*. In the process, we will discover the *Mandelbrot set* and its intricate self-similar designs as an added reward.

We begin our excursion into complexity with a recursive exercise in the discrete world of the real integers.

Self-Similarity from the Integers

The self-similar properties that can be squeezed from the integers are far from exhausted by the Fibonacci numbers and Pascal's triangle (see Chapter 17, on cellular automata). Consider the sequence of the nonnegative integers 0, 1, 2, 3, 4, 5, 6, 7, . . . in binary notation:

$$0, \quad 1, \quad 10, \quad 11, \quad 100, \quad 101, \quad 110, \quad 111, \ldots$$

and take the "digital root" (i.e., the sum of the digits modulo 2) of each binary number. This yields the sequence

$$0, \quad 1, \quad 1, \quad 0, \quad 1, \quad 0, \quad 0, \quad 1, \ldots$$

which is called the *Morse-Thue* (MT) *sequence*, in honor of the Norwegian mathematician Axel Thue (1863–1922), who introduced it in 1906 as an example of an aperiodic, recursively computable string of symbols, and after Marston Morse of Princeton (1892–1977), who discovered its significance in the symbolic dynamics in the phase-space description of certain nonlinear physical systems [Thu 06, Mor 21].

Interestingly, the MT sequence can also be generated by iterating the mapping $0 \rightarrow 01$ and $1 \rightarrow 10$, that is, the mapping in which each term is followed by its complement. Starting with a single 0, we get the following successive "generations":

$$0$$
$$0 \quad 1$$
$$0 \quad 1 \quad 1 \quad 0$$
$$0 \quad 1 \quad 1 \quad 0 \quad 1 \quad 0 \quad 0 \quad 1$$

and so on. A sequence generated in this manner is called a self-generating sequence [Slo 73].

Alternatively, each generation is obtained from the preceding one by appending its complement:

0

0 1

0 1 1 0

0 1 1 0 1 0 0 1

and so on. This is simply a consequence of the fact that the mapping $0 \rightarrow 01$ and $1 \rightarrow 10$ immediately gives rise to the mapping $01 \rightarrow 0110$ and $10 \rightarrow 1001$, and so on, where each higher-order map mimics the original generating rule "copy and append the complement." In other words, the original mapping rule is *inherited* by all successive generations. This kind of inheritance is an important consequence of iterated mappings and often leads to self-similar structures. Such generating processes are also called *inflation*, a term which (in its noneconomic and noncosmological sense) is associated with Penrose tilings and their fascinating scaling properties [GS 87].

The infinite sequence obtained from the iterated map $0 \rightarrow 01$ and $1 \rightarrow 10$ is invariant under this mapping; inflation leaves it untouched. The MT sequence is in fact *self-similar*: retaining only every other term of the infinite sequence (indicated by underlining), beginning with the first term, reproduces the sequence:

$$\underline{0} \quad 1 \quad \underline{1} \quad 0 \quad \underline{1} \quad 0 \quad \underline{0} \quad 1 \dots$$

Similarly, retaining every other *pair* also reproduces the sequence:

$$\underline{01} \quad 1 \quad 0 \quad \underline{10} \quad 0 \quad 1 \dots$$

as does the "renaming" of each pair, quadruplet, octet, and so on, by its left most digit (i.e., regenerating it from its "amputated" first digit). This "skipping" process is equivalent to what has been called *deflation* in tiling or *block renaming* in renormalization theories. These schemes follow simply from the inverse, $01 \rightarrow 0$ and $10 \rightarrow 1$, of the original mapping $0 \rightarrow 10$ and $1 \rightarrow 10$. Naturally, if inflation reproduces a given infinite sequence, so does the corresponding deflation. The fact that proper subsets can be equivalent to the entire set is, of course, a well-known property of infinite sets.

The self-similarity of the MT sequence is very easy to understand. Retaining only every other term of the infinite sequence is equivalent to multiplying the original numbers in the underlying integer sequence by 2. Since, in the binary representation, this means a left shift of the digits, the digital roots are not changed—by definition this *is* the MT sequence. (If we retain every other term beginning with the *second* term, the MT sequence, for similar reasons, turns into its own complement.)

Of course, there is nothing magic about the number 2 and the binary number system. The successive integers written in ternary notation:

$$0, \quad 1, \quad 2, \quad 10, \quad 11, \quad 12, \quad 20, \quad 21, \ldots$$

have digital roots (sums of digits modulo 3) that form a self-similar sequence with a similarity factor of 3:

$$p_k = 0, \quad 1, \quad 2, \quad 1, \quad 2, \quad 0, \quad 2, \quad 0, \ldots$$

Indeed, p_{3k} equals p_k because the ternary representation of $3k$ is the same as that of k except for a left shift. What iteration generates the p_k? (And what are the self-similarities of the sequences $\{p_{3k+1}\}$ and $\{p_{3k+2}\}$?)

Another interesting property of the MT sequence is its *aperiodicity*. We leave the proof, which is not too difficult, to the reader. Although aperiodic, the sequence is anything but random; it in fact has strong short-range and long-range structures. For example, there can never be more than two adjacent terms that are identical. And of course, any terms whose indices (beginning with the index 0 for the initial 0 in the sequence) differ by a factor of 2^n are identical.

This strong internal structure is reflected in the Fourier spectrum of the sequence (see Figure 1), which shows pronounced peaks, in spite of the fact that the sequence is aperiodic. The reader may want to show that the two strongest peaks occur at the frequencies one-third and two-thirds of the "sampling frequency."

A particularly convenient starting point for deriving the Fourier transform of the MT sequence in the ± 1 alphabet ($m_k = 1, -1, -1, 1, 1, -1, \ldots$) is its generating function

$$H(z) := \sum_{k=0}^{\infty} m_k z^k$$

The invariance of the MT sequence m_k under the substitution $1 \rightarrow 1, -1$ and $-1 \rightarrow -1, 1$ implies the functional equation

$$H(z) = (1 - z)H(z^2)$$

which in turn yields the generating function

$$H(z) = (1 - z)(1 - z^2)(1 - z^4)(1 - z^8) \cdots$$

Except for replacing z by z^{-1}, the generating function is identical with the "z transform," which is commonly used by electrical engineers to describe the transfer functions of digital filters.

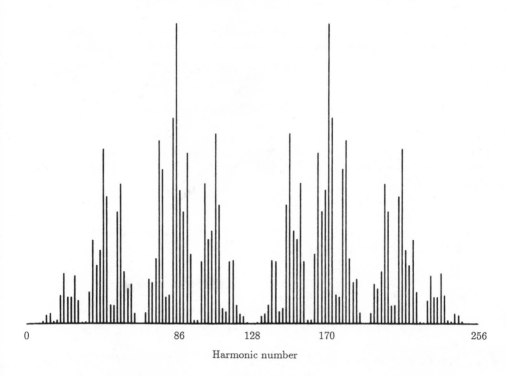

0 86 128 170 256

Harmonic number

Figure 1 Fourier amplitude spectrum of the Morse-Thue sequence (first 256 terms repeated periodically) [Schr 90].

By setting $z = \exp(i\omega)$, where ω is the radian frequency, the generating function yields the Fourier transform

$$M(\omega) = \prod_{k=0}^{\infty} [1 - \exp(i\omega 2^k)]$$

which obeys the scaling law $M(\omega) = [1 - \exp(i\omega)]M(2\omega)$. This scaling law, together with the symmetries $M(-\omega) = M^*(\omega)$ and $M(\omega + 2\pi) = M(\omega)$, determines the self-similar structure of the spectrum.

In physics, the MT sequence occurred originally in the symbolic dynamics for certain nonlinear dynamic systems. Marston Morse proved that the trajectories of dynamic systems whose phase spaces have a negative curvature everywhere can be completely characterized by a *discrete* sequence of 0s and 1s—a stunning discovery. This means that some complicated curve in $\mathbb{R}^n$, which, after all, represents an uncountably infinite set in a high-dimensional space, can be mapped into a discrete binary sequence! With the help of the MT sequence, Morse also proved the existence (under certain rules) of infinitely long chess games.

In the following section we shall study a particularly simple and instructive case of symbolic dynamics and its relation to the Morse-Thue sequence.

The Logistic Parabola and Period Doubling

Suppose in an ecological, economic, or other growth process the measure x_{n+1} of the next generation (the number of animals, for example) is a linear function of the present measure x_n:

$$x_{n+1} = rx_n$$

where $r > 0$ is the *growth parameter*. If unchecked, the growth will follow a geometric ("exponential") law:

$$x_n = r^n x_0$$

which for $r > 1$ will tend to infinity.

But growth is often limited by limited resources. In other words, the larger x_n, the smaller the growth factor r. The simplest way to model the decline in the growth factor is to replace r by $r(1 - x_n)$, so that, as x_n approaches some limit (1, in our case), the growth factor goes to 0. Thus, we get the growth law

$$x_{n+1} = f(x_n) = r(1 - x_n)x_n \tag{1}$$

which is called the *quadratic map* or, because of its use in logistics and its parabolic shape, the *logistic parabola* (see Figure 2).

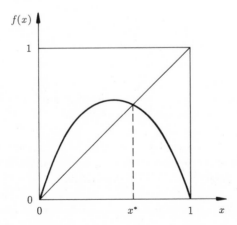

Figure 2   Quadratic map, also known as the *logistic parabola*. Note fixed point at $x = x^*$.

The logistic equation was introduced in 1845 by the Belgian sociologist and mathematician Pierre-François Verhulst (1804–1849) to model the growth of populations limited by finite resources [Ver 1845]. (The designation *logistic*, however, did not come into general use until 1875. It is derived from the French *logistique*, referring to the *lodgment* of troops.)

In its original form, the logistic equation was written as $x(t) = K/[1 + \exp(a - bt)]$, where x is the population whose growth pattern is being studied as a function of time t. The constants a and b set the origin and the scale of the time variable. Depending on these constants, the initial growth of x is approximately exponential, the growth rate reaching a maximum for $t = a/b$ and then tapering off to zero. The constant K determines the asymptotic value of x.

In another form, the Verhulst equation for the growth rate is $dx/dt = rx(K - x)/K$, which for $x \ll K$ leads to an exponential growth of x. But as x approaches K, the growth rate drops down to zero. Equation 1 is a recursive form of this equation in which time t has been replaced by the discrete variable n. The most important attribute of the Verhulst equation and its corollaries is their *nonlinearity*, which allows the modeling of nonlinearities and their consequences, such as chaotic dynamics, in many fields.

The quadratic map (equation 1) has two fixed points: $x = 0$ and, for $r > 1$, $x = x^* = 1 - 1/r$ (look again at Figure 2). The derivative of this map is

$$f'(x) = r(1 - 2x)$$

which equals r for $x = 0$ and $2 - r$ for the other fixed point, $x^* = 1 - 1/r$. Fixed points are stable as long as $|f'| < 1$. Thus the fixed point $x = 0$ is stable for $r < 1$. The fixed point $x = 1 - 1/r$ exists and is stable in the range $1 < r < 3$, because $|f'(x = 1 - 1/r)| < 1$ there.

In fact, for $r = 2$, $f'(x) = 0$ at the fixed point $x^* = 1 - 1/r = \frac{1}{2}$. Such fixed points are said to be *superstable*, because convergence to the fixed point is very rapid, as can be observed on any pocket calculator. In general, superstable *orbits* occur whenever $x = \frac{1}{2}$, for which $f'(x) = 0$, is a member of the orbit. (*Orbit* is the technical term for a succession of iterates x_n.) The parameter values for the superstable orbits of period length 2^k are called R_k; $r = R_0$ gives the superstable fixed point (period length 1).

For $r = 3$, the slope of equation 1 is -1 at the fixed point $x = \frac{2}{3}$. This fixed point is "indifferent," meaning that nearby values are neither attracted nor repelled. What happens for $r > 3$? The fixed point becomes unstable, splitting or *bifurcating* into an orbit of period length 2: $x_0, x_1, x_2 = x_0$ (see Figure 3). For example for $r = R_1 = 3.2360679775$, there is a stable (in fact, superstable) orbit of period length 2: $0.5 \to 0.8090169943 \ldots \to 0.5$, and so on (see Figure 4).

The value of R_1 is obtained from setting $f(f(0.5)) = 0.5$ as the solution of the cubic equation $R_1^3 = 4R_1^2 - 8$, which happens to have a *quadratic* irrational solution related to the golden mean $\gamma = 0.618 \ldots : R_1 = 2/\gamma = \sqrt{5} + 1$.

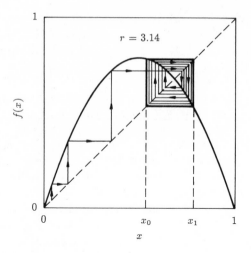

Figure 3   Fixed point turned unstable, leading to orbit of period length 2.

An orbit having a period length of 2, or a 2-orbit, for short, means, of course, that $f(f(x))$—abbreviated $f^{(2)}(x)$—has a fixed point (and because $f^{(2)'}(x) = 0$, the orbit is superstable). This is indeed the case (see Figure 4B). In fact, $f^{(2)}(x)$ has *two* fixed points, both superstable, at $x_0 = 0.5$ and $x_1 = 0.809 \ldots$.

If r is increased further, then these two fixed points of $f^{(2)}(x)$ will in turn become unstable. Indeed, they will become unstable at precisely the same value of r. Is this a coincidence? No, because, according to the chain rule of differentiating.

$$\frac{d}{dx} f(f(x)) \bigg|_{x=x_0} = f'(f(x)) \bigg|_{x=x_0} \cdot f'(x) \bigg|_{x=x_0}$$

or, with $f(x_0) = x_1$,

$$\frac{d}{dx} f(f(x_0)) = f'(x_1) \cdot f'(x_0)$$

Hence:

$$f^{(2)'}(x_0) = f^{(2)'}(x_1)$$

As a consequence of this equality, if x_0 becomes unstable because $\left| f^{(2)'}(x_0) \right| > 1$, so does x_1 at precisely the same value of the parameter r. This means that both fixed points of $f^{(2)}(x)$ will bifurcate at the same r value, leading to an orbit of period 4. In other words, now $f^{(2)}(f^{(2)}(x)) = f^{(4)}(x) := f(f(f(f(x))))$ will have a fixed point—in fact, four fixed points. For $r = R_2 = 3.498561699 \ldots$, the four fixed points of $f^{(4)}(x)$ are $x_0 = 0.500$, $x_1 = 0.874 \ldots$, $x_2 = 0.383 \ldots$, and

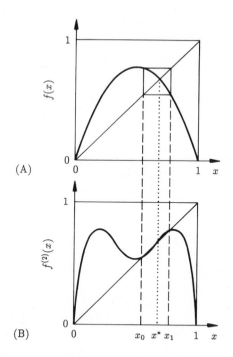

Figure 4   (A) The quadratic map for an orbit of period length 2. (B) The once-iterated map $f^{(2)}(x)$ has two stable fixed points, x_0 and x_1.

$x_3 = 0.827 \ldots$. These also form a superstable orbit of period length 4 of $f(x)$: $x_0 \to x_1 \to x_2 \to x_3 \to x_0$, and so on.

Again, because of the chain rule of differentiation, the four derivatives are the same at all four points of the orbit. Thus if, for a given value of r, the magnitude of one of the derivatives exceeds 1, then the magnitude of all four derivatives will. Hence, all four iterates will bifurcate at the same value of r, leading to an orbit of period 8. This bifurcation scenario will repeat again and again as the growth parameter r is increased, yielding orbits of period length 16, 32, 64, and so on *ad infinitum*, ending up in a "chaotic" orbit of infinite period length for $r = r_\infty = 3.5699 \ldots$.

These period-doubling bifurcations are also called *pitchfork bifurcations*, because of the resemblance to a pitchfork when the values of the iterates are plotted as a function of the parameter (see Figure 5). Two prongs of the fork are the new iterates after bifurcation, and the central prong (shown as a dashed line in Figure 5) is the old (now unstable) iterate, which has turned from attractor to *repellor*.

Period doubling is a very common phenomenon, encountered in a wide variety of physical, ecological, and economic systems. Think of predators and

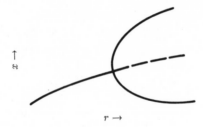

Figure 5   Pitchfork bifurcation: outer "prongs" are the new iterates after bifurcation, and dashed "prong" is the old iterate, now unstable.

their prey, such as foxes and rabbits. For sufficiently low fox reproduction rates, corresponding to $r < 3$ in equation 1, there may be enough rabbits to satiate the foxes' appetites, enabling a stable equilibrium between the number of foxes and rabbits to ensue. But if the fox reproduction rate is increased above a certain limit, corresponding to $r > 3$, the foxes will devour so many rabbits that, in the next season, there will not be enough rabbits to go around, so that the fox population will decrease. This will give the rabbits a chance to recover and become more plentiful again, allowing the number of foxes to increase too, giving rise to a two-season cycle.

   At which values r_n of the growth parameter r do these bifurcations, from period length 2^{n-1} to 2^n, take place? And for which values R_n do we get superstable orbits of equation 1 with period length 2^n? What happens to the iterates x_0, $x_1, \ldots, x_{2^n-1}$ as n goes to infinity? To answer these questions we have to exploit the self-similarities that *must* be hiding somewhere in the iterated quadratic map.

Self-Similarity in the Logistic Parabola

Let us consider the superstable orbits of equation 1 with period lengths $P = 1$, 2, 4, 8, and so on. The parameter values $r = R_n$ that give superstable orbits of period length 2^n are much better defined, both experimentally and theoretically, than the points of bifurcation, $r = r_n$. The fast convergence to the final orbit gives better numerical estimates, and one always knows one member of the orbit a priori: $x_0 = 0.5$. By contrast, numerical determination of r_n, a bifurcation value, is somewhat trickier.

   The period-doubling process is characterized by self-similarities that facilitate its analysis. To demonstrate one of these self-similarities, compare $f(x)$ for the parameter value $r = R_1$ for the superstable orbit with period length $P = 2$ (see Figure 6A) with the function $f^{(2)}(x)$ for $r = R_2$ for the superstable orbit with $P = 4$ (Figure 6B). The resemblance between the dashed square and its contents in

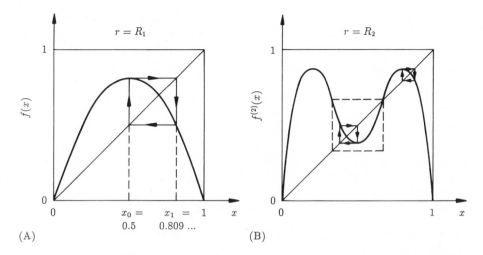

Figure 6   (A) The quadratic map $f(x)$ for the superstable orbit of period length 2. (B) The once-iterated map $f^{(2)}(x)$ for period length 4. Note the similarity between the contents of the small dashed square and part A. This self-similarity is characteristic for period doubling and facilitates its analysis [Schu 84].

Figure 6B and the large square and *its* contents in Figure 6A is striking. The discrepancy between the parabola in Figure 6A and the fourth-order curve inside the dashed square in Figure 6B is in fact quite small, as can be seen in Figure 7, which shows both the parabola (the solid curve) and the rescaled and inverted fourth-order curve (dashed curve).

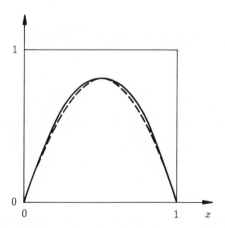

Figure 7   The difference between the quadratic map (solid curve) and the rescaled iterated quadratic map (dashed curve).

The rescaling factor for this transition from period length 2 to period length 4 is easily calculated to equal $-(2 + 2/R_1) = -(2 + \gamma) = -2.618\ldots$, where γ is the golden mean. (The minus signs account for the fact that period doubling engenders an upside-down orientation.) In going from the superstable orbit of period length $P = 4$ ($r = R_2$) to that for $P = 8$ ($r = R_2$), almost precisely the same "scenario" is repeated for $f^{(2)}(x)$ and $f^{(4)}(x)$ as that shown for $f(x)$ and $f^{(2)}(x)$ in Figure 6, only the scaling factor is slightly different from -2.618. In fact, the same scenario (a popular buzzword in this context) is repeated every time the parameter value is changed from $r = R_n$ to $r = R_{n+1}$. As $n \to \infty$, the scaling factor quickly converges on its asymptotic value of $-2.5029\ldots$, which is not very far from its initial value of $-2.618\ldots$. In the limit, the initial parabola of the quadratic map becomes a transcendental function, given by an infinite power series $g(x)$ originally derived by Mitchell Feigenbaum:

$$g(x) \approx 1 - 1.52763x^2 + 0.104815x^4 - 0.0267057x^6 + \cdots$$

Here the x coordinate has been shifted so that the maximum of $g(x)$ is at $x = 0$, instead of $x = 0.5$, and has a value of 1. The function $g(x)$ is the *fixed-point function* of the period-doubling transformation for quadratic maps. It obeys the scaling law $g(x) = \alpha g(g(x/\alpha))$, which also determines $\alpha = 1/g(1)$. The derivation of $g(x)$ as a universal function for *all* maps with a quadratic maximum by Feigenbaum, via a renormalization theory, is instructive but not exactly easy [Fei 79].

Numerically, the scaling parameter α can be obtained from any of the numerous self-similarities of the iterates $x_m^{(n)}$ generated by the quadratic map. A particularly attractive method is to calculate the value of the iterate $x_{P/2}^{(n)}$ at the half period for a superstable orbit of period length $P = 2^n$, starting with $x_0 = 0.5$. For the parameter value $r = R_{n-1}$, $x_{P/2}^{(n-1)} = x_0$; but for $r = R_n$, $x_{P/2}^{(n)}$ misses x_0 by a small amount, the difference $|x_{P/2}^{(n)} - x_0|$ scaling asymptotically with α as n is increased to $n + 1$. More precisely,

$$\alpha_n := \frac{x_{P/2}^{(n)} - x_0}{x_{P/2}^{(n+1)} - x_0} \to \alpha \qquad \text{for } n \to \infty$$

With a programmable calculator, one first determines R_n and R_{n+1} (by adjusting R until $x_P = x_0 = 0.5$, for $P = 2^n$ and $P = 2^{n+1}$) and then reads out the value of $x_{P/2}$. In this manner one quickly obtains $\alpha_7 \approx -2.502905$, with a relative deviation from α of about 10^{-6}.

The Scaling of the Growth Parameter

We have just learned that period doubling is asymptotically self-similar, with a scaling factor for the variable x equal to $-2.5029\ldots$. How do the parameter values r—say, those for the superstable orbits R_n—scale? Numerical evidence suggests that the differences $R_{n+1} - R_n$ become smaller and smaller according to the following geometric law:

$$R_{n+1} - R_n \approx \frac{R_n - R_{n-1}}{\delta} \qquad \text{for } n \to \infty$$

where δ is a *universal* constant, the famous (and probably transcendental) Feigenbaum constant (originally found by S. Grossmann and S. Thomae [GT 77]). This magic number has earned the epithet "universal" because it applies, as Feigenbaum has shown, to many different nonlinear maps, independent of the details of the mapping, as long as the absolute maximum of the mapping is quadratic. The convergence of $\delta_n := (R_n - R_{n-1})/(R_{n+1} - R_n)$ to δ is very rapid: $\delta_1 \approx 4.7$, $\delta_2 \approx 4.68$, ..., $\delta_6 \approx 4.66918$. The asymptotic value is

$$\delta = 4.6692016091029 \ldots$$

and the accumulation point of the growth parameter for the period doublings is

$$R_\infty = 3.5699456 \ldots$$

With these two values and another constant, one has $R_n \approx R_\infty - 1.542\delta^{-n}$.

The two scaling parameters α and δ are related to each other. A simplified theory yields $\delta \approx \alpha^2 + \alpha + 1 \approx 4.76$ [Fei 79].

At the critical value of the growth parameter $r = R_\infty$ the period has become infinite. In other words, the orbit is now *aperiodic*, comprising a point set of infinitely many values of x that never repeat. However, other x values are attracted to this point set, which is in fact a Cantor set (see Figure 8). Note the approximate

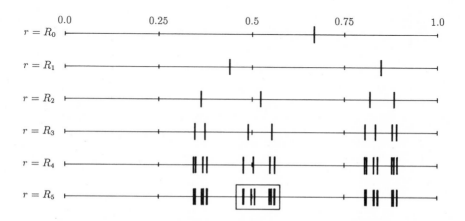

Figure 8　Self-similar Cantor set of iterates for period doubling. The iterates in the small box for $r = R_5$ are a scaled-down version of all the iterates for $r = R_3$.

self-similarity of this point set: the left half of the bottom line is the mirror image of the line above it compressed by a factor of about 2.5, and the right half of the bottom line is the line above it compressed by a factor of 2.5^2. The Hausdorff dimension $D = 0.538 \ldots$ of this set was derived analytically and numerically by P. Grassberger [Gra 81]. This and similar attractors in higher dimensions have been called *strange*, although once one knows about Cantor sets they are really not so strange after all.

Assuming that the limit set is strictly self-similar, we can use the well-known formula (see Appendix A) for the Hausdorff dimension D of a self-similar Cantor set with two different remainders, s_1 and s_2,

$$s_1^D + s_2^D = 1$$

to calculate a good approximation to the Hausdorff dimension of the strange attractor of the logistic parabola at the period-doubling accumulation point. With $s_1 = 1/2.5 = 0.4$ and $s_2 = s_1^2$, and setting $z = 0.4^D$, we obtain from

$$z + z^2 = 1$$

$z = \gamma \approx 0.618$ and $D \approx \log \gamma / \log 0.4 \approx 0.525$, a surprisingly good approximation to the more precise value $0.538 \ldots$. As so often, self-similarity may be only approximate, but ignoring the lack of exact scaling still gives good results that can be bettered only by more involved computation.

The Fourier spectrum of the periodic sequence x_m, too, shows pronounced self-similarity (see Figure 9). Let a_k^n be the Fourier coefficients of the x_m for a period length $P = 2^n$. In going from an orbit of period length $P = 2^n$ by a period-doubling bifurcation to an orbit of length 2^{n+1}, the new Fourier coefficients with

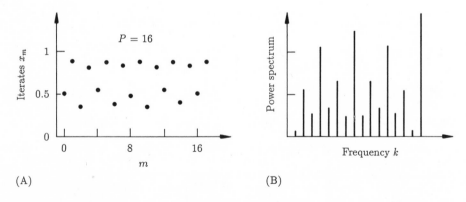

(A)   (B)

Figure 9   (A) Iterates x_m for the quadratic map at period length 16. (B) Fourier power spectrum (on a logarithmic scale) of the iterates x_m [Schu 84].

an even index a_{2k}^{n+1} are approximately equal to the old Fourier coefficients: $a_{2k}^{n+1} \approx a_k$ (because $x_{n+P} \approx x_n$). The odd-index coefficients a_{2k+1}^{n+1}, which describe the subharmonics that appear in the spectrum as a result of the period doubling, are determined by the difference $x_{n+P} - x_n$. A detailed analysis shows that the squared magnitudes of the Fourier coefficients, $|a_{2k+1}^{n+1}|^2$, are roughly equal to an adjacent component from the previous orbit scaled down by a factor of $8\alpha^4/(1 + \alpha^2) \approx 40$, corresponding to 16 decibels (dB) in logarithmic units [Fei 79]. (The number of decibels is, by definition, 20 $\log_{10}$ of a magnitude ratio or 10 $\log_{10}$ of a squared magnitude ratio, such as a spectral *power* ratio.)

An early confirmation of period-doubling bifurcations occurred in a hydro-dynamic ("Rayleigh-Bénard") experiment by Libchaber and Maurer in which the Reynolds number played the role of the growth parameter r [LM 80]. In studying the forced nonlinear oscillations of bubbles in water, Lauterborn and Cramer found a similar behavior: the appearance of more and more subharmonics until the onset of chaos, called *cavitation noise* in this context [LC 81]. As a result of such experiments, the destructive mechanism of cavitation, a much dreaded source of failure in ship propellers, is now well understood.

Self-Similar Symbolic Dynamics

Instead of listing the sequence of iterates x_n themselves, it often suffices to state whether they fall to the left (L) or the right (R) or *on* the maximum or center (C) of the map. The sequence of symbols L, R, C is then called the *symbolic dynamics* for a given orbit. Thus, the superstable orbit of period 2 has the symbolic dynamics or "kneading sequence" $CRCRCR$ .... Restricting the notation to a single period, we write simply CR.

It is not too difficult to show that the next superstable orbit, the one with period length 4, is obtained as follows. First one writes two periods of the orbit with period 2, $CRCR$, and then changes the second C to L if the number of R's to the left of it is odd. Otherwise the second C is changed to R. Thus, the superstable orbits have the following symbolic dynamics:

		Period 1:	C
Period 1	$\rightarrow$	Period 2:	$CC \rightarrow CR$
Period 2	$\rightarrow$	Period 4:	$CRCR \rightarrow CRLR$
Period 4	$\rightarrow$	Period 8:	$CRLRCRLR \rightarrow CRLRRRLR$

and so on. The orbit of period 8 is often more conveniently written as $CRLR^3LR$. In the same vein the superstable orbit of period 16 is written as $CRLR^3LRLRLR^3LR$.

This algorithm of counting the number of previous R's and checking whether it is even or odd is directly related to the fact that the slope of the quadratic map is *negative* for the right half of the map. Thus, each time the iterate x_n falls into the right half ($x_n > 0.5$), there is a sign change in how small differences in x_n are propagated, and an *odd* number of sign changes *is* a sign change (while an even number is not). This is one of the most important properties not only of the quadratic map but of all unimodal ("one-hump") maps. As a result these maps have a "universal" ordering of their symbolic dynamics as the growth parameter is changed.

More specifically, for $r = R_n$, that is, period length $P = 2^n$, the iterate $x_P^{(n)}$ equals x_0 by definition. In changing the growth parameter r from R_n to R_{n+1}, $x_P^{(n)} - x_0$ will be positive (negative) if $x_m^{(n)} - x_0$ was positive an even (odd) number of times for $m = 1, 2, \ldots, P - 1$. This is the reason for the aforementioned rule $C \to R$ (or L) for an odd (even) number of preceding R's.

These sequences are self-similar in the following sense. Retaining only every other symbol (starting with C) reproduces the sequence for the superstable orbit with half the period, except that L and R are interchanged. Thus, "pruning" the symbolic dynamics we derived for the orbit with period 16 results in $CLRL^3RL$, which is the complement of $CRLR^3LR$ that describes the orbit of period 8.

As in the Morse-Thue sequence, retaining every other term produces a similar, albeit complemented, sequence. Is there a closer connection between the Morse-Thue sequence and the symbolic dynamics of the superstable orbits? There is indeed. To see this, let us replace R by 1 and C and L by 0. With this notation, the superstable orbits of periods 1, 2, 4, and 8 are

$$0$$
$$0 \quad 1$$
$$0 \quad 1 \quad 0 \quad 1$$
$$0 \quad 1 \quad 0 \quad 1 \quad 1 \quad 1 \quad 0 \quad 1$$

which unfortunately does *not* look like the beginning of the Morse-Thue sequence. However, the running partial sums modulo 2 of these orbits (beginning at the left) do reproduce the Morse-Thue sequence:

$$0$$
$$0 \quad 1$$
$$0 \quad 1 \quad 1 \quad 0$$
$$0 \quad 1 \quad 1 \quad 0 \quad 1 \quad 0 \quad 0 \quad 1$$

and so on. Computing running partial sums modulo 2 is, of course, equivalent to keeping track of the number of preceding R's.

Conversely, the kneading sequences for unimodal maps in the binary notation, 01011101 . . . , are obtained from the Morse-Thue sequence by taking sums (or differences) modulo 2 of adjacent elements.

This connection allows us to write down directly (without iteration) the kneading sequence of period 2^m for any m. The rule—I encourage readers to derive it for themselves—is simply this: the kth term in every such symbolic sequence is obtained by writing $k = 2^q \cdot j$, where j is odd. Then the parity of q determines the choice between L and R: for odd q the symbol is L; for even q it is R (and of course for $k = 0$ the symbol is C by definition). Thus, for example, the ninety-sixth symbol in the kneading sequence is L, because $96 = 3 \cdot 2^5$, and 5 is odd.

One general result of this rule is that all terms with an odd k are R (including the last term of each periodic orbit with $P = 2^n$). All terms whose k is a power of 2 ($k = 2^q$) alternate between L and R.

The irrational number constructed with the help of the Morse-Thue sequence interpreted as a binary fraction, $0.0110100110010110 \ldots = 0.4124 \ldots$, which may be called the *Morse-Thue constant*, is intimately related to the period-doubling bifurcation scenario and the Mandelbrot set (see pages 295–299).

Periodic Windows Embedded in Chaos

Eyesight should learn from reason.
—JOHANNES KEPLER

Figure 10A shows the "behavior" of the iterated logistic parabola, that is, the values of its iterates x_n as the parameter r is increased from 3 to 4. There is a cascade of period-doubling bifurcations followed by chaos (the dense bands) interleaved with periodic "windows." In fact, the r values for periodic windows are dense, but one sees only a few in the treelike plot (Figure 10A), sometimes called a "Feigenbaum" ("figtree") plot. Most prominent is the period-3 window starting at $r = \sqrt{8} + 1$ (see Figure 10A). Once the period length 3 has been observed, we know from the work of Li and Yorke that all possible periods appear [LY 75].

Note that in the period-3 window the period doubling occurs again, leading to orbits of period length 6, 12, 24, and so on, and renewed chaos in which another period-3 window is embedded, and so forth *ad infinitum* in another self-similar cascade (see Figure 10B).

Interestingly, the chaos bands also show bifurcation, called *reverse bifurcation*, as we decrease r from its highest value $r = 4$. For r somewhat below 3.68, the single chaos band splits into two (see Figure 11), and for r near 3.6 the two chaos bands split into four, and so forth, mimicking the period-doubling bifurcations observed for increasing r. In fact, each chaos band is torn asunder by the surviving "ghost" of the corresponding period-doubling bifurcation.

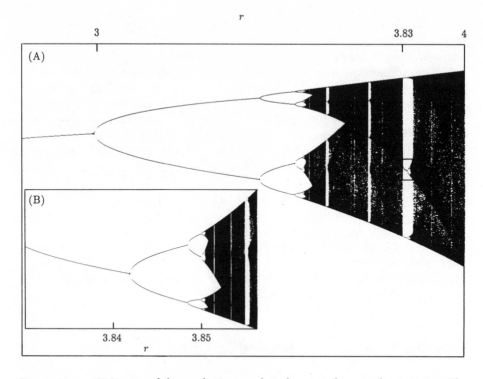

Figure 10   (A) Iterates of the quadratic map plotted against the growth parameters. The period-doubling bifurcation cascade is on the left, followed by chaos bands, on the right, in which the iterates are chaotic. The chaos bands are interspersed with "period windows" in which the iterates are periodic again. Most prominent is the window for period length 3 starting at $r \approx 3.83$. (B) An enlargement of the central portion of the period-3 window reveals another period-doubling cascade, followed by a second period-3 window, in which the period length is 9.

The reason for the bifurcation of chaos bands as the parameter r is decreased is easy enough to see and analogous to the bifurcation of periodic orbits. Consider the r value $\tilde{r}_1$, for which the third iterate x_3 of $x_0 = 0.5$ falls on the unstable fixed point $x^* = 1 - 1/\tilde{r}_1$. This yields the equation $\tilde{r}_1^3(4 - \tilde{r}_1) = 16$ for $\tilde{r}_1$, which yields $\tilde{r}_1 = 3.678573510 \ldots$. For r values slightly smaller than $\tilde{r}_1$, x_3 will fall just below x^* and x_4 slightly above, creating a gap around x^* into which no iterates can fall.

Similarly, for r values just below $\tilde{r}_2 = 3.5925721841 \ldots$, for which iterates of $x_0 = 0.5$ fall on the unstable period-2 orbit, bifurcation of the two chaos bands into four bands takes place. In fact, the dark sinusoidal contours visible in the Feigenbaum diagram (Figure 11), including upper and lower edges of chaos bands, are images of the stationary point $x = 0.5$. Chaos bands merge where these

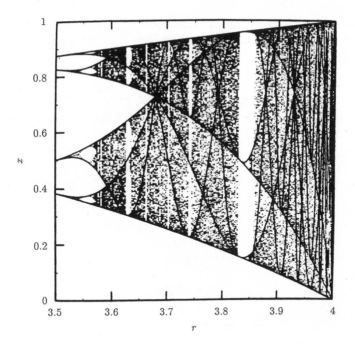

Figure 11   Reverse bifurcation of chaos bands as the growth parameter is decreased below 3.68 [JM 85].

contours cross each other, and periodic windows open where contours touch upper and lower edges [Lor 80, JM 85].

Interestingly, the parameter values $\tilde{r}_m$ at which 2^{m-1} chaos bands join 2^m chaos bands form a descending, asymptotically geometric progression with the same accumulation point, $r = 3.5699\ldots$, as that of the period-doubling bifurcations. And the scaling factor, too, is the same, namely, the Feigenbaum constant $\delta = 4.6692\ldots$.

The $\tilde{r}_m$ also correspond to the parameter values for the accumulation points of successive rows of the "Sharkovskii ordering" of orbits (see page 285), because the symbolic dynamics are the same. For example, the symbolic dynamics of the two chaos bands (starting with $x_0 = 0.5$), $CRLR^2(RL)^\infty$, are the same as those for the accumulation point of the orbits of period length $4 \cdot 3, 4 \cdot 5, 4 \cdot 9, \ldots$ (see pages 287–288).

The ordering with which the iterates x_n fall into the 2^m different chaos bands is also the same as the ordering of the iterates in a stable orbit of period length $P = 2^m$. For example, for both the period-4 orbit and the four chaos bands, the iterates, starting with the largest iterate x_1, are ordered as follows:

$x_1 > x_3 > x_4 > x_2$. This ordering is obtained from the ordering $x_1 > x_2$ for $P = 2$ by splitting each iterate x_i into the two iterates of x_i and x_{i+P} and inverting the order of every other pair of iterates. In this manner, the ordering for $P = 8$, $x_1 > x_5 > x_7 > x_3 > x_4 > x_8 > x_6 > x_2$, is easily derived from the ordering $x_1 > x_3 > x_4 > x_2$ for $P = 4$.

The ordering of the iterates x_n according to their values can also be obtained from a Gray code (which in turn is related to the Hilbert space-filling curve; see pages 10–13 in Chapter 1). For example, to deduce the correct order of the eight iterates x_n, $n = 1, 2, \ldots, 8$, interpret the three-digit Gray code (in which one digit at each step is changed, starting from the left) as ordinary binary numbers and add 1 to obtain the index n of x_n:

$$0 = 000 = x_1$$
$$1 = 100 = x_5$$
$$2 = 110 = x_7$$
$$3 = 010 = x_3$$
$$4 = 011 = x_4$$
$$5 = 111 = x_8$$
$$6 = 101 = x_6$$
$$7 = 001 = x_2$$

The close correspondence between period-doubling bifurcations and the reverse bifurcations of the chaos bands that we have glimpsed here is one of the many fascinating features of the quadratic map.

The Parenting of New Orbits

The algorithm described in pages 277–278 for constructing the symbolic dynamics of period-doubled superstable orbits can be considerably generalized. Thus the rule "append the string $CR \ldots$ to itself and change the second C to L (R) if the number of R's in the original string is odd (even)" applies not only to fundamental periods of length $P = 2^n$ but to any orbital length. In this manner the superstable orbit of length 3 with symbolic dynamics CRL is doubled via $CRLCRL$ to $CRLLRL$, which in turn is doubled to $CRL^2RLR^2L^2RL$ and so on to an infinite cascade of orbits of lengths $3 \cdot 2^n$.

More generally, one can derive orbits of lengths $k \cdot m^n$ from an orbit P of length k and an orbit Q of length m. For the orbit of length $k \cdot m$, one copies the symbolic dynamics of P m times and replaces each of the $(m-1)$ C's (except the first C) by one after another of the $(m-1)$ symbols of Q, interchanging L and R of Q if the number of R's in P is odd. (The initial C in Q is ignored.) For example, the 2-orbit CR is tripled by means of the 3-orbit CRL by first copying the 2-orbit three times—$CRCRCR$—and then replacing the second and third C's by the complement of the second and third symbols of $C\underline{RL}$. This yields the orbit $CR\underline{LRR}$ of period 6, distinct from the previously derived orbit CRL^2RL (The period-doubling algorithm described in the preceding paragraph is but a special case of this composition law with $Q = CR$.)

Which of the two orbits of period 6 just described "dominates" the other? By definition, this depends on the first symbol in which they differ. If the number of R's in the initial, equal portion of the strings is odd (even), then the orbit with L or C (R) as the first distinct symbol dominates the other. Thus, the orbit CRL^2RL dominates the orbit $CRLR^3$. The superstable r values of an orbit that dominates another orbit is the larger of the two superstable r values. Indeed, the approximate r values corresponding to CRL^2RL and $CRLR^3$ are 3.8445688 and 3.6275575, respectively.

A particularly attractive equivalent algorithm for "multiplying" two orbits of period lengths p_1 and p_2 to yield an orbit of length p_1p_2 proceeds as follows. Write the symbolic dynamics of the orbit as a sequence of plus or minus signs, $+$ corresponding to either of the letters L or C and $-$ to the letter R. (This reflects the fact that the slope of unimodal maps is *negative* to the *right* of the maximum.) Thus, the superstable $p_1 = 3$ orbit CRL is written as $+ - +$. Next form the running product of these signs starting from the left. This transforms $+ - +$ into $+ - -$, which we shall call the σ *sequence* of the orbit. Similarly, the σ sequence of the superstable orbit CR with period length $p_2 = 2$ is $+ -$.

Using these σ sequences, period multiplication becomes a simple appending process. For example, period doubling of any orbit is realized by appending its σ sequence to itself with the opposite signs, corresponding to the period-doubling "operator" $+ -$, that is, the σ sequence of the period-2 orbit. Using this rule, the period-doubling cascade now looks as follows. Starting with the fixed point C, which has period length $p = 1$ and thus corresponds to a σ sequence consisting of a single sign, $\sigma_0 = +$, we have

$$
\begin{array}{lll}
p = 1 & + & C \\
p = 2 & + - & CR \\
p = 3 & + - - + & CRLR \\
p = 4 & + - - + - + + - & CRLR^3LR
\end{array}
$$

and so on. This iterative construction of the σ sequence of orbits with period lengths 2^n ("append the complement") leads, of course, to the previously discussed self-similar Morse-Thue sequence, which is generated by the same rule.

To recover the symbolic dynamics, one writes C followed by R's and L's, depending on whether the corresponding sign in the σ sequence is different from or the same as the preceding sign, as was just shown for $p = 2, 3,$ and 4.

To triple the period length of an orbit, one appends its σ sequence two times to itself with opposite signs, corresponding to the σ sequence $+--$ of the period-3 orbit. The period-tripling cascade thus has the following σ sequences, starting with $\sigma_0 = +$:

$$p = 1 \quad + \qquad\qquad\qquad\qquad C$$

$$p = 3 \quad +-- \qquad\qquad\qquad CRL$$

$$p = 9 \quad +---++-++ \qquad CRL^2RLR^2L$$

and so on. The resulting infinite self-similar σ sequence with $\sigma_{3k} = \sigma_k$ is a generalization of the Morse-Thue sequence.

The kth symbol s_k, $k > 0$, in the period-tripled symbolic dynamics is given by writing the index $k = 3^m \cdot q$, where $q = 1$ or 2 modulo 3: $s_k = L$ if m and $\langle q \rangle_3 := q$ modulo 3 are both even or both odd. If the parities of m and $\langle q \rangle_3$ are different, then $s_k = R$. Writing -1 for R and $+1$ for L, we have $s_k = (-1)^{m + \langle q \rangle_3}$. Thus, with $405 = 3^4 \cdot 5$, s_k equals $(-1)^{4+2} = 1 = L$.

The "second harmonic" of the period-3 orbit CRL, with the σ sequence $+--$, has the σ sequence $+----++$, which corresponds to a period-6 orbit with the symbolic dynamics CRL^2RL. It is obtained by appending $+--$ to itself with the opposite sign. Similarly, the σ sequence for the tripled period-2 orbit CR, the σ sequence $+-$, is obtained by appending $+-$ twice to itself with opposite signs, corresponding to the period-tripling operator $+--$. This gives $+--+-+$, which corresponds to $CRLR^3$, as we derived before by a less elegant rule.

We shall encounter the σ sequence again in the next section, where it is used to calculate the growth parameter of a linearized logistic map, the so-called tent map.

Another algorithm *interpolates* a new orbit between two known orbits P and Q by taking the intersection of *harmonic* $H(P)$ of P and an *antiharmonic* $A(Q)$ of Q [MSS 73]. The harmonic of an orbit P is formed, as before, by appending P to itself and changing the second C to L (or R) if the number of R's in P is odd (even). The antiharmonic of an orbit Q, which in general is not a possible periodic orbit, is defined just like the harmonic except that R and L are interchanged in the replacement of the second C. The σ sequence of an antiharmonic is obtained by appending the original σ sequence to itself without sign change. (The reader may wish to derive the rules that distinguish possible orbits from impossible strings.)

For example, the harmonic of $P = CR$ is the orbit $H(P) = CRLR$, and the antiharmonic of $Q = CRL^\infty$ is the string $A(Q) = CRL^\infty RRL^\infty$. The intersection of these two strings, meaning the string in which the initial symbols of the two strings $H(P)$ and $A(Q)$ agree, is the "daughter" orbit CRL of period length 3. With this rule, the parameter value r for the daughter orbit always lies between those of the two parent orbits; hence the designation *orbit interpolation*. By repeated forming of harmonics and interpolation, *all* orbits of the map can be constructed from the "first" orbit (C) and the "last" orbit (CRL^∞).

The different period lengths p of stable periodic orbits of unimodal maps appear in a *universal order*. If r_p is the value of the growth parameter r at which a stable period of length p first appears as r is increased, then $r_p > r_q$ if $p \succ q$ (read p precedes q) in the following "Sharkovskii order":

$$3 \succ 5 \succ 7 \succ 9 \succ \cdots$$

$$2 \cdot 3 \succ 2 \cdot 5 \succ 2 \cdot 7 \succ \cdots$$

$$\cdots$$

$$2^n \cdot 3 \succ 2^n \cdot 5 \succ 2^n \cdot 7 \succ \cdots$$

$$\cdots$$

$$\cdots \succ 2^m \succ \cdots \succ 4 \succ 2 \succ 1$$

Thus, for example, the minimal r value for an orbit with $p = 10 = 2 \cdot 5$ is larger than the minimal r value for $p = 12 = 4 \cdot 3$ because $10 \succ 12$ in this witchcraft algebra.

Some of the consequences of this ordering are the following:

• The existence of period length $p = 3$ guarantees the existence of any other period length q for some $r_q < r_p$.

• If only a finite number of period lengths occur, their lengths must be powers of 2—that is, $p = 2^k, 2^{k-1}, \ldots, 4, 2, 1$, for some k.

• If a period length p exists that is not a power of 2, then there are infinitely many different periods.

The superstable orbits for the smallest parameter value r_p have the symbolic dynamics $CRLR^{p+3}$ for odd period lengths $p \geq 3$. The "last" superstable orbit of period length p, that is, the orbit with the *largest* value of r_p, has the symbolic dynamics CRL^{p-2} [CE 80]. For example, the last period-6 superstable orbit is CRL^4 and has $r \approx 3.9975831$.

These results are a consequence of Sharkovskii's theorem, which concerns the existence and ordering of orbits according to Sharkovskii's dominance definition given previously for a *fixed* value of r [Sha 64]. However, most of these orbits are unstable. (They are the remaining "ghosts" of orbits that were stable for smaller r values. The fixed point $x^* = 1 - 1/r$, for example, persists even for values of r exceeding 3, where it becomes unstable.) In fact, one-hump maps

$f(x)$ with a "Schwartzian derivative"

$$\frac{f'''(x)}{f'(x)} - \frac{3}{2}\left(\frac{f''(x)}{f'(x)}\right)^2$$

that is negative were shown by D. Singer to have at most one stable periodic orbit [Sin 78]. (Here $f(x)$ is assumed to be three times continuously differentiable and to map the unit interval into itself.)

For example, for $r = 3.83187405529$ and $x_0 = 0.5$ we obtain the superstable orbit *CRL* of period length 3. But by choosing $x_0 = 1 - 1/r = 0.73903108882$, the unstable orbit of period length 1 is initiated. And for $x_0 = 0.89208905218$, the unstable orbit of period length 2 is obtained. Other initial values lead to an unstable period-4 orbit that is descended from the orbit *CRLR*. However, the period-4 orbit based on *CRLL* cannot be realized for $r = 3.83 \ldots$, because *CRLL* dominates *CRL*. The reader may want to find initial values for other period lengths, *all* of which are possible according to Sharkovskii's theorem for $r = 3.83 \ldots$ because the period length 3 is possible.

The Calculation of the Growth Parameters for Different Orbits

Given a superstable orbit with the symbolic dynamics $Q = CRL \ldots$, what is the corresponding value of the growth parameter r? One method is to adjust r in $f(x) = rx(1 - x)$ iteratively until $f^{(p)}(0.5) = 0.5$, where p is the period length of the orbit and $f^{(p)}(x)$ is the pth iteration of $f(x)$. However, this method is likely to fail in regions where the r values for "similar" orbits are crowded. *Similar* here refers to orbits with equal parity (number of R's) and period lengths that divide the given period p. Also, of course, there must be a good initial guess of r.

The confusion with other orbits can be eliminated if the symbolic dynamics are actually used in the calculation of r—not just its parity and period length. For such a method it is advantageous to transform the variable x linearly to yield another, often-used form of the quadratic map: $g(x) = 1 - \mu x^2$, in which the growth parameter μ is related to r by the equation $\mu = r(r - 2)/4$ or $r = 1 + \sqrt{(1 + 4\mu)}$. For this form of the quadratic map, the maximum occurs for $x = 0$. Hence superstable orbits contain the value $x = 0$.

Let us take as an example the period-5 orbit with the smallest r value, which has symbolic dynamics $CRLR^2$. Set $x_0 = x_5 = 0$; then, because $g(0) = 1$, $g_R(g_R(g_L(g_R(1)))) = 0$. In this equation the subscripts (R or L) remind us which branch of $g(x)$ comes into play at each iteration; we need to know this in order to be able to invert the equation. Inverting yields

$$1 = g_R^{-1}(g_L^{-1}(g_R^{-1}(g_R^{-1}(0)))) \tag{2}$$

where

$$g_R^{-1}(x) = +\sqrt{\frac{1-x}{\mu}}$$

and

$$g_L^{-1}(x) = -\sqrt{\frac{1-x}{\mu}}$$

Multiplying equation 2 by μ gives

$$\mu = \sqrt{\mu + \sqrt{(\mu - \sqrt{(\mu - \sqrt{\mu})})}} \tag{3}$$

from which μ may be determined by iteration, as suggested by H. Kaplan. Starting with $\mu_0 = 2$, one obtains quickly and without ambiguity the correct parameter value for the orbit $CRLR^2$: $\mu = 1.625413725\ldots$, which corresponds to $r \approx 3.738914913$.

For an arbitrary allowed orbit $CRL\ldots$, the plus or minus signs appearing in equation 3 are determined by the symbolic dynamics with the letter R corresponding to a minus sign and L to a plus sign. The first sign under the square root $(+)$ corresponds to the first letter (L) after the initial CR in $CRL\ldots$.

Equation 3 is particularly useful for the calculation of the parameter values for accumulation points of certain orbits. For example, looking at the Sharkovskii order, we may want to know at which μ value the orbits with odd period lengths $p = 3, 5, 7, 9, \ldots$ accumulate. These orbits have the symbolic dynamics $CRLR^{p-3}$, as already noted. For $p \to \infty$, equation 3 therefore becomes

$$\mu = \sqrt{\mu + \sqrt{(\mu - \sqrt{(\mu - \sqrt{(\mu \ldots)})})}}$$

with an infinite sequence of minus signs. Setting $\mu - \sqrt{(\mu - \sqrt{(\mu \cdots)})} = x$, we have $x = \mu - \sqrt{x}$. Eliminating x leads to the cubic equation for μ, namely, $\mu^3 - 2\mu^2 + 2\mu - 2 = 0$, with the solution $\mu = 1.543689012\ldots$, which corresponds to $\tilde{r}_1 = 3.67857351\ldots (= \mu^3)$. This is also the parameter value at which the last two chaos bands merge, because they have the same symbolic dynamics (see pages 279–280).

Because the orbits of *even* period length of the form $p = 2 \cdot 3, 2 \cdot 5, 2 \cdot 7,$ $\ldots$ have the same symbolic dynamics, $CRLR^{p-3}$, as the odd orbits, they accumulate at the same parameter value. But they approach it from below.

Similarly, one determines the parameter value $\tilde{r}_2$ of the accumulation point of the orbits with period lengths $p = 4 \cdot 3, 4 \cdot 5, 4 \cdot 7, 4 \cdot 9, \ldots$ with the symbolic dynamics $CRLR^3(LR)^{p/2-3}$. This gives $\tilde{r}_2 = 3.5925721841\ldots$, which is also the point at which the four chaos bands merge into two bands.

The orbits CL^n, $n \to \infty$, lead to $\mu = \sqrt{(\mu + \sqrt{(\mu \cdots)})}$, that is, $\mu = \sqrt{(\mu + \mu)}$, for the accumulation point, with the solution $\mu = 2$ (corresponding to $r = 4$).

Another "one-hump" map, the piecewise linear "tent" map

$$f(x) = \lambda(1 - |x - 1|) \qquad 0 \le x \le 2 \tag{4}$$

is much easier to analyze than the quadratic map [DGP 78]. Yet the tent map in equation 4 shares many properties with the quadratic map, such as the ordering of orbits, and their symbolic dynamics as the parameter λ is increased from 1 to 2. (However, some orbits of the quadratic map are missing in the tent map, such as those that have resulted from period doubling. Also, because $|f'(x)| > 1$ for $\lambda > 1$, there are *no* stable orbits.)

The determination of the parameter value λ for a given orbit $CRL \ldots$ is particularly simple. The value of λ is given as the solution of the equation

$$\sum_{k=1}^{P} \frac{\sigma_k}{\lambda^k} = 0 \tag{5}$$

For purposes of iteration, starting with $\lambda = 1.5$, say, the following form is more convenient (it also uses the fact that $\sigma_1 = 1$ and $\sigma_2 = -1$):

$$\lambda = 1 - \sum_{k=3}^{P} \frac{\sigma_k}{\lambda^{k-2}} \qquad 1 < \lambda < 2 \tag{6}$$

Here the σ_k equal $+1$ (or -1), depending on whether the number of R's in the orbit $CRL \ldots$ to the left of and including the kth symbol is even (or odd). For example, for the orbit CRL of period length $P = 3$, equation 6 reads $\lambda = 1 + 1/\lambda$, which has the solution $\lambda = (\sqrt{5} + 1)/2 = 1.618 \ldots$. For the "lowest" of the three orbits of period length 5, $CRLRR$, equation 6 yields $\lambda = 1 + 1/\lambda - 1/\lambda^2 + 1/\lambda^3 \approx 1.5128763968$. (Note that σ_3 always equals -1. In fact, there are no solutions to equation 6 for $1 < \lambda < 2$ otherwise.)

The σ_k are the same as the "σ sequence" that we introduced on page 283 to facilitate the generating of new orbits from known orbits. In a sense, the σ sequences are the most useful form of the quantized orbital dynamics of unimodal maps.

The common accumulation point for orbits with odd period lengths p and orbits with period lengths p of the form $2 \cdot 3, 2 \cdot 5, 2 \cdot 7, \ldots$ is particularly easy to find with equation 6. As noted before, the common symbolic dynamics of these orbits are $CRLR^{p-3}$. The corresponding σ sequences are $+ - - + - + - + - \cdots$; that is, $\sigma_k = (-1)^k$ for $k \ge 3$. For $p \to \infty$, we therefore have $\lambda = 1 + 1/(\lambda + 1)$, with the solution $\lambda = \sqrt{2}$.

For finite p, the solution is given by

$$\lambda^2 = 2 - \frac{(-1)^p}{\lambda^{p-2}}$$

or, asymptotically,

$$\lambda \approx \sqrt{2}\left[1 - \frac{(-1)^p}{2\sqrt{2^p}}\right]$$

Thus, the differences between the λ values for successive odd (or even) orbits and their accumulation point $\lambda = \sqrt{2}$ form an asymptotically geometric progression with a factor of $\frac{1}{2}$ as the period length is increased by 2. This compares with the factor $1/4.669\ldots$ for the period-*doubling* sequence of the quadratic map. And while the period-doubled orbits have *adjacent* parameter intervals, many other orbits intervene between the parameter values for period lengths 3 and 5 or those of other adjacent odd period lengths.

Another helpful property of the σ sequence is that equation 5 can be factored for orbits generated by parent orbits, yielding interesting relationships between the λ values of parents and offspring. For example, for the period-6 orbit with σ sequence $+--+-+$ (the tripled period-2 orbit), $\lambda = \lambda_3^{1/2}$. Here λ_3 is the parameter for the period-3 orbit $+--$.

Tangent Bifurcations, Intermittency, and $1/f$ Noise

Something strange happens for $r = 1 + \sqrt{8}$: a so-called *tangent bifurcation*. For r just below $1 + \sqrt{8}$ (see Figure 12), the iterates become "trapped" for a long time between the logistic parabola and the straight line $x_{n+1} = x_n$. Figure 13 shows intermittent period-3 pulses for $r = 1 + \sqrt{8} - 10^{-4}$. The power spectrum of this process decreases as the reciprocal of frequency f. This phenomenon, called *intermittency*, is one of the main mechanisms of $1/f$ noise in nature.

For r slightly above $1 + \sqrt{8}$, the thrice-iterated quadratic map $f^{(3)}(x)$ (see Figure 14) acquires six additional fixed points: three with an absolute slope larger than 1, which belong to an unstable orbit of period length 3, and three with a slope less than 1, which are the three points belonging to the stable orbit with period length 3. This is the famous period-3 orbit, which guarantees that all other period lengths exist, albeit as unstable orbits, at the same parameter value. This coexistence of an infinite number of unstable orbits has been called *chaotic* by Li and Yorke. The fact that "period three implies chaos," the title of their paper, was enunciated by these authors in 1975 [LY 75].

As r is decreased below $1 + \sqrt{8}$, all these orbits become stable at small but finite intervals of the growth parameter. All orbital periods except $p = 2$

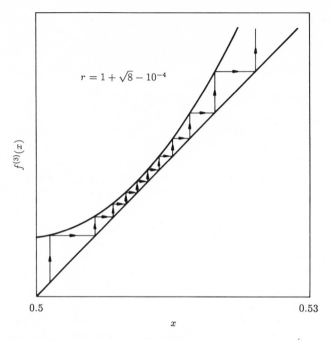

$$r = 1 + \sqrt{8} - 10^{-4}$$

0.5

0.53

x

Figure 12 Trapped iterates near a tangent bifurcation for $r = 1 + \sqrt{8} - 10^{-4}$.

and 3 are stable at more than one r interval. For p an odd prime, the number of such intervals with different orbits equals $(2^{p-1} - 1)/p$. Together with the period-doubling pitchfork bifurcation, the tangent bifurcation is the main source of new orbits.

The r intervals for stable orbits are dense; that is, the parameter values for which no stable periodic orbits exist form no intervals. Nevertheless, they have

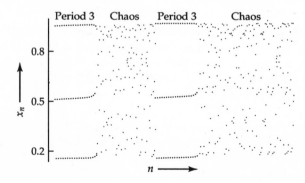

Figure 13 Intermittency for growth parameter just below tangent bifurcation: period-3 pulses alternate with random pulses.

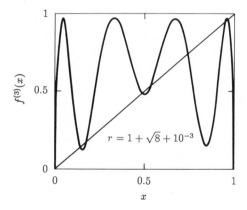

Figure 14 The thrice-iterated quadratic map for the growth parameter slightly above $1 + \sqrt{8}$. This iterated map has acquired six additional fixed points by "tangent bifurcation." Three of these six fixed points are stable (absolute slope smaller than 1) and are members of the stable period-3 orbit visible in Figures 10 and 11.

positive Lebesgue measure. This means that a random choice of the growth parameter has a nonvanishing probability of leading to an aperiodic orbit. This behavior is reminiscent of the irrational numbers, which, too, have positive Lebesgue measure although they form no intervals. Of course, with a finite-state automaton such as a digital computer—not to mention analog machines—an aperiodic orbit can never be proved as such.

Apart from the period-doubling bifurcations starting with the stable fixed point and ending at $r = 3.5699\ldots$, the range of r values for the period length 3 is larger than that for any other period length. The period-3 "window" in the Feigenbaum diagram (see Figure 10A) is thus the most prominent among all the periodic windows and one of the few that are actually visible without a "magnifying glass" (i.e., a computer program that enlarges a small interval of the growth parameter for better visibility). At sufficient magnification one can see period-doubling bifurcations, as in Figure 10B, in each of these periodic windows, each governed by the same Feigenbaum constant $\delta = 4.6692\ldots$.

A Case of Complete Chaos

A particularly interesting value of the growth parameter for the quadratic map is $r = 4$. The transformation

$$y = \frac{1}{2} - \frac{1}{\pi} \arcsin\,(1 - 2x) \qquad 0 \le y \le 1 \tag{7}$$

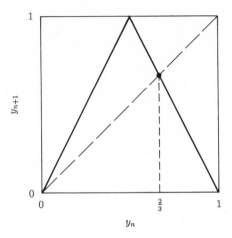

Figure 15   A chaotic tent map: $y_{n+1} = 1 - |2y_n - 1|$.

turns the quadratic map into another tent map:

$$y_{n+1} = 1 - |2y_n - 1|$$

which consists of two straight-line segments with slopes $+2$ and -2, and a maximum at $y_n = 0.5$ (see Figure 15).

By "flipping" the right half of the tent map ($y_{n+1} \rightarrow 1 - y_{n+1}$ for $y_n > 0.5$), we obtain the exceedingly simple *binary-shift map*:

$$y_{n+1} = 2y_n \bmod 1 \tag{8}$$

If we express y_n as a binary fraction, then this map is nothing but a left shift of the digits, with any 1s protruding to the left of the binary point dropped. As a consequence, a value of y that has a terminating binary fraction is mapped eventually into 0. In general, rational values of y, which have periodic binary fractions, lead to periodic orbits. By contrast, irrational y—that is, almost all y in the interval (0, 1)—give rise to nonperiodic orbits.

Although for $r = 4$ almost all initial values y in (0, 1) entail aperiodic orbits, such aperiodic values of y form no intervals. In fact, the initial values for periodic orbits are dense in (0, 1). To see this we truncate a given value of y after an arbitrarily large number of binary places and repeat the remaining bits periodically. Thus, for example, within less than 2^{-5} of $y = 0.10110001\ldots$ we find an initial value with period length 5, namely, $y_0 = 0.\overline{10110}$ (or any longer period length).

Excepting "nonnormal" binary numbers (see pages 241–243 in Chapter 11) as starting values y_0, the iterates of irrational y_0 fill the unit interval with a uniform density. Such a distribution is called the *invariant distribution of the mapping*, because a random variable starting out with it stays with it. As a consequence of the uniform distribution of y, the invariant distribution $p(x)$, which is related to y by equation 7, is the U-shaped distribution well known from the arcsine law of random walk theory:

$$p(x) = \left|\frac{dy}{dx}\right| = \frac{1}{\pi}[x(1-x)]^{-1/2} \qquad 0 < x < 1$$

See Figure 16, which, after proper scaling and shifting, also approximates the invariant distributions in the chaos bands for other parameter values.

The map in equation 8 also illustrates very nicely what is meant by *deterministic chaos*. Suppose the initial condition of a physical dynamic system is represented by a value of y_0 with some *finite* precision. For example, for an eight-place binary precision, we would have

$$y_0 = 0.10110101\genfrac{}{}{0pt}{}{000}{111}\cdots$$

where the double entries reflect the fact that we do not know whether those digits beyond the eighth are 0 or 1. Because of this *unavoidable* lack of perfect precision, the ninth and all higher iterates of y_0 are

$$y_k = 0.\genfrac{}{}{0pt}{}{000}{111}\cdots$$

that is, they assume any possible values in the unit interval in a completely

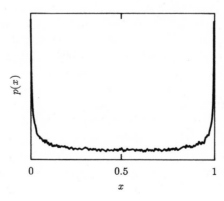

Figure 16 Invariant distribution of quadratic iterates for $r = 4$. Most iterates cluster near $x = 0$ and $x = 1$ [CE 80].

unpredictable succession. It is in this manner that a deterministic law, including the simple equation 8, produces chaotic results, called, appropriately, *deterministic chaos*. Even for completely deterministic dynamic laws, initial conditions with finite precision will—under proper magnifying circumstances—eventually produce totally unpredictable results.

For this to happen, the iteration law must be nonlinear and, in fact, not uniquely invertible. Maps that meet this requirement include those with a maximum (quadratic or otherwise) or with a remainder (mod) operation, as in equation 8. In addition, the nonlinearity must be strong enough that any initial uncertainty will grow exponentially. For example, for the binary-shift map (equation 8), the uncertainty grows by a factor of 2 with each iteration.

By contrast, for the logistic parabola with $r = 2$, for example, any initial value (other than 0 or 1) will lead to the fixed point $x^* = 0.5$ and the differences ε_n between successive iterates x_n and x^* will decrease (asymptotically) by a factor of $-2\varepsilon_n^2$ with each iteration. This is often referred to as *quadratic convergence*. For instance, starting with $x_0 = 0.45$ ($\varepsilon = -0.05$), successive differences will be approximately -0.0005, -0.00005, -0.000000005, and so on. In other words, the "error" becomes rapidly smaller and smaller; the distance of the 5 from the decimal point is doubled with every iteration—as opposed to the one-digit shift per iteration for the map in equation 8.

The convergence is not as rapid for values of r that do not correspond to superstable orbits (as $r = 2$ does, because the orbit includes the "flat top" of the parabola at $x = 0.5$). For example, for $r = 2.5$, the fixed point is $x^* = 1 - 1/2.5 = 0.6$, which differs from 0.5 and is therefore not superstable. Differentiation at the fixed point will show us how the iteration will converge. For $r = 2.5$, we obtain

$$f'(x^*) = 2 - r = -0.5$$

Thus, the differences between x_n and x^* will decrease (asymptotically) by a factor of -0.5. For example, $x_0 = 0.61$ will lead to the successive differences $x_n - x^* = 0.01000$, -0.00525, 0.00256, -0.00129, 0.00064, and so on. For this nonsuperstable orbit the convergence is much slower and is described as linear. In the example, it corresponds to an asymptotic left shift of $|x_n - x^*|$ by (an average) $\log_{10} 2 = 0.3$ decimal places—as opposed to the *doubling* of the left shift with each iteration for a superstable orbit.

Thus, superstable orbits have two major advantages:

1 They converge much faster and are more stable in the presence of small perturbations—hence the name *superstable*—and are therefore more easily measured experimentally.

2 They have simpler theoretical descriptions, such as the symbolic dynamics discussed in this chapter.

Many of the properties of the quadratic map are paradigmatic not only for other unimodal maps but for different nonlinear mappings as well. These laws in turn model a broad range of contemporary problems in which nonlinearities play an essential role.

The Mandelbrot Set

The shortest path between two truths in
the real domain passes through the
complex domain.
—JACQUES HADAMARD

As we saw in the preceding section, the quadratic map for $r = 4$ can be transformed into a simple tent map, which can be further simplified to the binary-shift map $y_{n+1} = 2y_n \bmod 1$. From this map the orbit of any initial point y_0 can be directly inferred by writing y_0 as a binary fraction: periodic binary fractions lead to periodic orbits, but irrational y_0 with normal aperiodic fractions lead to chaotic orbits.

Unfortunately, this simple mapping is not applicable for other values of the growth parameter of the real quadratic map that we have studied so far. However, if we "complexify" both the growth parameter and the variable, the quadratic map becomes considerably more transparent and amenable to analysis. As in many other branches of mathematics (number theory, for example) the introduction of *complex* variables makes many proofs and relations much simpler (as, for example, the proof of the prime number theorem, which dictates the distribution of primes). Hence the paradoxical mathematical motto "complexify to simplify."

In its complexified version, the quadratic map is often rendered in the form

$$z_{n+1} = z_n^2 + c \tag{9}$$

where both the variable z and the growth parameter c are allowed to assume complex values, graphically represented by points in the complex z plane and the complex c plane. For real c, the relation with the previously used parameters is $c = -\mu = -r(r-2)/4$.

One of the first questions that comes to mind when looking at equation 9 is, For what values of the parameter c do the z_n stay bounded as the iteration is continued indefinitely? Obviously, for $c = 0$ and z_0 in the unit disk $|z_0| \leq 1$, z_n stays within the unit disk for $n \to \infty$. However, even for $c = 0$, the initial value $z_0 = 2$, for example, gives $z_7 = 2^{2^7} > 10^{38}$; that is, the seventh iteration already exceeds the diameter of the universe measured in atomic units.

What about $c \neq 0$? For $c = -2$, for example, and $z_0 = 0$, equation 9 gives $z_1 = -2$, $z_2 = 2$, $z_3 = 2$, and so on. Thus, $c = -2$ leads to a preperiodic orbit, with period length 1—in other words, a fixed point, $z = 2$.

In general, the set of all points c for which the iteration $z_{n+1} = z_n^2 + c$, with $z_0 = 0$, stays bounded as $n \to \infty$ is called the *Mandelbrot set*, or **M** set for short, after Benoit Mandelbrot, who discovered it and analyzed many of its intricate details [Man 80]. The **M** set, shown in black in Color Plate 8A, consists of a large heart-shaped ("cardioid") area to which smaller disks are attached, to which even smaller disks are attached and so forth *ad infinitum* in a roughly self-similar progression. The same cardioid shape festooned with a proliferation of disks, also called *Apfelmännchen*, can be discovered in many other regions of the complex parameter plane if it is sufficiently magnified (see Color Plate 8B and C). But other characteristic shapes, too, are revealed by the computer "microscope": dendrites, whorls, and "sea horse" tails; see the color plates. In these illustrations, black areas belong to the **M** set, and the different colors signify different rates of escape to infinity of z_n for values of c outside the **M** set. (The individual colors were selected from a digital "palette" for distinctiveness and aesthetic appeal.)

Although the **M** set is not self-similar as a whole, it possesses many approximately self-similar substructures, such as cardioids and disks, and sea horse tails and whorls within whorls within whorls with infinitely fine filigree. To think that a resplendent structure such as the **M** set and its surround results from a simple quadratic equation is indeed astonishing; its mathematical gossamer continues to inspire awe even in the hardened professional. The complexity of the **M** set is also a vivid reminder that the complexity that we observe in many natural phenomena, including *Life Itself* (the title of a well-known book by Crick), can result from relatively simple laws [Cri 81]. Clearly, complex behavior does not necessitate complex laws.

Although parts of the **M** set look rather like isolated spots (in fact they are known to have been obliterated by some overzealous art editors), the set is actually a connected set, as proved by Douady and Hubbard [DH 82, 85]. However, it is not known whether the **M** set is everywhere *locally* connected. (A circle from which a single point has been removed is still connected, but is no longer everywhere locally connected: points on different sides of the gap, no matter how close, are connected only through a long circular arc.)

The large cardioid area of the **M** set and each circular disk correspond to a particular periodic orbit: the cardioid to period 1, the largest disk to period 2, and the other horizontally attached disks to periods 2, 4, 8, . . . , terminating in the Feigenbaum accumulation point of period-doubling bifurcations. The largest remaining cardioid on the real axis corresponds to orbits with period 3. Each of the infinitely many disks sprouting on the cardioids in the complex plane corresponds to a periodic orbit with a particular period related to that of its cardioid, and each of these disks has infinitely many smaller disks attached to it that all look similar to each other. In fact, the only deviation from self-similarity apparent to the naked eye in this succession of disks is the cleavage in the "rear" of the mother cardioid.

The Julia Sets of the Complex Quadratic Map

The fact that a given parameter value lies in the **M** set may, of course, not be all we want to know. We are eager to discern how the iterates z_n behave for different z_0. For which values of z_0 are the z_n bounded? For a given parameter c, the set of initial values z_0 for which the z_n are bounded form the so-called *filled-in Julia set J_c.* (The Julia set proper consists of the boundary points of J_c.)

(A)

(B)

Figure 17 (A) Filled-in Julia set, defined as the set of all z_0 for which the iteration $z_{n+1} = z_n^2 + c$ is bounded. (B) Another, barely connected Julia set, illustrating the great variety of shapes obtained when the parameter c is changed [Man 83].

Computer experiments have shown that different values of c can lead to a stunning variety of Julia sets, minute changes in c often causing enormous metamorphoses in J_c (see Figure 17A and B).

Some Julia sets are connected; others are just "floating-dust" Cantor sets. Interestingly, those values of the parameter c for which J_c is connected are precisely all the members of the **M** set, so that the latter can also be defined as the set of all c values for which J_c is a connected set. This equivalence is a consequence of a theorem proved independently in 1918 by Gaston Julia and Pierre Fatou, a fact that was rediscovered jointly by Douady and Hubbard, who added many more insights to our store of—still sporadic—knowledge of the deceptively simple iteration $z \rightarrow z^2 + c$ [DH 82].

One of the most consequential discoveries of Douady and Hubbard is that the boundary of the Mandelbrot set can be mapped conformally to the unit circle and that the iteration $z_{n+1} = z_n^2 + c$ corresponds simply to doubling the angle on the unit circle. Thus, measuring angles α in multiples of 2π, the complex quadratic map corresponds to $\alpha_{n+1} = 2\alpha_n \mod 1$. If the "external angle," as it is called, is expressed by a binary fraction, then the iteration is a left shift of the binary digits modulo 1. A c value with an external angle of $\alpha = \frac{13}{31} = 0.\overline{11001}$, for example, will lead to a periodic orbit of period length 5. The individual digits tell us which of the iterates z_n will fall into the upper (0) or lower (1) half plane.

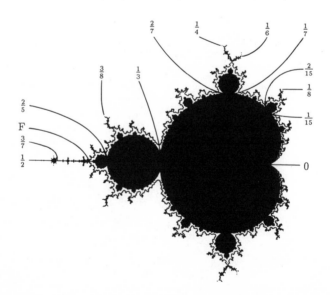

Figure 18 External angles for the Mandelbrot set. The fractions determine the period lengths of the iterates z_n for a given choice of the parameter c. The point "F" marks the accumulation point of the period-doubling cascade [Dou 86].

Figure 18 shows the **M** set and some rational external angles. The accumulation point of period-doubling bifurcations $c = -1.4011\ldots$ (marked "F," for Feigenbaum) has as its external angle the Morse-Thue constant $0.0110100110010110\ldots = 0.412\ldots$, whose binary digits are the Morse-Thue sequence.

The conformal mapping from the boundary of the **M** set to the unit circle may be visualized physically as a problem in electrostatics as follows. Consider an infinitely long conducting bar whose cross section is the **M** set, surround it with a distant electrode in the shape of a circular cylinder, and apply a voltage difference between bar and cylinder. Then the electric field lines from a point on the circle at an angle α with the real axis end on a point c on the boundary of the **M** set with external angle α. This is so because electric field lines obey the laws of conformal mapping. And the equipotential lines, which are orthogonal to the field lines, correspond to c values with equal rates of divergence to infinity for z_n with $z_0 = 0$. (Remember that values of c outside the **M** set lead to unbounded iterates z_n.)

The interested reader will find more fascinating details in H.-O. Peitgen and P. H. Richter's *The Beauty of Fractals*, which also includes an illuminating essay by Douady [PR 86a].

13

C H A P T E R

A Forbidden Symmetry, Fibonacci's Rabbits, and a New State of Matter

The above proposition $[1 + 1 = 2]$ *is occasionally useful.*

—A. N. WHITEHEAD and B. RUSSELL,
in *Principia Mathematica*

Modern theoretical physics is a luxuriant . . . world of ideas and a mathematician can find in it everything to satiate himself except the order to which he is accustomed.

—YURI MANIN

In this chapter we shall taste some of the forbidden fruits that self-similarity breeds: a new solid state of matter, namely, a "quasicrystal" with a fivefold axis of rotational symmetry (like that of a five-legged starfish and many flowers). Curiously, the new matter is related to a simple iterated map that was itself bred by multiplying rabbits—rather rare rabbits, that is, of the famous Fibonacci family. Said simple iterated map is in turn intimately entangled with the continued fraction for the golden mean, easily abased to the "silver means," which predict more forbidden symmetries—some of which have since been seen in actual quasicrystals.

The Forbidden Fivefold Symmetry

From snowflakes to gemstones, people have forever prized crystals, formations in which the individual atoms are arranged in orderly periodic lattices. But we are also familiar with *disordered* substances, such as most liquids, in which the atoms are randomly distributed. Likewise, most solid substances encountered in nature are disordered, or *amorphous,* just like a liquid, except that they are solidified. Glasses are transparent examples of amorphous solids. In fact, among physicists the designation *glass* has become the generic term for disordered systems. Thus, a *metallic glass* does not mean a pewter cup, nor does it have much else to do with glass: it is simply a metal in which the individual atoms are arranged in disorderly fashion. And a *spin glass* is not spun glass, nor is there much spinning going on. Rather, a spin glass is a disordered arrangement of magnetic spins or, by extension, the values of any other physical variable that has two[1] preferred states, such as on-off neurons in a neural network.

And then there are some showy states of matter, such as *liquid crystals,* which are now ubiquitous as alphanumeric displays (LCDs) in watches and calculators. In a liquid crystal, the molecules are randomly located but their orientations are well ordered, under the control of an external voltage, which permits the displayed information to be changed.

Until recently, few if any people suspected that there could be another state of matter sharing important aspects with both crystalline *and* amorphous substances. Yet, this is precisely what D. Shechtman and his collaborators discovered when they recorded electron diffraction patterns (see Figure 1) of a rapidly cooled aluminum-manganese alloy (Al_6Mn), now called a *quasicrystal* [SBGC 84]. The diffraction pattern (essentially a two-dimensional Fourier transform) of their quasicrystals showed *sharp peaks,* implying long-range order, just as for periodic crystal lattices. But the pattern also showed a *fivefold* symmetry that is forbidden for periodic crystals; see Figure 2 for a simple proof. Fivefold symmetry means that the lattice can be brought into coincidence with itself by a rotation through $360°/5 = 72°$. But the only allowed symmetry axes are two-, three-, four-, and sixfold; all other rotational symmetries conflict with the translational symmetry of a periodic crystal.

What then *is* going on in these new substances? Several other quasicrystals with other forbidden symmetries have been identified since the original discovery. Thus, quasicrystals are not an isolated quirk; they represent a new solid state of matter—Linus Pauling's tenacious doubts notwithstanding [Pau 85]. And as we shall shortly see, the explanation of their existence is rooted in self-similarity.

---

1. There is a strong link between the physicists' *spin* and the number 2 (also known as the "oddest prime" because it is the only even prime). Because elementary spin is a two-valued variable ("up" or "down"), two electrons are allowed in the same atomic orbit, thereby explaining (together with Pauli's exclusion principle) the periodic table of elements. Einstein, in the only experiment he ever performed himself (with W. J. de Haas), on the gyromagnetic ratio of the electron, got a result that was off by a factor of 2 (an error of 100 percent!). However, this did not bother the great theorist in the least—*close enough,* he is said to have remarked. Subsequently, in turned out that his result was quite accurate and the factor of 2 had to do with the spin of the electron. Later on, many occasions arose in physics and chemistry where one had to "multiply by 2 because of the spin." (However, the fact that the *proof* figure on a liquor bottle is the alcohol percentage multiplied by 2 is probably unrelated to the spin induced in some imbibers by the liquid potion.)

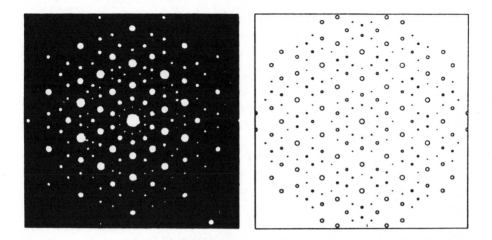

Figure 1   Electron diffraction pattern of crystal with forbidden fivefold symmetry [SBGC 84].

Figure 2   Simple proof that fivefold symmetry is impossible in a periodic crystal. In a periodic crystal there is a smallest distance between two atoms. Let the segment $\overline{AB}$ be one of these shortest distances. If the crystal has fivefold symmetry, then, in addition to the points A and B, the points C and D (obtained by $360°/5 = 72°$ rotations) should also be occupied by atoms. But the distance between C and D is smaller (by a factor of $0.382\ldots$, equal to the golden mean squared) than $\overline{AB}$—contradicting the claim that $\overline{AB}$ was the smallest distance.

Long-Range Order from Neighborly Interactions

As we know from the number-theoretic Morse-Thue sequence (see Chapter 12), sharp spectral peaks and aperiodicity are no contradiction, as long as *long-range order* prevails. In fact, the simplest example of aperiodicity and long-range order leading to sharp spectral peaks is furnished by the superposition of two sine waves with incommensurate frequencies, for example,

$$s(t) = \sin(\omega_0 t) + \sin(\alpha \omega_0 t)$$

where the frequency ratio α is an irrational number. There is no nonzero value T for which $s(t) = s(t + T)$ for all t. Yet Fourier-analyzing $s(t)$ (properly windowed to make the Fourier transform converge) will, of course, show sharp peaks at the incommensurate radian frequencies $\omega = \omega_0$ and $\omega = \alpha \omega_0$.

Periodicity in crystals is easy to explain. For example, in a crystal of table salt (sodium chloride, chemically speaking), sodium atoms (Na) prefer chlorine atoms (Cl) as neighbors and vice versa: chlorine atoms like to surround themselves with sodium atoms. Thus, going along one of the crystal axes, sodium and chlorine alternate: Na-Cl-Na-Cl-Na-Cl-, and so on. The result is perfect periodicity and long-range order.

But how can we explain long-range order in an aperiodic quasicrystal? That is not so easy. If there is no simple mutual attraction between different kinds of atoms or molecules (or if high temperature overcomes this), the usual result is *no* long-range order: a random structure, as in liquids (or "frozen" liquids, such as window glass).

Perhaps the only way to produce long-range order from the short-range interactions that dominate solid structures *without* resulting in a periodic lattice (as in our table salt example) is to rely on iterated maps. Iterated maps are models of short-range interactions. For example, a 0 attracts a 1, which engenders the mapping $0 \rightarrow 01$; and a 1 attracts a 0, or $1 \rightarrow 10$. Yet, as we know from the Morse-Thue sequence, iterated maps can also produce aperiodic long-range order. Since iterated maps often lead to self-similarities, an explanation of quasicrystals by this approach means that the crystals (and their diffraction patterns) must exhibit scaling invariances. This is indeed the case, as a closer inspection of the diffraction pattern in Figure 1 shows. The most prominent scaling factor in Figure 1 turns out to be the golden mean $\gamma = (\sqrt{5} - 1)/2 = 0.618 \ldots$. (*Note:* Some authors—including the present one, in another book—call the *reciprocal* of the golden mean, $1/\gamma = 1.618 \ldots$, the golden section or golden mean.)

Thus, there is an odds-on chance that quasicrystals might be modeled by an iterative map related to the golden mean. To find such a map, we have to turn the clock back a bit.

Around the year 1200, Leonardo da Pisa (ca. 1175–1250)—better known as "Fibonacci," that is, son of Bonacci—was considering the problem of how

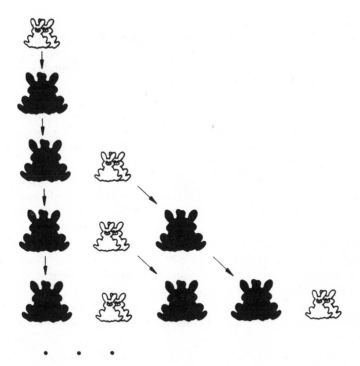

Figure 3   Multiplying rabbits, as Fibonacci saw them. Small, "white" rabbit symbols: immature pairs; large, "filled" symbols: mature pairs.

rabbits multiply, something rabbits were obviously very good at even then [Fib 1202]. In best modern style he postulated a highly simplified model of the procreation process: each season every adult pair of rabbits begets a young pair, which will be mature one generation later. Starting with one immature pair of rabbits and assuming that rabbits go on living forever, the rabbit population grows rapidly, as shown in Figure 3.

More formally, Fibonacci was considering the iterated map $0 \to 1$ and $1 \to 10$, where 0 stands for an immature rabbit pair and 1 for a mature pair. Thus, the first six generations are represented by the binary sequences

$$
\begin{array}{l}
0 \\
1 \\
10 \\
101 \\
10110 \\
10110101 \\
\cdots
\end{array}
\tag{1}
$$

which, like the Morse-Thue-sequence, is a self-generating sequence. The nth generation has precisely F_n pairs of rabbits, where F_n is the nth Fibonacci number defined by $F_1 = F_2 = 1$ and the recursion $F_{n+2} = F_{n+1} + F_n$. This yields the well-known Fibonacci sequence 1, 1, 2, 3, 5, 8, 13, ... A simple formula for generating the F_n for $n > 0$ is obtained by rounding the value of $\gamma^{-n}/\sqrt{5}$ to the nearest integer, where $\gamma = 0.618$ . . . is the aforementioned golden mean. Thus, the ratio of two successive Fibonacci numbers asymptotically approaches the golden mean, as does the ratio of 0s to 1s in each line of pattern 1. In fact, the numbers of 0s and 1s in the nth line are precisely F_{n-2} and F_{n-1}, respectively. (Note that, by backward recursion, $F_0 = 0$ and $F_{-1} = 1$.)

Another law for constructing the infinite sequence whose beginning is shown in pattern 1, and which I have called the *rabbit sequence*, is quite apparent: after the first two rows, simply append to each row the previous row to form the next one. This property is a direct consequence of iterating the mapping. The first iteration of the map $0 \to 1$, $1 \to 10$ gives $0 \to 10$, $1 \to 101$, and iterating the iterated map results in $0 \to 101\ 10$, $1 \to 101\ 10\ 101$, and so on. Thus, the fifth line (101 10) in pattern 1 can be considered to have been generated from the third line (10) by using the once iterated map $1 \to 101$ and appending to it the result of $0 \to 10$. But 101 is, of course, the fourth line, and the appended 10 is in fact the third line. Thus, each sequence in pattern 1 can be obtained by appending to the predecessor sequence the *prepredecessor*, an "inflation" rule which mirrors the original map $0 \to 1$ and $1 \to 10$. It is this kind of structure that causes *long*-range order to occur in the rabbit sequence although it was defined on the basis of only a *short*-range law ($0 \to 1$, $1 \to 10$) involving only next neighbor symbols.

As we saw in Chapter 11, iterated maps often lead to self-similarity, and the rabbit sequence is no exception: it abounds with self-similarities. One self-similarity of the rabbit sequence can be demonstrated by retaining the first two out of every three symbols for every 1 in the sequence and retaining the first one out of every two symbols for every 0. This decimation indeed reproduces the infinite rabbit sequence, as indicated in the following by underlining:

$$\underline{10}1\underline{1}0\underline{10}1\underline{1}0\underline{1}1 \ldots$$

This property reflects the fact that the rabbit sequence reproduces itself upon *reverse* mapping (also called *block renaming* or "deflation" in renormalization theories in physics) according to the law $10 \to 1$, $1 \to 0$. (Note that the non-underlined bits also mimic the 1s and 0s of the rabbit sequence—there is no escaping from those foxy rabbits.)

Let us try to get some useful work out of our rabbits. Consider the following "synchronization" problem (with potential applications to keeping digital transmission channels in step, as in picture transmission from distant space vehicles). How many steps do we have to move to the right in the rabbit sequence

$$101101011011010110101 \ldots$$

to find a given subsequence (e.g., 10) again? We find the answer by inspection. First we have to move 3 places, then 2, then 3 again, and so forth: 3, 2, 3, 3, 2, 3, 2, . . . , a sequence (of Fibonacci numbers, incidentally) that mimics the rabbit sequence, which is in fact reproduced by the substitution $2 \rightarrow 0$ ($F_3 \rightarrow F_0$) and $3 \rightarrow 1$ ($F_4 \rightarrow F_1$). To cope with large synchronization errors, one has to focus on long subsequences. In general, a subsequence of length $F_n - 1$ does not reoccur before F_{n-1} steps. It will, however, reoccur after at most F_{n+1} steps.

There is a simple formula (first encountered on pages 53−54) for calculating the indices k for which the rabbit sequence symbol r_k equals 1:

$$k = \left\lfloor \frac{n}{\gamma} \right\rfloor \qquad n = 1, 2, 3, \ldots$$

while the indices for which $r_k = 0$ are given by

$$k = \left\lfloor \frac{n}{\gamma^2} \right\rfloor \qquad n = 1, 2, 3, \ldots$$

where the "floor function" $\lfloor x \rfloor$ means the largest integer not exceeding x.

These two equations can be interpreted as formulas for generation of 1s and 0s by two incommensurate frequencies: γ and γ^2, respectively. Note that $\gamma + \gamma^2 = 1$, which is the frequency of occurrence of either 1 *or* 0, also called the "sampling" frequency by engineers. If γ is replaced by any positive irrational number $w < 1$ and γ^2 by $1 - w$, then the resulting two sequences, which together cover all the positive integers, are called a pair of *Beatty sequences*. Because of this covering property, Beatty sequences are useful as index sequences [Slo 73].

With the rabbit sequence being generated by the frequencies $\gamma \approx 0.618$ and $\gamma^2 \approx 0.382$, it is small wonder that the spectrum (i.e., the magnitude of the Fourier transform) should show pronounced peaks at these two frequencies; see Figure 4, which was obtained by truncating the rabbit sequence after 144 terms and taking the Fourier transform [Schr 90]. The two main peaks are located at the harmonic numbers 55 and 89 corresponding to the frequencies $\frac{89}{144} \approx \gamma$ and $\frac{55}{144} \approx \gamma^2$. The spectrum also reflects the self-similarity of the rabbit sequence. In fact, the peaks occur at frequencies that scale with the golden mean γ (actually, the ratio of successive Fibonacci numbers for the truncated sequence), and the amplitudes scale approximately as γ^2.

Generation of the Rabbit Sequence from the Fibonacci Number System

For most purposes we do not need the *index* sequences for the 1s or 0s of the rabbit sequence but we need the sequence itself and a *direct* formula to generate it. Here is a first stab:

Figure 4 Fourier amplitude spectrum of binary rabbit sequence (the first 144 terms repeated periodically) [Schr 90].

Consider

$$r_k = m_r + 1 \bmod 2$$

where m_r is the *index* of the least significant term in the representation of k in the Fibonacci number system [Schr 90]. In this sytem, n is represented as the unique sum of Fibonacci numbers with descending indices, starting with the largest-index Fibonacci number not exceeding n:

$$n = F_{m_1} + F_{m_2} + \cdots + F_{m_r}$$

where $m_1 > m_2 > \cdots > m_r$. For example, $12 = 8 + 3 + 1 = F_6 + F_4 + F_2$. Thus, since the index (2) of the last term (F_2) is even, $r_{12} = 1$. (Note that in this representation no two adjacent indices can appear; i.e., $m_{i+1} \geq 2 + m_i$. Also, by convention, $1 = F_2$, so that $r_1 = 1$.) However, while this approach does address itself to the r_k themselves, the calculation via the Fibonacci number system can hardly be called direct.

The Self-Similar Spectrum of the Rabbit Sequence

A more direct representation of the rabbit sequence r_k is the following:

$$r_k = \begin{cases} 1 & \text{if } \langle (k+1)\gamma \rangle_1 < \gamma \\ 0 & \text{otherwise} \end{cases} \tag{2}$$

where, as before, $\langle x \rangle_1$ is the fractional part of x. This formula for r_k has an attractive geometric representation; see Figure 5.

Two direct formulas for generating the rabbit sequence, which also put its long-range order *and* aperiodicity into direct evidence, are given by

$$r_k = \lfloor (k + 1)\gamma \rfloor - \lfloor k\gamma \rfloor$$

and, rewriting equation 2,

$$r_k = \frac{1}{2} + \frac{1}{2} \operatorname{sgn} [\gamma - \langle (k + 1)\gamma \rangle_1] \tag{3}$$

where sgn $[x]$ is the algebraic sign of x ($+1$ or -1) for $x \neq 0$ (sgn $[0]$ is defined as 0).

Interpreted in engineering terms, equation 3, with k considered a continuous variable ("time"), says that r_k is a square wave (jumping between the values 1 and 0) with a fundamental frequency γ (and a "duty cycle" of γ). However, k is *not* a continuous variable; it is discrete, increasing in steps of 1. This means that the square wave is *sampled* with a sampling frequency of 1. Since the frequencies γ and 1 are incommensurate, the resulting sequence of samples is aperiodic, while retaining a perfectly rigid long-range order. For example, setting $k = 144$, we find that $r_{144} = 1$, which is quickly confirmed by noting that $144 = F_{12}$ and by applying the general rule $r_{F_n} = (1 + (-1)^n)/2$. (*Note:* For $F_n = 1$, one has to take $n = 2$.)

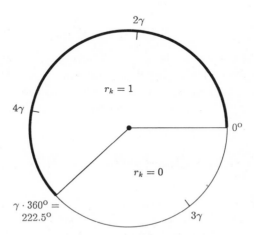

Figure 5   The rabbit sequence, generated geometrically. The sequence term r_k equals 1 if the angle $\langle (k + 1)\gamma \rangle_1 \cdot 360°$ falls within the heavy circular arc. Otherwise r_k equals 0.

Since the period γ is an *irrational* number, the samples r_k taken at sampling intervals are aperiodic. As a result, the spectrum of r_k shows strong peaks at certain preferred frequencies, namely, the sampling frequency multiplied by γ^2, γ^3, γ^4, γ^5, and so on, and the corresponding mirror frequencies ($1 - \gamma^2 = \gamma$, $1 - \gamma^3, \ldots$).

A mathematical expression of the spectrum of r_k is obtained by Fourier-transforming r_k as given by equation 3. This yields

$$R_m = \operatorname{sinc} m\gamma \qquad \text{for frequencies } f_{nm} = n + m\gamma$$

where the "sinc function" $\operatorname{sinc} x$ is defined as $(\sin \pi x)/\pi x$.

Self-Similarity in the Rabbit Sequence

Where do these spectral self-similarities come from? Obviously, they must be hiding already in the sequence itself. Indeed, if we look at the index sequence $\lfloor n/\gamma \rfloor$ for the 1s, we see that scaling n by a factor of γ, that is, by substituting n/γ for n, will give the index sequence $\lfloor n/\gamma^2 \rfloor$ for the 0s. Thus, the magnitude of the Fourier transform of r_k will remain unchanged under this rescaling, except for the constant scaling factor.

We can also observe the self-similarity in r_k itself. Since the self-similarity factor is $1/\gamma$, we have to "hop along" $1.618\ldots$ places on average to effect the scaling by $1/\gamma$. Since the asymptotic ratio of 1s to 0s is precisely γ, we might try to skip two terms in r_k if we encounter a 1 and skip only one term every time we encounter a 0 in the original sequence. This indeed reproduces the series:

$$1 \quad 0 \quad \underline{1} \quad 1 \quad \underline{0} \quad 1 \quad 0 \quad \underline{1} \quad 1 \quad 0 \quad \underline{1} \quad 1 \quad \underline{0} \ldots = 1 \quad 0 \quad 1 \quad 1 \quad 0 \ldots$$

as the reader may be tempted to show. This decimation process is the complement of the "deflation" or *block renaming* that we mentioned before. Here we have renamed each 101 block 1 and each 10 block 0. The block renaming is the inverse of the generating map ($0 \to 1$, $1 \to 10$) iterated once, that is, $0 \to 10$, $1 \to 101$.

A One-Dimensional Quasiperiodic Lattice

How do we convert our discoveries about self-similar sequences producing aperiodic long-range order into something more physical, such as a one-dimensional (1D) lattice, say, as a precursor to a full-blown 3D quasicrystal? A simple method is illustrated in Figure 6. We place atoms on a straight line according to the following rule. For each 1 in the rabbit sequence r_k we take an

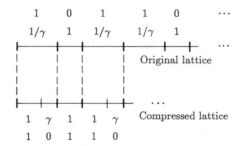

Figure 6   One-dimensional quasiperiodic "crystal," obtained from the rabbit sequence. The self-similarity factor is the golden mean γ.

interatomic distance of $1/\gamma = 1.618 \ldots$ units (say, Ångström units), and for each 0 we take a distance of 1 unit. The lower part of Figure 6 shows the same 1D lattice compressed by a factor of $1.618 \ldots$, which demonstrates that every atom in the original lattice coincides precisely with an atom in the compressed lattice. (The compressed lattice, having a higher atomic density, will of course have some extra atoms with no partners in the original lattice.) Thus, the 1D lattice, so constructed, has self-similarity. (The reader may want to generalize this result to self-similar 1D lattices based on suitable irrational numbers other than the golden section γ.)

If every atom in the "rabbit lattice" is represented by a Dirac delta function, we obtain the Fourier transform [ZD 85]:

$$S_{nm} = \text{sinc}\left(\frac{f}{\sqrt{5}} + m\right) \quad \text{at frequencies } f_{nm} = \frac{n}{\sqrt{5}} - m$$

Self-Similarity from Projections

An alternative method of constructing the 1D rabbit lattice is illustrated in Figure 7 [de B 81]. It shows the square 2D integer lattice, called $\mathbb{Z}^2$, and a straight line with a slope (the tangent of the angle between it and the abscissa) equal to $1/\gamma$. For each unit square that this straight line enters, the upper left corner of the square is projected normally onto the line. And lo and behold, the footprints of these projections generate the previously defined 1D rabbit lattice. The rabbit sequence is recovered by designating the larger intervals by 1s and the shorter intervals by 0s. This geometric construction is a direct consequence of the arithmetic description of the rabbit sequence (equation 2), as the reader may want to show.

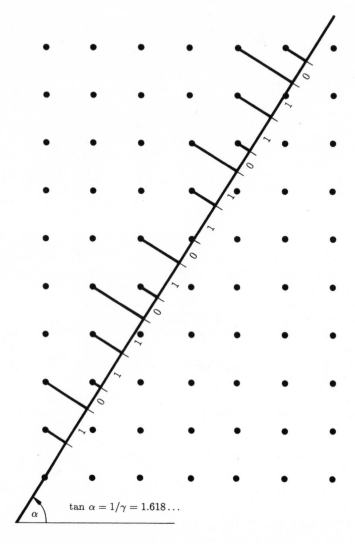

Figure 7 The one-dimensional crystal of Figure 6, obtained by projections from a square lattice [deB 81].

The projection method, being based on a perfectly periodic square lattice, also demonstrates again both the long-range order in the 1D rabbit lattice and its aperiodicity (because of the irrational slope of the straight line). But most important, the projection method can easily be generalized to generate quasiperiodic lattices in two and three dimensions that mimic real quasicrystals [Mac 82, DK 85, Els 85, Jan 89].

Figure 8   Two-dimensional quasicrystal, obtained by projections from a five-dimensional hypercubic lattice.

To construct a 2D quasiperiodic lattice, one needs *four* rationally independent vectors; see Levin and Steinhardt [LS 84]. It is more convenient, however, to project a region of the five-dimensional "cubic" lattice onto an appropriately inclined plain. See Figure 8—which was, of course, generated by a computer, since five-dimensional lattices are still out of reach in the tangible world. It is interesting to note that the points in this image correspond to the vertices of an *aperiodic* tiling of the plane by *two* different tiles, the famous Penrose tiling (see Figure 9)—a feat that had long been considered impossible [Pen 74, Mac 82]. (For an aperiodic tiling with just *one* tile, see Figure 10 [Gar 77].)

When a photographic slide with the point pattern of Figure 8 is placed into a laser beam, the diffraction pattern of Figure 11A results, which shows the puzzling fivefold symmetry (which looks like a tenfold symmetry because we cannot see the signs of the scattered amplitudes in the *intensities* recorded photographically). In fact, Figure 11A resembles the diffraction pattern from an actual quasicrystal. Note particularly the self-similarities with the scaling factor γ in

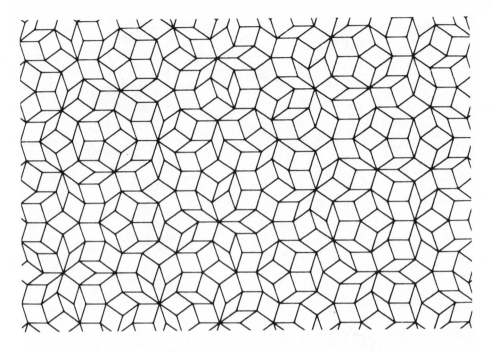

Figure 9   Penrose tiling: a tiling of the plane with only two different tiles [Mac 82].

Figure 10   An aperiodic tiling of the plane by Heinz Voderberg with one tile shown alternately in black or white [Gar 77].

many of the details of the pattern such as the numerous regular pentagons of different sizes.

The experiment just described is amenable to a "live" demonstration before sizable audiences using large television monitors for the display and a lensless TV camera for capturing the diffraction pattern.[2] By increasing the brightness of the diffraction pattern (by turning up the laser intensity), more and more diffraction spots can be made visible until the entire monitor screen is filled. Parts A, B, and C of Figure 11 show the diffraction pattern at three progressively higher intensities. In this manner, one of the crucial differences between quasicrystal diffraction patterns and those of periodic crystals can be demonstrated most convincingly: diffraction patterns of quasicrystals, while consisting of delta functions just like those of periodic crystals, consist of a (countably) *infinite* collection of delta functions that are everywhere dense—in contrast to the isolated diffraction spots of periodic crystals! The original quasicrystal patterns looked like those of periodic crystals only because they were obtained with a relative low incident intensity.

To generate a three-dimensional quasi-periodic lattice, one projects a *six*-dimensional cubic lattice onto three dimensions [KD 86].

More Forbidden Symmetries

Having tasted a first forbidden fruit of fivefold symmetry, we might ask whether there are quasicrystals with other outlawed symmetry axes that can be distilled from self-similar iterated maps. There are indeed.

The mapping $0 \to 1$, $1 \to 10$, which generates the rabbit sequence, is intimately related to the continued fraction for the golden mean γ:

$$\gamma = \frac{1}{1+} \frac{1}{1+} \frac{1}{1+} \cdots$$

which is customarily written as $[1, 1, 1, \ldots]$ or, since the continued fraction is periodic, simply as $[\bar{1}]$. Note that the period length equals 1. The continued fraction expansion for γ follows immediately from its definition as the positive root of the quadratic equation $x^2 + x = 1$, which can also be written as $x = 1/(1 + x)$. Using this form of the defining equation for γ recursively results in the foregoing continued fraction expansion for γ. (Note that, because $|1 + \gamma| > 1$, the recursion converges.)

---

2.   I am grateful to Hans Werner Strube for the computer-generated "quasicrystal" and to Heinrich Henze for the brilliant laser diffraction patterns.

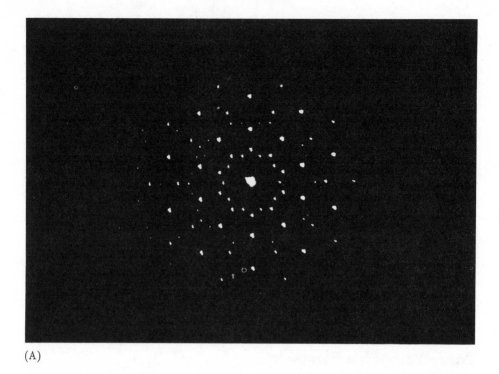

(A)

Figure 11   (A) Laser diffraction pattern with fivefold symmetry of quasicrystal shown in Figure 8. (B) Laser diffraction pattern at higher laser intensity showing an increased number of diffraction spots. (C) Diffraction pattern at still higher intensity showing a nearly dense pattern of diffracted energy.

Could it be that iterated mappings related to the lesser "*silver* means," $\tau_N^{\pm}$, can be pressed into service to generate self-similar lattices? The silver means $\tau_N^{\pm}$ are defined by the equation $1/\tau_N^{\pm} = N \pm \tau_N^{\pm}$; that is, they are all those quadratic irrational numbers that can be expressed by periodic continued fractions with period length 1 and ± 1 as the numerator.[3] It can be shown that

3. The *noble means*, another generalization of the golden mean, are defined as all those numbers whose continued fraction expansions end in infinitely many 1s. They distinguish themselves both in the present case and in the quasiperiodic route to chaos of nonlinear dynamic systems. In this nomenclature the golden mean is but the noblest of the noble means.

Cassini's divisions in the rings of Saturn are a manifestation of what happens when, instead of noble numbers, simple rational numbers reign: rocks and ice particles constituting the rings, whose orbital periods are in simple rational relation with the periods of the moons of Saturn, are simply swept out of their paths by the resonance effects between commensurate orbital periods. In fact, the very stability of the solar system depends on the nobility of the orbital period ratios.

(B)

(C)

$\tau_2^+ = [2, 2, 2, \ldots] = \sqrt{2} - 1$ generates a quasicrystal with the crystallographically forbidden eightfold symmetry, while $\tau_4^- = 2 - \sqrt{3}$ underlies a likewise forbidden twelvefold symmetry axis. Both eight- and twelvefold symmetry have recently been observed experimentally [INF 85, INF 88, CLK 88].

In addition to the golden mean $\gamma = \tau_1^+$, all $\tau_N^\pm$, where N is the nth Lucas number and the sign superscript in $\tau_N^\pm$ equals $(-1)^n$, generate quasicrystals with a fivefold symmetry [Schr 90]. The Lucas numbers L_n obey the same recursion as the Fibonacci numbers, but start with $L_1 = 1$ and $L_2 = 3$. The Lucas numbers 1, 3, 4, 7, 11, 18, . . . , too, are related to the golden mean γ. In fact, for $n \geq 2$, L_n is given by γ^{-n} rounded to the nearest integer.

14

Periodic and Quasiperiodic Structures in Space—The Route to Spatial Chaos

> *A great truth is a truth whose opposite is also a great truth.*
>
> —NIELS BOHR

In the preceding chapter we saw that, in addition to crystals with perfectly periodic lattices, there are *quasicrystals* whose spatial structure is *quasi*periodic. In one-dimensional models, this quasiperiodicity can be described by two *incommensurate* frequencies involving the golden mean or other quadratic irrational numbers whose continued fraction expansions have a short period length.[1] And, of course *amorphous* substances with no discernible periodic structure have always been known.

At the time of their discovery in 1984, quasicrystals came as a real surprise: only a very sparse sampling of scientists had foreseen the possibility of a spatial structure—other than liquid crystals—intermediate between amorphous glasses and regular crystals. Yet, the surprises are not over.[2] There is still another spatial structure lurking between quasiperiodicity and amorphous disorder: spatial chaos.

The existence of spatial chaos should not really come as a surprise to anyone familiar with temporal chaos, considering that space and time are mere components of a unified space-time. Whatever can happen in time could also happen in space

---

1. Periodic continued fractions with long periods involving large integers would give rise to unrealistic quasicrystals with molecular interactions much too complicated to be believable.

2. "It's not over until the fat lady sings," as P. W. Anderson remarked on St. Patrick's Day in 1987 in New York City at the "Woodstock" meeting of the American Physical Society on the new high-T_c ("room temperature in Alaska") superconductors.

(and vice versa). In the time domain, we have long been acquainted with periodic, quasiperiodic, and random phenomena:

• The motion of a swing is periodic.

• The phases of the moon at midnight every Sunday, say, are quasiperiodic; they are governed by two (as yet) incommensurate frequencies: the moon's orbit around the earth and the earth's elliptic orbit around the sun.

• The hiss of air escaping from a punctured bicycle tire can be considered a random noise. Thermal motion is another example of a random process.[3]

Yet, more recently, physicists have come to appreciate a fourth kind of temporal behavior: *deterministic chaos*, which is aperiodic, just like random noise, but distinct from the latter because it is the result of deterministic equations. In dynamic systems such chaos is often characterized by small fractal dimensions because a chaotic process in phase space typically fills only a small part of the entire, energetically available space.

Taking a cue from chaos in the time domain, we should expect to find chaos entrenched also in the space domain. In fact, *turbulence* is now considered a case of spatial chaos, albeit a very complicated one. In this chapter we first focus on a particularly simple case of spatial quasiperiodicity and chaos, with important analogies in the time domain: a one-dimensional Ising spin model of magnetism.

Periodicity and Quasiperiodicity in Space

Imagine a one-dimensional system of electron spins $s_i = +1$ or $s_i = -1$ positioned at equal intervals along a single spatial dimension, as considered by Bak and Bruinsma [BB 82]. In the presence of an external magnetic field H, the energy E of the system is given by

$$E = -\sum_i H \, s_i + \sum_{i \neq j} J_{ij} \, s_i \, s_j \tag{1}$$

where J_{ij} is an antiferromagnetic interaction ($J_{ij} > 0$) between spins s_i and s_j that

---

3. If our ears were a bit more sensitive and if there were no distracting sounds, we would hear the thermal motion of the air molecules: a constant hissing, the perception of which would offer no added survival value to ourselves or the species. In fact, such a super-ear would be an evolutionary liability, considering the extra expense to the species to develop, protect, and maintain it.

decays with increasing spatial distance $|i - j|$ according to the power law

$$J_{ij} = |i - j|^{-\alpha} \tag{2}$$

with $\alpha = 2$, for example.

The fact that J_{ij} is positive means that adjacent spins would like to have *opposite* sign (to minimize the energy E). This is why an interaction like that in equation 2 is said to be *anti*ferromagnetic. (In a "ferromagnet," adjacent spins prefer to align themselves in the *same* direction, creating a strong external magnetic field, such as that of a horseshoe magnet.)

With adjacent spins having opposite signs, spins at alternating locations, of course, have the same value, giving a positive, though smaller, contribution to the energy E (for $\alpha > 0$). Thus, without an external field H, the minimum energy is obtained by a fraction $w = \frac{1}{2}$ of spins pointing up. Setting $s_0 = +1$ as an initial condition, we have

$$s_{2k} = +1 \qquad \text{and} \qquad s_{2k+1} = -1 \tag{3}$$

which is a perfectly periodic antiferromagnetic arrangement.

The Devil's Staircase for Ising Spins

For nonzero values of the external magnetic field H, $w = \frac{1}{2}$ may no longer give the minimum energy E. In fact, for $H \to \infty$ all spins would turn up, so that w, the fraction of up-spins, would go to 1. But how?

For small changes of H (and zero temperature), no spins will flip; they are *locked* into their given configuration. In fact, for each rational $w = p/q$ there is a *range* of H values, $\Delta H(p/q)$, for which w remains fixed. As a result, the plot of w versus H looks like a devil's staircase; see Figure 1 [BB 82, BB 83]. Indeed, the staircase is "complete," like the one we shall encounter later in this chapter in connection with the mode locking of two oscillators. *Complete* means that the rational plateaus in Figure 1 add up to the entire H interval. Irrational values of w occur at values of H that form a thin Cantor set, whose fractal dimension D can be determined analytically for power-law interactions of the form $J_{ij} = |i - j|^{-\alpha}$:

$$D = \frac{2}{1 + \alpha} \tag{4}$$

The plateau for $w = \frac{1}{2}$ (see Figure 1) has a relative length of 0.44, and the two intervals for $w \neq \frac{1}{2}$ are equal: $r_1 = r_2 = 0.28$. If the devil's staircase for the one-

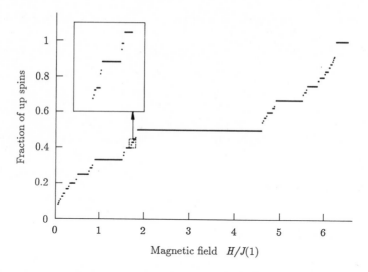

Figure 1   Fraction of up-spins as a fraction of magnetic field for an Ising spin glass [BB 82].

dimensional Ising "antiferromagnet" were exactly self-similar, the Hausdorff dimension would equal

$$D = \frac{\log N}{\log (1/r)} = \frac{\log 2}{\log (1/0.28)} = 0.54$$

Since, according to equation 4, $D = 0.\bar{6}$ for $\alpha = 2$, the staircase cannot be exactly self-similar. However, the plateau at $w = \frac{1}{3}$ (and the one at $w = \frac{2}{3}$), with $r_1 = 0.47$ and $r_2 = 0.18$, implies a fractal dimension $D = 0.59$, which is obtained from the formula for unequal remainders, $r_1 \neq r_2$:

$$r_1^D + r_2^D = 1$$

Thus, it is not unreasonable to expect the devil's staircase of Figure 1 to be asymptotically self-similar or self-affine. This is also suggested by the 10-times magnified portion of staircase shown in the insert in Figure 1. The largest plateau in the insert corresponds to $w = \frac{3}{7}$.

Quasiperiodic Spatial Distributions

The devil's staircase for Ising spins, not being precisely self-similar, is at best asymptotically self-similar. How do we approach such a staircase to test its self-similarity? Figure 1 suggests that the larger the denominator q in $w = p/q$, the smaller the locked plateau.

The physical situation would lead to the same conclusion. After all, q is the period length of the spin configuration, and intuition tells us that the larger the period length, the more tenuous the configuration, and hence the smaller the plateau.

How are the spins arranged for $w \neq \frac{1}{2}$? The answer is about the simplest formula imaginable to satisfy basic symmetries. Suppose $w = p/q$, where p and q are *coprime* integers (i.e., their largest common divisor equals 1).[4] Then p of every q spins should point up. More precisely, p of every q consecutive spins should be up. Thus, the spin pattern must be periodic with period length q. Within each period, precisely p spins are up and $q - p$ spins are down—but which are up and which are down? Obviously, for $w = \frac{3}{7}$, for example, having three adjacent spins point up and the next four down is not a minimum-energy solution. For a lower energy, the up and down spins must be better intermingled. But how? Detailed theoretical analysis shows that, with the initial condition $s_0 = +1$, the locations u_k of the up spins are given by the simple formula

$$u_k = \left\lfloor \frac{k}{w} \right\rfloor \qquad k = \ldots, -2, -1, 0, 1, 2, 3, 4, 5, \ldots \qquad (5)$$

where the floor function $\lfloor a \rfloor$ is the largest integer not exceeding a.

With $w = \frac{1}{2}$, equation 5 tells us that the position of the kth up spin is $2k$, in agreement with equation 3.

The corresponding locations d_k of the down spins are given by the set of integers complementary to the set u_k:

$$d_k = \left\lceil \frac{k}{1 - w} \right\rceil - 1 \qquad (6)$$

where the ceiling ("gallows") function $\lceil a \rceil$ means the smallest integer not smaller than a.

For $w = p/q = \frac{3}{7}$, for example, equation 5 tells us that the up spins are at location

$$u_k = \ldots, -5, -3; 0, 2, 4; 7, 9, 11; \ldots \qquad (7)$$

while the down spins, according to equation 6, are to be found at locations

$$d_k = \ldots, -4, -2; -1, 1, 3, 5; 6, 8, 10, 12; \ldots \qquad (8)$$

As can be seen, both sets u_k and d_k are periodic with period $q = 7$, in the sense

---

4. In number theory, this frequent condition is rendered as $(p, q) = 1$. In general, $(p, q) = m$ means that m is the largest common divisor of p and q.

that if $u_k = n$ for some k, then, for some other k' ($= k + 3$), $u_{k'} = u_k + 7$. Similarly, for $k' = k + 4$, $d_{k'} = d_k + 7$. The locations u_k and d_k together cover all the integers, each integer exactly once. (A simple proof for irrational w is given in my book on number theory [Schr 90].)

Equations 5 and 6 distribute the up- and down-spins as uniformly as possible under the given constraints—something we already guessed for a minimum-energy "antiferromagnet." In fact, equation 7 shows that for $w = \frac{3}{7}$, the three spacings per period between the up-spins are 2, 2, 3. The spacing pattern for the four down-spins per period for $w = \frac{3}{7}$ is—according to equation 8—2, 2, 2, 1. In general, it can be shown that, since the spacings have to be integers and add up to the period length q for both up- and down-spins, the spacings generated by equations 5 and 6 have in fact the least variance.

There is a close connection between these stable spin patterns and the motion in phase space of simple conservative dynamic systems with two degrees of freedom, such as two coupled oscillators. Depending on the nonlinear coupling strength, the motion will be periodic, quasiperiodic, or chaotic. If the two oscillator frequencies are commensurate, that is, if they are locked into a rational frequency ratio p/q, then the phase space trajectory of the system on the surface of a torus (an "inner tube") will be periodic with period q. The trajectory will close after p cycles around the torus's second dimension. Thus, a plane cut (called a *Poincaré section*) normal to the first dimension of the torus will be pierced by q distinct points with angles $\theta_1, \theta_2, \ldots, \theta_q$.

To simplify the description even further, we can replace the angles θ_k by their signs: a plus sign if $0 \leq \theta_k < w 2\pi$, say, and a minus sign for $w 2\pi \leq \theta_k < 2\pi$. Setting $\theta_0 = 0$, the successive five angles (modulo 2π) are $\theta_k/2\pi = 0$, $\frac{3}{5}, \frac{1}{5}, \frac{4}{5}, \frac{2}{5}$. The corresponding sign sequence, also called the *symbolic dynamics*, is $+ - + - +$. This is precisely the spin pattern of our antiferromagnetic Ising spin system locked in an up-spin ratio of $w = \frac{3}{5}$. Indeed, equation 5 gives the locations for the up-spins for $k = 1, 2, 3$ at $u_k = 1, 3,$ and 5. Thus, the spin pattern is $+ - + - +$. Because of the close analogy between toroidal trajectories winding around a torus in phase space and (quasi) periodic or chaotic spatial patterns (spins and quasicrystals, for example), people often refer to the ratio w as a *winding number*.

For irrational w, the spin pattern will be quasiperiodic rather than periodic. For example, for $w = (\sqrt{5} - 1)/2$, the golden mean, the up-spins, according to equation 5, will be at locations

$$u_k = 1, 3, 4, 6, 8, \ldots \qquad \text{for } k = 1, 2, 3, \ldots$$

and the down spins, according to equation 6, will be at

$$d_k = 2, 5, 7, 10, 13, \ldots$$

Note that the differences $d_k - u_k = k$. Each pair (u_k, d_k) forms a so-called Beatty pair, a winning combination in a game of Fibonacci nim (see page 53).

Beatty Sequence Spins

If the energetically favorable proportion of up-spins equals w, which in general will be different from $\frac{1}{2}$, the locations u_k of the individual up-spins are given by the *Beatty* sequence (equation 5). The locations of the down-spins are given by the *complementary* Beatty sequence (equation 6).

Instead of the Beatty sequences for the *locations* of the up- and down-spins, we can give a simple formula for the spin values themselves. For irrational w we have

$$s_m = \text{sgn}\,[w - \langle (m + 1)w \rangle_1] \tag{9}$$

where sgn is the algebraic sign function and $\langle \; \rangle_1$ stands for the fractional part.

For the golden mean, $w = (\sqrt{5} - 1)/2$, the sequence of spin signs is $+ - + + - + - + + - + + - \cdots$, which can also be obtained from the iterated "rabbit" map (see Chapter 13)

$$
\begin{aligned}
- &\to + \\
+ &\to + -
\end{aligned}
\tag{10}
$$

Starting with a single minus sign yields the following successive generations and their lengths L_i:

$-$	$L_1 = 1$
$+$	$L_2 = 1$
$+ -$	$L_3 = 2$
$+ - +$	$L_4 = 3$
$+ - + + -$	$L_5 = 5$

and so on. An equivalent rule to generate generation n is to append generation $n - 2$ to generation $n - 1$. Note that, as a consequence, the length L_n of the nth generation obeys the recursion $L_n = L_{n-1} + L_{n-2}$. With $L_1 = L_2 = 1$, this results in the Fibonacci numbers $L_n = 1, 1, 2, 3, 5, 8, 13, \ldots$.

The relative number of up-spins in generation n equals L_{n-1}/L_n, which approaches the golden mean $(\sqrt{5} - 1)/2 = 0.618\ldots$, as it should.

The Beatty sequence for $w = (\sqrt{5} - 1)/2$ leads to a one-dimensional analogue of a *quasicrystal* with a fivefold rotational symmetry (see Chapter 13). Such a symmetry is forbidden for *periodic* crystals, but was observed in 1984 when the first quasicrystal was discovered

The mapping in equation 10 is closely related to the continued fraction (CF) expansion of the golden mean $(\sqrt{5} - 1)/2 = [1, 1, 1, \ldots] = [\overline{1}]$. For another simple continued fraction with period length equal to 1, that of $w = \sqrt{2} - 1 = [\overline{2}]$, the corresponding mapping that generates the spins according to equation 9, namely, $- + - + - - + - + - - + \cdots$, is

$$- \rightarrow - +$$
$$+ \rightarrow - + - \qquad (11)$$

which, starting with a single minus sign, gives the following successive generations:

$$- \qquad\qquad L_1 = 1$$
$$- + \qquad\qquad L_2 = 2$$
$$- + - + - \qquad\qquad L_3 = 5$$
$$- + - + - - + - + - - + \qquad L_4 = 12$$

and so on. The alternate rule for generating generation n is to repeat generation $n - 1$ twice and append generation $n - 2$. Beginning with $-$ and $- +$, this yields the generations just shown, which grow in length L_n according to the recursion $L_{n+1} = 2L_n + L_{n-1}$. With $L_1 = 1$ and $L_2 = 2$, this gives the successive lengths 1, 2, 5, 12, 29, 70, 169. Note that L_{n-1}/L_n approaches the value of $w = \sqrt{2} - 1$. Also, the relative number of up-spins approaches, as it should, $w = \sqrt{2} - 1$. In fact, of the L_n spins in generation n, precisely L_{n-1} are up and $L_n - L_{n-1} = L_{n-1} + L_{n-2}$ are down. The ratio L_{n-1}/L_n approaches w as quickly as possible (for given bounds on the denominators). Note that, according to these lengths, the ratio $70/(29 + 70) = 70/99 = 0.70707\ldots$, for example, should be a good approximation to $1/\sqrt{2} = 0.70710\ldots$, as it is indeed.

When $w = \sqrt{2} - 1$ is used in equation 9, it will generate the same sequence as the iterated mapping in equation 11. Equations 5 and 6, with $w = \sqrt{2} - 1$, will generate the corresponding *locations* of the up- and down-spins.

In the context of quasicrystals, the quadratic irrational number $\sqrt{2} - 1 = [\overline{2}]$ leads to a one-dimensional model of a quasicrystal with a "forbidden" eightfold rotational symmetry, first described by Wang, Chen, and Kuo [WCK 87].

Similarly, the spin sequence for $w = [\overline{3}] = (\sqrt{13} - 3)/2$ in equation 9,

$$- - + - - + - - + - \ldots$$

can also be generated by the iterated mapping

$$- \rightarrow - - +$$
$$+ \rightarrow - - + - \qquad (12)$$

Equivalently, starting with the two initial generations $-$ and $- - +$, generation $n > 2$ is generated by repeating generation $n - 1$ three times and appending generation $n - 2$. It follows directly that the length L_n of the nth generation obeys the recursion $L_n = 3L_{n-1} + L_{n-2} = 1, 3, 10, 33, 109, 360, \ldots$ and L_{n-1}/L_n approaches $w = [\overline{3}]$.

Does the mapping in equation 12 tell us that the number of up-spins in generation n, m_n^+, divided by the total number of spins, $L_n = m_n^+ + m_n^-$, approaches the desired ratio $w = [\overline{3}]$? It does indeed. First we observe from mapping 12 that each spin (either up or down) in generation $n - 1$ generates precisely one up-spin in generation n. Therefore $m_n^+ = L_{n-1}$. Thus, the relative number of up-spins equals L_{n-1}/L_n, which, as we already saw, approaches

$$[\overline{3}] = \frac{(\sqrt{13} - 3)}{2} = 0.3027756\ldots$$

The convergence is quite rapid, too. For example,

$$\frac{L_5}{L_6} = \frac{109}{360} = 0.3027777\ldots$$

In general, the spin pattern for w equal to a periodic CF with period length 1, $w = [\overline{n}]$, is given by the iteration

$$\begin{aligned} - &\to (-)^{n-1} + \\ + &\to (-)^{n-1} + - \end{aligned} \tag{13}$$

where $(-)^{n-1}$ means a sequence of $n - 1$ minus signs.

For $w = [\overline{n}]$, we have the relation

$$\frac{1}{w} = n + w$$

The positive solution of this quadratic equation yields

$$\tau_n^+ = \frac{\sqrt{n^2 + 4} - n}{2}$$

which is called a *silver mean* because τ_n^+, like the golden mean, has a periodic continued fraction with period length equal to 1. For the special case of $n = 1$, silver turns into gold and we get the golden mean $w = \tau_n^1 = \gamma = (\sqrt{5} - 1)/2$.

If we relax the condition that the terms of the continued fraction have to be positive, then we get a second family of silver means, defined by

$$\frac{1}{\tau_n^-} = n - \tau_n^- \qquad n = 2, 3, \ldots$$

with the unique root in the interval [0, 1]

$$\tau_n^- = \frac{n - \sqrt{n^2 - 4}}{2}$$

and the continued fraction expansion

$$\tau_n^- = [n-, n-, n-, \ldots]$$

Spin patterns for these silver means, too, can be generated by simple iterated mappings. For example, for $n = 4$, we have $\tau_4^- = 2 - \sqrt{3} = 0.268\ldots$. With this τ_4^- in equation 5, we find the up-spins at $u_k = 3, 7, 11, 14, 18, \ldots$. Thus, the pattern of spins for τ_4^- is

$$- \quad - + - \quad - - + - - - + - - + -, \quad - - + - - - + \ldots \quad (14)$$

To find the spin-mapping law, we have to compute the approximants of the continued fraction: $\frac{1}{4}$, $1/(4 - \frac{1}{4}) = \frac{4}{15}$, $\frac{15}{56}$, and so on, where the denominators $(4, 15, 56, \ldots)$ are the period lengths. The length of the nth generation L_n $(n > 1)$ is given by $4L_{n-1} - L_{n-2}$, starting with $L_0 = 0$ and $L_1 = 1$. Marking off sub-sequences of these lengths by commas (see spin pattern 14) reveals the iteration law

$$- \rightarrow - - + -$$
$$+ \rightarrow - - - +$$

which, starting with a single minus sign, produces the following successive generations:

$-$	$L_1 = 1$
$- - + -$	$L_2 = 4$
$- - + - - - + - - - + - - + -$	$L_3 = 15$

and so on, in uncanny agreement with spin pattern 14.

The silver mean $\tau_4^- = 2 - \sqrt{3}$ is the basis for generating quasicrystals with a forbidden twelvefold symmetry, which were discovered in 1988 [CLK 88].

All silver means τ_N^+ (τ_N^-), where N equals an even- (odd-) index Lucas number, belong to the irrational number field $\mathbb{Q}(\sqrt{5})$ and lead to fivefold symmetric quasicrystals. The Lucas numbers, $L_n = 2, 1, 3, 4, 7, 11, 18, \ldots$, are defined by the same recursion as the Fibonacci numbers: $L_n = L_{n-1} + L_{n-2}$, but with the initial condition $L_0 = 2$, $L_1 = 1$. For $n > 1$, L_n can be obtained by rounding γ^{-n} to the nearest integer.

It is a fair guess that the general mapping law for τ_n^- is given by

$$
\begin{aligned}
- &\to (-)^{n-2} + - \\
+ &\to (-)^{n-2} +
\end{aligned}
\tag{15}
$$

where $(-)^{n-2}$ stands for a sequence of $n-2$ minuses. The reader may wish to prove this and the following equivalent law: beginning with the first two generations $-$ and $(-)^{n-2} + -$, generation k is obtained by repeating the previous generation $(n-1)$ times and appending generation $k-1$, from whose beginning generation $k-2$ has been eliminated.

The Scaling Laws for Quasiperiodic Spins

Equation 9 allows us to calculate the sign of any spin directly without recursion. For example, for $w = (\sqrt{13} - 3)/2$, the 1000th spin is $+$. On the other hand, the very fact that, for w equal to a periodic continued fraction of the form $[\bar{n}]$, our antiferromagnetic Ising spins can be calculated recursively by an iterated mapping suggests that these spin patterns must have some scaling invariance. And indeed they have—in fact, they enjoy numerous self-similarities.

Let us focus on the spin pattern for the winding number w equal to the golden mean. In $\{0, 1\}$ notation this is the "rabbit"sequence:

$$
\underline{1 \quad 0} \quad 1 \quad \underline{1} \quad 0 \quad \underline{1 \quad 0} \quad 1 \quad \underline{1 \quad 0} \quad 1 \quad \underline{1} \quad 0 \ldots
\tag{16}
$$

The infinite rabbit sequence reproduces itself if, for every 1 we encounter (beginning on the left), we hop ahead two places and strike out the third bit. For every 0, we hop only one place and eliminate the second bit. The digits retained by this mad hopping and striking-out scheme are underlined in sequence 16, and the decimated sequence does reproduce the original rabbit sequence. The pattern of long and short underscores also corresponds to the rabbit sequence, albeit by construction. And, maddeningly, the *struck-out* (nonunderlined) digits *also* reproduce the original rabbit sequence. Can the reader show why?

What the hopping scheme really does is to map $101 \to 10$ and $10 \to 1$, as can be seen from sequence 16. This "block-renaming" scheme, which we have encountered before, is in effect the reverse of the mapping $1 \to 10 \to 101$, which is the next iteration of the original mapping $(0 \to 1 \to 10)$.

The block renaming $101 \to 10$ and $10 \to 1$ corresponds to a simple scaling of the index k by a factor w. Specifically, in the formula for the up-spins (equation 5), k is replaced by $k' = k/w$ and the spin pattern is laterally transposed by one unit:

$$
u_{k'} = u_{k/w} - 1 = \left\lfloor \frac{k}{w^2} \right\rfloor - 1
$$

However, according to equation 6, $\lfloor k/w^2 \rfloor = \lfloor k/(1-w) \rfloor = \lceil k/(1-w) \rceil - 1$ are the locations of the original *down*-spins d_k. Thus, we see that in the scaled spin pattern, obtained by the block-renaming renormalization, the retained up-spins are one site to the left of the original down-spins. Indeed, as we can see from expression 16, the surviving up-spins are precisely those that have a 0 as a right neighbor. All other up-spins "die" in the construction.

We leave it to the reader to show that the surviving *down*-spins are to be found one site to the left of the original up-spin doublets 11 (whose density, like that of the surviving down-spins, is $w^3 = \sqrt{5} - 2 = 0.236 \ldots$).

Self-Similar Winding Numbers

The scaling law for the antiferromagnetic Ising spins can also be derived from the formula (equation 9) that generates the spins in the ± 1 notation:

$$s_m = \text{sgn} \left[w - \langle (m+1)w \rangle_1 \right]$$

Obviously, there is no change in the spin s_m if we add any integer—for example, $n(m+1)$—to the contents of the fractional-part brackets $\langle \ \rangle_1$:

$$s_m = \text{sgn} \left[w - \langle (m+1)(w+n) \rangle_1 \right] \tag{17}$$

For w equal to the golden mean, we have $w + 1 = 1/w$. Hence, with $n = 1$,

$$s_m = \text{sgn} \left[w - \left\langle (m+1)\frac{1}{w} \right\rangle_1 \right]$$

which tells us that the index $m + 1$ can be formally scaled by the factor $1/w^2$ without changing the spin pattern.

What other winding numbers show this kind of self-similarity? Equation (17) shows that for all w for which $(w + n) = 1/w$ we can scale $m + 1$ by a factor of $1/w^2$. Here n can be a positive or negative integer. (Note that $\langle \alpha \rangle_1 := \alpha - \lfloor \alpha \rfloor$ lies in the interval $[0, 1)$, so that for $\alpha = -4.7$, for example, $\langle \alpha \rangle_1 = \langle -4.7 \rangle_1 = 0.3$.) For positive n, these winding numbers are precisely those whose continued fraction expansion is periodic and has period length 1:

$$w = \tau_n^+ := [\bar{n}] \qquad n > 0$$

In fact, this equation can be written as

$$w = \frac{1}{n+w} \tag{18}$$

which is the relationship we need for the scaling invariance of equation 17.

The solution of the quadratic equation 18 for positive w is

$$w = \tau_n^+ = \frac{\sqrt{n^2 + 4} - n}{2} \qquad n > 0$$

as already stated.

For negative n, the two roots of this equation are outside the "legal" interval $(0, 1)$ for w. However, we can still get a self-similar solution, namely, for

$$w = \tau_n^- := [\overline{n-}] \qquad n < -1$$

The solution of this equation that lies in $(0, 1)$ is

$$w = \tau_n^- = -\frac{\sqrt{n^2 - 4} + n}{2} \qquad n < -2 \tag{19}$$

The numbers τ_n^+ and τ_n^- are the *silver means*, another generalization of the golden mean that we have previously encountered.

How does the scaling law for the spin-generating formula in equation 5 translate into a scaling law for the up-spins? In other words, what change in equation 5 is necessary so that it gives the locations of the up-spins surviving the block-decimation process for $n > 1$?

Circle Maps and Arnold Tongues

Next to the quadratic map, discussed in Chapter 12, another nonlinear law plays an important role for modeling a great many natural phenomena, the famous *circle map*:

$$\theta_{n+1} = \theta_n + \Omega - \frac{K}{2\pi} \sin(2\pi\theta_n) \tag{20}$$

Here K is a "coupling constant," which regulates the degree of nonlinearity; in fact, for $K = 0$, equation 20 is linear. The variable θ_n represents an angle, usually in the phase space of a dynamic system. The average increment per iteration of θ_n is called the *dressed winding number*, defined as

$$w := \lim_{n \to \infty} \frac{\theta_n - \theta_0}{n} \tag{21}$$

The parameter Ω in equation 20 is called the *bare winding number*. This curious nomenclature stems from the fact that the phase space in question is often a torus around which the trajectory *winds* its way. In a typical application, the *bare* winding number represents a frequency ratio, such as the resonant frequency of an oscillator, a swing, say, divided by the frequency of a periodic force acting on the resonator. The *dressed* winding number w represents the frequency ratio, usually a rational number $w = P/Q$, into which the system has been "locked" by some nonlinear coupling. Of course, for $K = 0$, $w = \Omega$ and there is no mode locking to rational winding numbers w. But for $K = 1$, yielding the *critical* circle map, the mode-locked regions cover the entire Ω interval (see Figure 7 in Chapter 7), leaving only a Cantor set of Ω values unlocked. The locked regions are called *Arnold tongues* after their discoverer, the Russian mathematician V. I. Arnol'd.

The critical circle map has a *cubic inflection point* for $\theta_n = 0$ and can be approximated by

$$\theta_{n+1} = \Omega + \frac{2\pi^2}{3}\,\theta_n^3 \tag{22}$$

for $|\theta_n| \leq 1$. Most of the results obtained for the critical circle map are in fact *universally* valid for all maps with a zero-slope cubic nonlinearity. This universality corresponds to the universality of the results for unimodal maps with a quadratic maximum. Together the quadratic map and the cubic map model many nonlinear phenomena, characterized by either a symmetric (even) nonlinearity or an *anti-symmetric* (odd) nonlinearity.

For $K > 1$, the circle map is nonmonotonic and the Arnold tongues overlap each other, giving rise to chaotic motion. Just as the transition to chaos in the quadratic map can be studied by period-doubling bifurcations, the route to chaos in the critical circle map too is analyzed in terms of orbits with increasing period lengths. But here the preferred period lengths are equal to the Fibonacci numbers F_n and the dressed winding numbers w (or $1 - w$) are ratios of adjacent Fibonacci numbers. With n going to infinity, these winding numbers approach the golden mean.

Even the most important symbolic dynamics of these two prototypical nonlinearities are similar to each other, as we shall see later in this chapter.

Figure 6 in Chapter 7 shows the dressed winding number $w = P/Q$ as a function of the bare winding number Ω for the critical circle map. It is a devil's staircase with horizontal plateaus at *all* rational values of w (as opposed to the devil's staircase based on the original Cantor set, which has plateaus only for $w = P/Q$ with $Q = 2^m$). Although not exactly self-affine, like the Cantor staircase, the mode-locking staircase shows an approximate self-affinity, as seen in the inset.

The widths of the plateaus obviously have a tendency to decrease with increasing value of Q, which is the period length of the frequency-locked motion. This is intuitively clear, because modes lock preferentially into frequency ratios involving small integers, such as the ratio of planet Mercury's orbital frequency around the sun to its spin frequency around itself, which equals $\frac{2}{3}$.

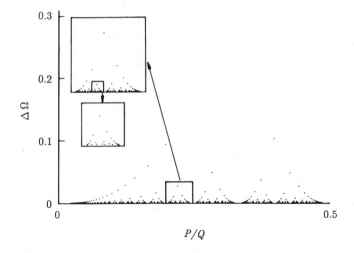

Figure 2 Width of frequency-locked intervals as a function of frequency ratio. Note self-similarity revealed by successive magnifications ratio [JBB 84].

Figure 2 shows the width of the plateaus $\Delta\Omega$ as a function of P/Q, calculated by Jensen, Bak, and Bohr [JBB 84]. Again there is a great deal of self-similarity, as shown in the insets. The scaling of $\Delta\Omega$ with Q found by these authors has an exponent $\log(\Delta\Omega)/\log Q \approx -2.292$.

The fractal defined by those values of Ω for which *no* mode locking takes place is in fact a *multi*fractal, with D_q ranging from $D_{-\infty} \approx 0.924$ down to $D_\infty = 0.5$. As explained in Chapter 9, $D_{-\infty}$ corresponds to the thinnest region of the fractal, which here is located around Ω equal to the golden mean γ, the most difficult frequency ratio to mode-lock. Shenker found the lengths r_d to scale asymptotically as $F_n^{-\delta} \sim \gamma^{n\delta}$, with $\delta = 2.1644\ldots$, for $\Omega_n = F_{n-1}/F_n$, $n \to \infty$, called the golden-mean route to chaos [She 82]. With the probabilities given by $p_n \sim \gamma^{2n}$ and $r_n \sim \gamma^{n\delta}$, one finds

$$D_{-\infty} = \lim_{n\to\infty} \frac{\log p_n}{\log r_n} = \frac{2}{\delta} \approx 0.924$$

The most concentrated region of the mode-locking fractal lies just to the right of the locked interval for the frequency ratio 0 (near $\Omega = 1/2\pi$) as it is approached by the frequency ratios $1/Q$, with $Q \to \infty$. This is the so-called harmonic series, which is comparatively easy to mode-lock. For the harmonic series, changes in dressed winding numbers are asymptotically proportional to the square root of the changes in the bare winding number—that is, $p_n \sim r_n^{1/2}$. Thus, $D_\infty = 0.5$ exactly.

The entire multifractal spectrum $f(\alpha)$ was calculated from 1024 mode-locked intervals by Cvitanović, Jensen, Kadanoff, and Procaccia [CJKP 85]. The maximum of $f(\alpha)$ equals approximately 0.868 and corresponds to the Hausdorff dimension D_0 of the multifractal underlying the mode-locking staircase.

Mediants, Farey Sequences, and the Farey Tree

In order to calculate the dimensions D_q of the mode-locking fractal and its multifractal spectrum $f(\alpha)$, some order has to be imposed on the rational numbers P/Q representing different frequency ratios. One such ordering is used in the standard proof that the rational numbers (as opposed to the irrational numbers) form a countable set. Here we need a different ordering, one that better reflects the physics of mode locking.

Suppose the parameter Ω in equation 20, the bare winding number, is such that the dressed winding number falls somewhere between $\frac{1}{2}$ and $\frac{2}{3}$ without actually locking into either one. What is the most likely locked-in frequency ratio for a nonlinear coupling strength just below the value that would cause mode locking at $\frac{1}{2}$ or $\frac{2}{3}$? It seems reasonable that it should be a frequency ratio P/Q in the interval $(\frac{1}{2}, \frac{2}{3})$ with Q as small as possible.

Indeed, this is precisely what happens in dynamic systems modeled by the circle map. Adjust the nonlinear coupling strength K and the bare winding number Ω to a point just below the crossing of the two Arnold tongues for the locked frequency ratios $\frac{1}{2}$ and $\frac{2}{3}$. The dressed winding number w for this point in the Ω-K plane must be rational because $K > 1$. In fact, the rational value P/Q that w assumes is given by $\frac{1}{2} < P/Q < \frac{2}{3}$ with Q as small as possible.

This raises an interesting mathematical question with a curious but simple answer: What *is* the ratio following $\frac{1}{2}$ and $\frac{2}{3}$ with the *smallest* denominator? If you ask a kindergartner to add $\frac{1}{2}$ and $\frac{2}{3}$, he or she may well add numerators and denominators separately and write

$$\frac{1}{2} + \frac{2}{3} = \frac{3}{5}$$

and in so doing will have discovered the looked-for "locked-in" fraction with the smallest denominator.[5]

---

5.   This is somewhat reminiscent of S. N. Bose (1894–1974), the celebrated Indian physicist, who, in deriving photon statistics, "forgot" to take account of the photon's (nonexistent) distinguishability. When *Nature* (not nature) turned his paper down, Bose wrote to Einstein, who saw the light and recognized Bose's "mistake" as the long-sought-after answer in the statistical physics of light. Bose's name has become enshrined ever since in the *Bose-Einstein* distribution, *bosons* (integer-spin particles, such as the photon) and *Bose condensation*, which gives us superconductivity and other macroscopic marvels of the microscopic quantum world.

What can such a strange strategy for forming intermediate fractions possibly mean? Physically, the frequency ratio 1/2 of two oscillators can be represented by a pulse (1) followed by a "nonpulse" (0) of the faster oscillator during every period of the slower oscillator. Thus, the frequency ratio 1/2 is represented by the sequence 101010 . . . or simply $\overline{10}$. Similarly, the frequency ratio 2/3 is represented by two 1s repeated with a period of three: $\overline{110}$.

Now, to form an intermediate frequency ratio, we simply *alternate* between the frequency ratios 1/2 (i.e., $\overline{10}$) and 2/3 (i.e., $\overline{110}$), yielding $\overline{10110}$, which represents the frequency ratio 3/5 (3 pulses during 5 clock times). So, in averaging frequency ratios, taking *mediants*, as this operation is called, is not such a strange thing after all.

In general, given two reduced fractions P/Q and P'/Q', the desired intermediate fraction is given by

$$\frac{P''}{Q''} = \frac{P + P'}{Q + Q'}$$

and is called the *mediant* by number theorists. In a penetrating analysis of Diophantine equations, John Horton Conway showed that numerators and denominators can be interpreted as the components of a two-dimensional *vector* and that the intermediate fraction with the lowest denominator is obtained by componentwise vector addition [unpublished, personal communication, 1989]. Thus, for example, the mediant of $\frac{5}{13}$ and $\frac{2}{5}$ equals $\frac{7}{18}$ (the revolutionary frequency ratio that Jupiter and Pallas selected for their gravitationally coupled orbits around the sun). (As it happens, there is not a single fraction between $\frac{5}{13}$ and $\frac{2}{5}$ with a denominator smaller than 18.) For this to be true, the two parent fractions must be sufficiently close. More precisely, they must be *unimodular*. The *modularity* of two reduced fractions P/Q and P'/Q', which measures their closeness for our purposes, is defined as the absolute difference $|QP' - PQ'|$, and unimodular fractions are those for which $|QP' - PQ'|$ equals 1.

The mediant of two fractions has the same modularity with its two parents as the parents have between them: modularity is another hereditary trait. Inheritance is a pivotal property, in self-similarity, including the self-similarities found in mode locking.

Mediants occur naturally in *Farey sequences*. A Farey sequence is defined as the sequence of fractions between 0 and 1 of a given largest denominator (called the *order* of the sequence). Thus, the Farey fractions of order 5 are (in increasing magnitude):

$$\frac{0}{1} \quad \frac{1}{5} \quad \frac{1}{4} \quad \frac{1}{3} \quad \frac{2}{5} \quad \frac{1}{2} \quad \frac{3}{5} \quad \frac{2}{3} \quad \frac{3}{4} \quad \frac{4}{5} \quad \frac{1}{1}$$

Notice that each fraction is the mediant of its two neighbors. The modularity between all adjacent fractions equals 1, but they are not uniformly spaced.

However, Riemann's famous hypothesis, concerning the zeros of his zeta function, guarantees that the spacings between adjacent fractions are relatively uniform [Schr 90].

While Farey sequences have many useful applications and nice properties, such as classifying the rational numbers according to the magnitudes of their denominators (in fact, there are entire books listing nothing but Farey fractions), they suffer from a great irregularity: the number of additional fractions in going from Farey sequences of order $n - 1$ to those of order n equals the highly fluctuating Euler's function $\phi(n)$, defined as the number of positive integers smaller than and coprime with n. For example, $\phi(5) = 4$, $\phi(6) = 2$, and $\phi(7) = 6$. A much more regular order is infused into the rational numbers by *Farey trees*, in which the number of fractions added with each generation is simply a power of 2.

Starting with two fractions, we can construct a Farey tree by repeatedly taking the mediants of all numerically adjacent fractions. For the interval [0, 1], we start with $\frac{0}{1}$ and $\frac{1}{1}$ as the initial fractions, or "seeds". The first five generations of the Farey tree then look as follows:

$$\frac{0}{1} \qquad\qquad\qquad\qquad\qquad\qquad\qquad\qquad\qquad\qquad \frac{1}{1}$$

$$\frac{1}{2}$$

$$\frac{1}{3} \qquad\qquad\qquad\qquad\qquad\qquad \frac{2}{3}$$

$$\frac{1}{4} \qquad\qquad \frac{2}{5} \qquad\qquad \frac{3}{5} \qquad\qquad \frac{3}{4}$$

$$\frac{1}{5} \quad \frac{2}{7} \quad \frac{3}{8} \quad \frac{3}{7} \quad \frac{4}{7} \quad \frac{5}{8} \quad \frac{5}{7} \quad \frac{4}{5}$$

Each rational number between 0 and 1 occurs exactly once somewhere in the infinite Farey tree. The tree's construction reflects precisely the interpolation of locked frequency intervals in the circle map by means of mediants. The Farey tree is therefore a kind of mathematical skeleton of the Arnold tongues.

The location of each fraction within the tree can be specified by a binary address, in which 0 stands for moving to the left in going from level n to level $n + 1$ and 1 stands for moving to the right. Thus, starting at $\frac{1}{2}$, the rational number $\frac{3}{7}$ has the binary address 011. The complement of $\frac{3}{7}$ with respect to 1 (i.e., $\frac{4}{7}$) has the complementary binary address: 100. This binary code for the rational numbers is useful in describing coupled oscillators.

Note that any two *numerically* adjacent fractions of the tree are unimodular. For example, for $\frac{4}{7}$ and $\frac{1}{2}$, we get $2 \cdot 4 - 1 \cdot 7 = 1$.

Some properties of the Farey tree are particularly easy to comprehend in terms of continued fractions, which for numbers w in the interval $[0, 1]$ look as follows:

$$w = \cfrac{1}{a_1 + \cfrac{1}{a_2 + \cfrac{1}{a_3} \cdots}}$$

but are more conveniently written as $w = [a_1, a_2, a_3, \ldots]$, where the a_k are positive integers. Irrational w have nonterminating continued fractions. For quadratic irrational numbers the a_k will (eventually) repeat periodically. For example, $1/\sqrt{3} = [1, 1, 2, 1, 2, 1, 2, \ldots] = [1, \overline{1, 2}]$ is preperiodic and has a period of length 2; $1/\sqrt{17} = [\overline{8}]$ has period length 1 and $1/\sqrt{61}$ has period length 11. (It is tantalizing that no simple rule is known that predicts period lengths in general.)

Interestingly, for any fraction on level n of the Farey tree, the sum over all its a_k equals n:

$$\sum_k a_k = n \qquad n = 2, 3, 4, \ldots$$

We leave it to the reader to prove this equation (by a simple combinatorial argument, for example).

There is also a direct way of calculating, from each fraction on level $n - 1$, its two neighbors or direct descendants on level n. First write the original fraction as a continued fraction in two different ways, which is always possible by splitting off a 1 from the final a_k. Thus, for example, $\frac{2}{5} = [2, 2] = [2, 1, 1]$. Then add 1 to the last term of each continued fraction; this yields $[2, 3] = \frac{3}{7}$ and $[2, 1, 2] = \frac{3}{8}$, which are indeed the two descendants of $\frac{2}{5}$.

Conversely, the close parent of any fraction (the one on the adjacent level) is found by subtracting 1 from its last term (in the form where the last term exceeds 1, because $a_k = 0$ is an illegal entry in a continued fraction). The other (distant) parent is found by simply *omitting* the last term. Thus, the two parents of $\frac{3}{7} = [2, 3]$ are the close parent $[2, 2] = \frac{2}{5}$ and the distant parent $[2] = \frac{1}{2}$. (But which parent is greater, in general—the close or the distant one? And how are mediants calculated using only continued fractions?)

Interestingly, if we zigzag down the Farey tree from its upper right ($\frac{1}{1} \rightarrow \frac{1}{2} \rightarrow \frac{2}{3} \rightarrow \frac{3}{5} \rightarrow \frac{5}{8}$, and so on), we land on fractions whose numerators and denominators are given by the Fibonacci numbers F_n, defined by $F_n = F_{n-1} + F_{n-2}$; $F_0 = 0$, $F_1 = 1$. In fact, on the nth zig or zag, starting at $\frac{1}{1}$, we reach the fraction F_{n+1}/F_{n+2}, which approaches the golden mean $\gamma = (\sqrt{5} - 1)/2 = 0.618 \ldots$ as $n \rightarrow \infty$ [Schr 90]. (Starting with $\frac{0}{1}$ we land on the fractions F_n/F_{n+2}, which converge on $\gamma^2 = 1 - \gamma$.) The binary address of γ in the Farey tree is $101010\ldots$.

The continued fraction expansions of these ratios F_n/F_{n+1} have a particularly simple form. For example,

$$\frac{F_3}{F_4} = \frac{2}{3} = [1,\, 1,\, 1]$$

and in general

$$\frac{F_n}{F_{n+1}} = [1,\, 1,\, \ldots,\, 1] \qquad \text{(with } n \text{ 1s)}$$

Obviously, continued fractions with small a_k converge relatively slowly to their final values, and continued fractions with only 1s are the slowest converging of all. Since

$$\gamma = \lim_{n \to \infty} \frac{F_n}{F_{n+1}} = [1,\, 1,\, 1,\, \ldots] = [\bar{1}]$$

where the bar over the 1 indicates infinitely many 1s, the golden mean γ has the most slowly converging continued fraction expansion of all irrational numbers. The golden mean γ is therefore sometimes called (by physicists and their ilk) "the most irrational of all irrational numbers"—a property of γ with momentous consequences in a wide selection of problems in nonlinear physics, from the double swing to the three-body problem.

Roughly speaking, if the frequency ratio of two coupled oscillators is a rational number P/Q, then the coupling between the driving force and the "slaved" oscillator is particularly effective because of a kind of a resonance: every Q cycles of the driver, the same physical situation prevails so that energy transfer effects have a chance to build up in resonancelike manner. This resonance effect is strong, of course, particularly if Q is a *small* integer. This is precisely what happened with our moon: resonant energy transfer between the moon and the earth by tidal forces slowed the moon's spinning motion until the spin period around its own axis locked into the 28-day cycle of its revolution around the earth. As a consequence the moon always shows us the same face, although it wiggles ("librates") a little.

Similarly, the frequency of Mercury's spin has locked into its orbital frequency at the rational number $\frac{3}{2}$. As a consequence, one day on Mercury lasts two Mercury *years*. (And one day—in the distant future, one hopes—something strange like that may happen to Mother Earth!)

The rings of Saturn, or rather the gaps between them, are another consequence of this resonance mechanism. The orbital periods of any material (flocks of ice and rocks) in these gaps would be in a rational resonance with some periodic force (such as the gravitational pull from one of Saturn's "shepherding" moons). As a consequence, even relatively weak forces have a cumulatively significant effect over long time intervals, accelerating any material out of the gaps.

For rational frequency ratios with large denominators Q, such a resonance effect would, of course, be relatively weak, and for *irrational* frequency ratios, resonance would be weaker still or absent.

For strong enough coupling, however, even irrational frequency ratios might be affected. But there is always one irrational frequency ratio that would be least disturbed: the golden mean, because, in a rational approximation to within a certain accuracy, it requires the largest denominators Q. This property is also reflected in the Farey tree: on each level n the two fractions with the largest denominators are the ones that equal F_{n-1}/F_{n+1} and F_n/F_{n+1}, which for $n \to \infty$ approach $\gamma^2 = 0.382\ldots$ and $\gamma = 0.618\ldots$, respectively. (Conversely, the fractions with the smallest Q on a given level of the Farey tree are from the harmonic series $1/Q$ and $1 - 1/Q$.)

Another way to demonstrate the unique position of the golden mean among all the irrational numbers is based on the theory of rational approximation, an important part of number theory. For a good rational approximation, one expands an irrational number w into a continued fraction and terminates it after n terms to yield a rational number $[a_1, a_2, \ldots, a_n] = p_n/q_n$. This rational approximation to w is in fact the best for a given maximum denominator q_n. For example, for $w = 1/\pi = [3, 7, 15, 1, 293, \ldots]$ and $n = 2$, we get $p_n/q_n = 7/22$, and there is no closer approximation to $1/\pi$ with a denominator smaller than 22.

Now, even with such an optimal approximation as afforded by continued fractions, the differences for the golden mean γ

$$\left| \gamma - \frac{p_n}{q_n} \right|$$

exceed c/q_n^2 (where c is a constant that is smaller than but arbitrarily close to $1/\sqrt{5}$) for *all* values of n above some n_0. And this is true only for the golden mean γ and the "noble numbers" (defined as irrational numbers whose continued fractions end in all 1s). Thus, in this precise sense, the golden mean (and the noble numbers) keep a greater distance from the rational numbers than does any other irrational number. Small wonder that the golden mean plays such an important role in synchronization problems.

The golden mean is also visible in visual perception (see Figure 3). For a computer-generated image of a "sunflower" using the *golden angle* $\Delta\phi = 360° \gamma \approx 222.5°$ as the angular increment in the placement (r_n, ϕ_n) of successive seeds, where

$$(r_n, \phi_n) = (c \cdot r_{n-1}, \phi_{n-1} + \Delta\phi)$$

we get a realistic image of the sun flower's seed pattern, which uses the golden

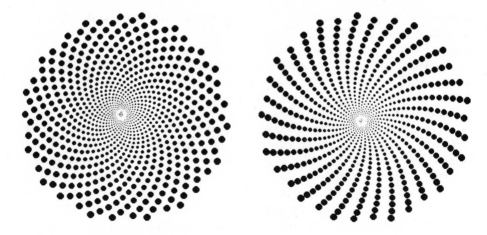

Figure 3   The golden angle in visual perception. (Courtesy T. Gramss, after [RS 87])

angle in its construction (the left part of Figure 3) [RS 87]. But for angular increments $\Delta\phi$ that differ by just 0.04 percent from the golden angle (222.4°), the human eye perceives pronounced spirals (the right part of Figure 3)—a psychovisual mode-locking phenomenon!

The Golden-Mean Route to Chaos

For the critical circle map

$$\theta_{n+1} = \theta_n + \Omega - \frac{1}{2\pi} \sin (2\pi\theta_n) \tag{23}$$

the sequence of the locked-in frequency ratios P/Q equal to the ratio of successive Fibonacci numbers $F_{n-1}/F_n = [1, 1, \ldots, 1]$ is in many respects the most interesting route to aperiodic behavior and deterministic chaos of the variable θ_n. In the transition to chaotic motion, these frequency ratios and equivalent ones, such as $F_{n-2}/F_n = [2, 1, 1, \ldots, 1]$, are usually the last to remain unaffected as the degree of nonlinear coupling is increased. *Chaotic* means, as always, that initially close values of θ will diverge exponentially so that all predictability is lost as the system evolves in time.

In the Farey-tree organization of the rational numbers, introduced in the previous section, the ratios F_{n-1}/F_n or F_{n-2}/F_n lie on a zigzag path approaching the golden mean γ or its square, $1 - \gamma = \gamma^2$, respectively. Each fraction is the mediant of its two predecessors. For example, the sequence F_{n-2}/F_n, beginning with $\frac{0}{1}$ and $\frac{1}{2}$, equals $\frac{1}{3}, \frac{2}{5}, \frac{3}{8}, \frac{5}{13}, \frac{8}{21}, \ldots$. The corresponding continued fractions, beginning with $\frac{1}{2}$, keep adding 1s: [2], [2, 1], [2, 1, 1], [2, 1, 1, 1], [2, 1, 1, 1, 1], and so on to $[2, \bar{1}] = \gamma^2$.

The parameter value Ω_n that gives a dressed winding number equal to the frequency ratio F_{n-2}/F_n has to be determined numerically. A simple calculator program that adjusts Ω so that, for $\theta_0 = 0$, $\theta_{F_n} = F_{n-2}$ yields the following approximate parameter values:

$$\Omega(\tfrac{1}{2}) = 0.5$$
$$\Omega(\tfrac{1}{3}) \approx 0.3516697$$
$$\Omega(\tfrac{2}{5}) \approx 0.4074762$$
$$\Omega(\tfrac{3}{8}) \approx 0.3882635$$
$$\Omega(\tfrac{5}{13}) \approx 0.3951174$$
$$\Omega(\tfrac{8}{21}) \approx 0.3927092$$
$$\Omega(\tfrac{13}{34}) \approx 0.3935608$$

and so on, converging to $\Omega_\infty \approx 0.3933377$.

These parameter values give rise to superstable orbits because the iterates θ_n include the value $\theta_n = 0$ for which the derivative of the critical circle map vanishes. These Ω values therefore correspond to the superstable values R_n of the quadratic map, and Ω_∞ corresponds to R_∞.

Is there a universal constant, corresponding to the Feigenbaum constant, which describes the rate of convergence of the parameter values $\Omega_n :=$ $\Omega(F_{n-2}/F_n)$ to Ω_∞ as n goes to infinity? Numerical evidence suggests that there is, and that the differences between successive values of Ω_n scale with an asymptotic factor:

$$\frac{\Omega_{n-1} - \Omega_n}{\Omega_n - \Omega_{n+1}} \to \delta$$

with $\delta = -2.8336\ldots$, which thus corresponds to the Feigenbaum constant $4.6692\ldots$. (The minus sign signifies that successive differences alternate in sign.)

Other self-similar scaling behaviors can be observed in the iterates of the variable θ_n. For example, for $\Omega = \Omega(F_{n-2}/F_n)$ the differences $\theta_{F_{n-1}} - F_{n-3}$ converge to 0 in an asymptotically geometric progression:

$$\theta_{F_{n-1}} - F_{n-3} \approx \alpha^n$$

with $\alpha = -1.288575\ldots$, which corresponds to the scaling parameter $-2.5029\ldots$ for the iterated variable of the quadratic map.

Both α and δ are universal for maps with a zero-slope cubic inflection point. This result follows from a renormalization theory for the golden-mean transition to chaos of such maps. The functional equation of the fixed-point function for the renormalized cubic map is

$$f(x) = \alpha f(\alpha f(x/\alpha^2))$$

(The corresponding functional equation for the period-doubling transition of the quadratic map looks rather similar: $g(x) = \alpha g(g(x/\alpha))$; see page 274).

Another similarity between these two prototypical transitions to chaos is in their symbolic dynamics. For $\Omega = \Omega(F_{n-2}/F_n)$ and $K = 0$,

$$\theta_n = \theta_0 + n \frac{F_{n-2}}{F_n}$$

We consider $\theta_n \bmod 1$ in the interval $(-0.5, 0.5]$ and write L for $\theta_n < 0$, C for $\theta_n = 0$, and R for $\theta_n > 0$. For $\theta_0 = 0$, we then get the following symbolic dynamics as $n \to \infty$:

$$CRLRRLRLRRLRRL \ldots$$

Following the initial C, this sequence is, of course, none other than the familiar rabbit sequence in which 1 has been replaced by R and 0 by L. For $K = 1$, the actual iterates are different from those for $K = 0$, but the symbolic dynamics are given by the same sequence.

As we know from the discussion of quasicrystals in Chapter 13, this sequence can be constructed from the iteration $0 \to 1$, $1 \to 10$ or, in our present alphabet, $L \to R$, $R \to RL$. Again there is a great similarity with the period-doubling transition for which the symbolic dynamics are generated from the iteration $L \to RR$, $R \to LR$. In fact, these two transitions were treated by a unified renormalization theory by Procaccia, Thomae, and Tressor [PTT 87].

For finite n, the symbolic dynamics can be obtained from the formula for Ising spin positions (see pages 323–324). Thus, for the dressed winding number $P/Q = 3/8$, the positions of the L's are given by equation 6 for the down-spins with $1 - w = 3/8$, namely, 2, 5, 7. Thus, the superstable orbit with frequency ratio 3/8 is $CRLRRLRL$, which is the initial eight-term segment of the infinite sequence.

For the winding number $P/Q = 2/5$, the dynamics computed in this manner are $CRLRL$, which differs in the last letter from the corresponding letter in the infinite sequence. In general, for $Q = F_{2k+1}$, the last letter is L and not R, as in the infinite sequence. However, this minor blemish is self-inflicted. It is easily removed by considering $\theta_n \bmod 1$ not in the interval $(-0.5, 0.5)$ but in the slightly shifted interval $(-x, 1 - x)$, where $x \approx 0.4461583$ is the solution of the transcendental equation $x = \Omega_\infty + (\sin 2\pi x)/2\pi$.

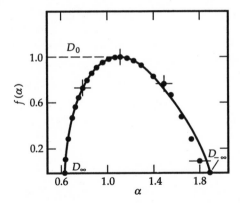

Figure 4 Multifractal spectrum of the critical circle map at the golden-mean winding number [JKLPS 85].

The aperiodic iterates θ_n of the critical circle map, for the dressed winding number equal to the golden mean γ, form a multifractal with a spectrum of singularities $f(\alpha)$ shown in Figure 4. The same spectrum obtains for all equivalent numbers, that is, all those irrational numbers whose continued fraction ends in all 1s, the so-called *noble numbers* (see Appendix B). This spectrum was computed from the periodic orbit with period length $Q = 2584 = F_{18}$. The distances $r_k = \theta_{k+F_{17}} - \theta_k \bmod 1$ were taken as the length scales of the multifractal, and p_k was set equal to $1/2854$. The dimension $D_{-\infty}$ for the most rarefied region was already obtained by Shenker from the abovementioned scaling exponent $\alpha = -1.288575\ldots$ as

$$D_{-\infty} = \frac{\log \gamma}{\log(-1/\alpha)} \approx 1.898$$

The most concentrated region of the multifractal scales as α^3, giving the smallest dimension D_{∞} as $D_{-\infty}/3 \approx 0.6327$ [She 82].

A selection of the applications related to spatial or temporal mode locking and chaos are listed in the references. These relate to acoustics [LP 88], solar system dynamics [Las 89, WPM 83, AS 81], physical chemistry [BB 79], astrophysics [HLSTLW 86], solid-state electronics [MM 86, BBJ 84, GW 87, CMP 87], cardiology [CJ 87], turbulence [JKLPS 85, FHG 85], and nonlinear mechanical oscillators [AL 85, Moo 84].

15
C H A P T E R

Percolation: From Forest Fires to Epidemics

Observation always involves theory.
—EDWIN HUBBLE

The function of an expert is not to be more right than other people, but to be wrong for more sophisticated reasons.
—DAVID BUTLER

Percolation permeates nature and man-made devices in many modes. In a coffee percolator, water seeps, or percolates, through the ground beans to emerge as drinkable coffee at the output spout.

To remain at the breakfast table, when you boil an egg long enough, its protein bonds will link up, and before too long the bonds will percolate through the entire egg to solidify it—so that you can eat it safely with a spoon and without spillage.

By contrast—and happily—an *epidemic* does not always percolate through an entire population. There is a *percolation threshold* below which the epidemic has died out before most of the *people* have. And an undercooked egg, too, is below the percolation threshold.

On a grander scale, percolation theory, eloquently expounded by Dietrich Stauffer [Sta 85], has something to offer for a better understanding of the formation of galaxies and clusters of galaxies. And at the other extreme, percolation has infiltrated even one of the tiniest scales: atomic nuclei. Their fragmentation is now being analyzed as a percolation process [Cam 86].

Another famous and preferred paradigm for percolation is a forest fire. If we ignite a few trees, will the whole forest burn down, or will most trees still be standing by the time the fire has stopped?

At the percolation threshold (clearly exceeded in the 1988 conflagrations in Yellowstone National Park), statistical self-similarities abound, and it is these self-similarities that make the mathematical treatment tractable and lead to simple scaling laws at or near the threshold.

Percolation theory is also a good preparation for the study of more complicated physical phenomena, such as phase transitions in magnetic materials and in thermodynamics in general. For example, the correlation length of the spin directions in a dilute "ferromagnet" becomes infinite at the percolation threshold, called the *critical* or *Curie* point in physics. This means that *clusters* of magnetic domains as large as the sample appear. In fact, clusters of *all* sizes, or length scales, arise, and these clusters are self-similar. In the entire range, from atomic distances to the size of the sample, clusters look similar and become stochastically indistinguishable when scaled to the same size.

Below the percolation threshold (*above* the Curie point for magnets), clusters of only finite size exist: the coffee does not drip through, and the chunk of iron is only "paramagnetic." But above the percolation threshold (below the Curie point), infinite clusters are common, with well-known consequences, depending on the application: a forest fire will spread to the other end of the forest, epidemics become pandemics, and iron (ironically?) becomes a ferromagnet. And near the percolation threshold, self-similarity reigns supreme!

Let us take a closer look at one of the hotter paradigms in percolation theory: forest fires, often fought but still ablaze.

Critical Conflagration on a Square Lattice

Assume, for simplicity, that a forest can be modeled as a square point lattice in which the lattice points are independently occupied by trees with probability $p < 1$ (see Figure 1). Now ignite the lowest row of trees and watch how the fire spreads as a digital clock ticks along in discrete time.

We assume that a burning tree will ignite all nearest-neighbor trees after one unit of time. After one more time unit of burning, a tree is burned out.

Elaborate computer simulations [Sta 85] have confirmed the obvious: below a critical tree density, p_c, the fire dies out before reaching the other edge of the forest, the uppermost rows in Figure 1. By contrast, for $p > p_c$, the fire will reach the far edge (and would threaten more trees if the forest were longer).

While $p \ll p_c$ would be a very safe forest and $p \approx 1$ a natural powder keg, the most interesting things happen *near the percolation threshold*, that is, for $p \approx p_c$ or for

$$\varepsilon := \frac{p - p_c}{p} \ll 1$$

It turns out that for $\varepsilon \ll 1$, the crucial variables obey simple scaling laws reflecting the self-similarity of percolation near the threshold, or *critical point* (to borrow a term from the physics of phase transitions).

Let us call the number of trees in the nth row of the lattice that have burned down at time t, divided by the average number of trees per row, $Z(n, t, \varepsilon)$. Here n and t are assumed to be large compared to 1. Numerous numerical analyses [Sta 85] have suggested that near the threshold (i.e., $\varepsilon \ll 1$), the *order parameter* Z is a (generalized) homogeneous function of its arguments:

$$Z(n, t, \varepsilon) = \frac{Z(\lambda^{a_n}n, \lambda^{a_t}t, \lambda^{a_\varepsilon}\varepsilon)}{\lambda} \qquad n \gg 1, t \gg 1, \varepsilon \ll 1 \qquad (1)$$

That is, Z is some *universal* function that scales as shown, with three *scaling exponents*: a_n, a_t, and a_ε [Gri 89]. What else can we say about this important function? If we wait long enough (i.e., as $t \to \infty$) and go far enough away (as $n \to \infty$), equation 1 becomes

$$Z(\infty, \infty, \varepsilon) = \frac{Z(\infty, \infty, \lambda^{a_\varepsilon}\varepsilon)}{\lambda} \qquad (2)$$

Postulating a power-law dependence of $Z(\infty, \infty, \varepsilon)$ on ε:

$$Z(\infty, \infty, \varepsilon) = \text{const} \cdot \varepsilon^\beta \qquad (3)$$

we obtain, from equation 2, $\varepsilon^\beta = (\lambda^{a_\varepsilon}\varepsilon)^\beta/\lambda$, or $a_\varepsilon = 1/\beta$. The exponent β is called a *critical exponent*, and we have just succeeded in relating it to one of the scaling exponents, a_ε.

Next, we introduce two more parameters: a characteristic length (e.g., the correlation length) ξ, and a characteristic time θ. Both ξ and θ are known to diverge to infinity as ε goes to 0 according to simple power laws:

$$\xi = \text{const} \cdot \varepsilon^{-\nu} \qquad (4)$$

and

$$\theta = \text{const} \cdot \varepsilon^{-\delta} \qquad (5)$$

We subsume the dependence of $Z(n, t, \varepsilon)$ on ε into the characteristic quantities and write, tentatively,

$$Z(n, t, \varepsilon) = n^x g\left(\frac{n}{\xi}, \frac{t}{\theta}\right) \qquad (6)$$

where the function g depends only on two variables. The exponent x must equal $-\beta/\nu$ for the original scaling law (equation 1) to hold.

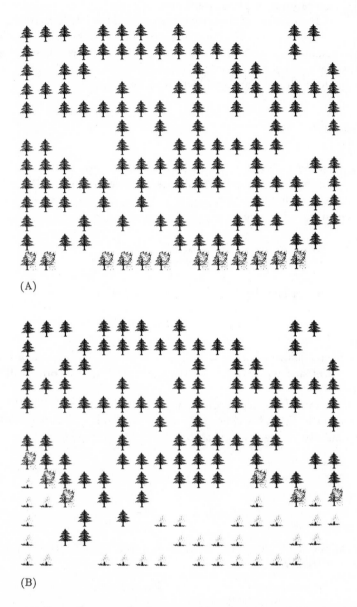

(A)

(B)

Figure 1 (A) Square lattice randomly occupied by trees at percolation threshold. Lowest row of trees has been ignited. (B) A little later: nearby trees have caught fire. (C) The conflagration has reached the upper edge of the forest. (D) The fire has died out, and so have most of the trees. (Courtesy of H. Behme).

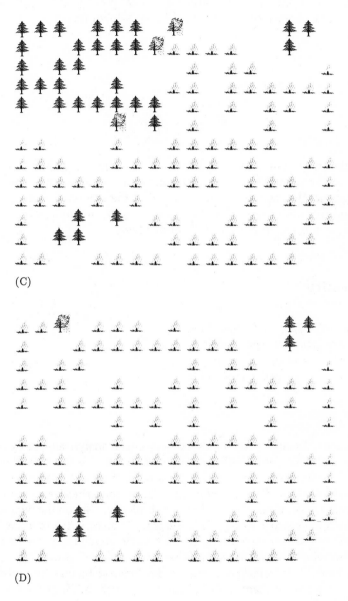

(C)

(D)

Figure 1 Continued

We have now related all three scaling exponents in equation 1 to three critical exponents: β, ν, δ, which describe how Z and the length and time parameters, ξ and δ, scale with ε near the critical point ($\varepsilon \ll 1$).

Defining a characteristic number of burned-down trees,

$$\zeta := n^{-\beta/\nu}$$

we can write equation (6) as

$$\frac{Z(n, t, \varepsilon)}{\zeta} = g\left(\frac{n}{\xi}, \frac{t}{\theta}\right) \tag{7}$$

which is the most symmetric and useful way of expressing the power-law dependence of Z on n and t and, with equations 3 to 5, on ε. These equations tell us, for example, that as we change p, that is, ε, the new values of $Z(n, t, \varepsilon)$ can be obtained from the *same* "universal" function g by multiplication with $n^{-\beta/\nu}$ and scaling n and t with $\varepsilon^{-\nu}$ and $\varepsilon^{-\delta}$, respectively.

Universality

For quite a while I have set for myself the
rule if a theoretician says "universal" it
just means pure nonsense.

        —WOLFGANG PAULI

The "critical exponents" ν and δ must be determined analytically, or by computer simulation, and the critical reader is invited to try this on his or her home computer. The surprising result is that, for a wide variety of problems in physics, chemistry, biology, and many other disciplines, the critical exponents do *not* depend upon the details of the situation but, typically, only on the dimensionality of the embedding space (e.g., two dimensions for a square lattice) and the "degrees of freedom" of the variable considered—for example, 2 in the case of a spin (or tree) system where spins (trees) are either up or (burned) down.

This kind of universality is one of the liveliest themes in contemporary physics, giving rise to many burning questions, such as, How does a specified random walk diverge on a *fractal* lattice, like the Sierpinski gasket, for example, near the critical point (percolation threshold)? How is electricity conducted on fractal networks? or, How does the *speed* of a forest fire depend on the density p of the trees?

According to equations 4, 5, and 7, Z/ζ has a fixed value if n changes with time t as

$$n = \text{const} \cdot t^{\nu/\delta}$$

Thus, the mean propagation speed of the fire defined, for example, by observing the progress in time of the fire's front scales as $t^{\nu/\delta - 1}$. By counting newly ignited trees in a computer simulation (to conserve forests and oxygen), the ratio of two of the critical exponents, ν/δ, can easily be determined [ASS 86].

Figure 2A shows one of the results of a simulation by Albinet and coworkers in which the mean position of the fire's cutting edge is plotted versus time in log-log coordinates. In this simulation, the lattice had a size of 200 times 200 points and the density of trees was at the critical value for a square lattice, $p_c \approx 0.593$. Thus, there were a total of 23,720 trees to burn or not to burn.

(A)

(B)

Figure 2 (A) Mean position of a fire's cutting edge as a function of burning time. Straight line, based on power-law postulate, has slope 0.87. (B) Number of burned trees as a function of time. Slope is 0.79 [ASS 86].

The straight line conforms to the power-law postulate, and its slope, v/δ, equals 0.87. Thus, not surprisingly, the fire spreads faster than a diffusion process (slope 0.5), but slower than on a fuse cord (slope 1.0).

The deviations from the straight line for very short times are residual effects of how the fires were laid, and those for very long times are size effects due to "saturation" (where all trees have burned down).

The total number of burned trees,

$$N_b(t, \varepsilon) = \text{const} \cdot \sum_n Z(n, t, \varepsilon)$$

scales as $t^{(v - \beta)/\delta}$. Thus simple unrestrained counting gives the ratio of $(v - \beta)/\delta$, or β/δ once v/δ has already been determined. Figure 2B shows the result of a computer experiment. Again, after initial effects have died out, the number of burned-down trees as a function of time is a straight line in a log-log plot with a slope of $(v - \beta)/\delta = 0.79$.

The third critical exponent, β, is determined by exploiting equation 3, that is, counting the dead trees after the fire has stopped. The simulation results turned (burned?) out to be rather prone to sampling error: $\beta = 0.12 \pm 0.03$; the theoretical value [Sta 85] is $\beta = \frac{5}{36} \approx 0.139$.

The exponent β is rather small, as would be expected if the fraction of trees that eventually burn down does not depend on ε as strongly as the speed of the fire does.

Another exponent that can be calculated analytically is v, which governs the correlation length (equation 4); it equals $\frac{4}{3}$ exactly. With $v = \frac{4}{3}$ and $v/\delta = 0.87$, the critical *time* exponent δ equals 1.533, close to the value found by Peter Grassberger [Gra 85]. Thus, as ε goes to zero, characteristic times diverge more rapidly than the correlation lengths. This makes sense, because for p smaller than but near p_c, that is, $\varepsilon \ll 1$, there may be a long period of time in which the fire keeps burning after its outer perimeter has stopped advancing much further. As firefighters know all too well, fires spread backward as well as forward.

It is interesting to note that all critical exponents, v, δ, and β, were found to be independent of the size of the interacting neighborhood; their values are indistinguishable for the following cases: 4 nearest neighbors on the square lattice, 8 nearest and next-nearest neighbors, and 24 neighbors in a 5 × 5 square.

This invariance illustrates what is meant by *universality*; the critical exponents depend only on the embedding dimension ($d = 2$) and the degrees of freedom (also 2) for all three coordination numbers studied.

However, the critical *densities* p_c do differ. Experimental values are $p_c = 0.592745, 0.407355$, and 0.168, respectively, for these three different co-ordination numbers. This is not surprising, because if a fire can jump not just to the nearest trees, but also to the second-, third-, fourth-, and fifth-nearest neighbors, wider gaps can exist in the forest without stopping the fire.

The Critical Density

In Figure 3, the average termination times of the fire t_∞ are plotted against the tree density p for a 300 by 300 square lattice. The divergence near $p = p_c \approx 0.593$ is quite pronounced and follows the theoretical expectation $t_\infty = \text{const} \cdot |\varepsilon|^{-\tau}$ with $\tau \approx 1.5$ [Sta 85].

Similar behavior, with the same value for τ, is found for triangular lattices, except that $p_c \approx 0.5$, in accordance with the exact theoretical prediction for the critical density, $p_c = 0.5$.

The triangular lattice is one of the lattices for which an analytical value is available. The Bethe lattice (see Chapter 16), called a Cayley tree in graph theory, is another instance: for z nearest neighbors, the percolation threshold p_c equals precisely $1/(z-1)$. It is clear from Figure 3 that computer experiments with varying p are a good way to determine p_c: at the critical point $p = p_c$, many parameters show a sharp peak. In physics such peaks (of specific heat or magnetic susceptibility, for example) as a function of temperature signal second-order phase transitions. Indeed, percolation *is* a phase transition, albeit much cleaner and clearer than the "average" thermodynamic phase transition, which can be very "mean" to treat indeed.

The Fractal Perimeters of Percolation

Does the forest fire advance in a straight front like a Greek phalanx, or is its cutting edge more fingerlike? The front is in fact fractal, complete with a Hausdorff

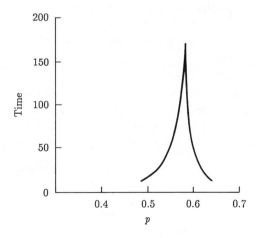

Figure 3   Average termination time of fires as a function of tree density. Note divergence near critical density $P \approx 0.593$. [Sta 85].

dimension D that lies between 1 and 2. A seemingly related phenomenon is *invasion percolation*, also known as "fingering" in oil exploration, where it has been much touted, especially when oil became "temporarily" scarce in the 1970s (see Chapter 9). However, fingering results from an instability at the interface of two liquids. By contrast, the fractal fire front stems from a connectivity paradigm: the adjacency of trees.

Finite-Size Scaling

What does computer simulation of forest fires teach us about the fractal dimension f of the perimeter? If we define the perimeter as the number F of burned sites which border an unburned site and plot this number as a function of the size of the lattice L, we find a simple power law, called *finite-size scaling* [Sta 85]:

$$F = \text{const} \cdot L^f$$

with $f = 1.75$ for all three neighborhoods—a value uncomfortably close to 2 for a *perimeter*, which, topologically, has only one dimension. (Note, though, that our definition of *perimeter* includes the boundaries of internal pockets of unburned trees.)

Another fractal dimension, $\bar{d}$, describes the total number M of burned trees. If all trees burned down, or even if just a fixed fraction of trees burned down, M would be proportional to the total number of trees or the area of the lattice: $M = \text{const} \cdot L^2$. But that is not what happens near the percolation threshold. In fact, we find another power law describing the finite-size scaling:

$$M = \text{const} \cdot L^{\bar{d}} \tag{8}$$

with $\bar{d} \approx 1.9$, somewhat less than 2 because pockets of unburned trees persist. After the fire has stopped, the *burned-down* trees form a kind of two-dimensional sponge (or Swiss cheese) with many holes on many scales.

Stauffer [Sta 85] also cites an interesting relationship for the difference between the fractal dimension $\bar{d}$ and the embedding dimension d, and the critical exponents β and v:

$$d - \bar{d} = \frac{\beta}{v} \tag{9}$$

For $d = 2$, $\beta = \frac{5}{36}$, and $v = \frac{4}{3}$, this relation gives $\bar{d} = \frac{91}{48}$, in excellent agreement with the experimental data; see Figure 4.

Figure 4   Size of largest cluster at the percolation threshold of the triangular lattice plotted against lattice size. The slope of the straight line corresponds to the theoretical fractal dimension $\frac{91}{48}$ [Sta 85].

Of course, far from the critical point there is little dependence of the number of burned-down trees on ε. Thus, the exponent β (see equation 3) would tend to zero so that, with equation 9, $\bar{d} \rightarrow d$, as expected, because forest fires are fractal only near the critical point $\varepsilon \ll 1$.

In fact, far from the critical point, the correlation length is much, much smaller than the forest (or the magnet). Thus, M does not vary as $L^{\bar{d}}$, as in equation 8, with a fractal exponent $\bar{d}$, but in a completely "Euclidean" manner: $M = \mathrm{const} \cdot L^d$, where d is the Euclidean dimension of the embedding space. In short: Percolation is a fractal phenomenon only near the critical point; above or below, it shows classical Euclidean behavior.

Still another fractal dimension, $\tilde{d}$, can be defined by the penetration time t_∞ as a function of forest size L;

$$t_\infty = \mathrm{const} \cdot L^{\tilde{d}}$$

Numerical evidence confirms this power law and gives $\tilde{d} \approx 1.16$, in close agreement with the expectation that characteristic times scale as lengths raised to the power $\nu/\delta = 1.159$.

While we have focused our attention on forest fires, much the same laws govern the spread of epidemics, the formation of galaxies, nuclear fragmentation, and countless other phenomena [Kes 87].

Percolation is a widespread paradigm. Percolation theory can therefore illuminate a great many seemingly diverse situations. Because of its basically geometric character, it facilitates the analysis of intricate patterns and textures without needless physical complications. And the self-similarity that prevails at critical points permits profitably mining the connection with scaling and fractals.

16

Phase Transitions and Renormalization

Is nature trying to tell us something by
using only renormalizable interactions?
—HEINZ PAGELS

The concepts of renormalization and self-similarity are closely related. In fact, renormalization is one of the most fruitful applications of self-similarity. In physics, renormalization theories have shed light on nonlinear dynamics and the mysteries of phase transitions in areas ranging from freezing to ferromagnetism, spin glasses, and self-organization [Hak 78].

Where do the puzzling fractional exponents, describing behavior near critical points, come from, and why are they so often identical in widely different situations? And what is the reason behind the small integers in these exponents? All this has been greatly clarified in the last two decades by Leo Kadanoff, Michael Fisher, and others, and especially by Kenneth Wilson, who has won the Nobel Prize in physics for his work. One of the more spectacular phase transitions that are now completely transparent is *critical opalescence*, in which a translucent medium, at the critical point, becomes optically opaque as a result of a "soft mode" that scatters light much as thick smoke does in a smoke-filled foyer.

Here we shall touch only very lightly on the subject, but hope, nevertheless, to convey some of the spirit by the sprinklings that follow.

A First-Order Markov Process

A *Markov process* is a stochastic process in which present events depend on the past only through some finite number of generations. In a *first-order* Markov process, the influential past is limited to a *single* earlier generation: the present can be fully accounted for by the *immediate* past.

Such processes are often represented by state diagrams, such as that shown in Figure 1, with various transition probabilities. Thus, in the simple first-order Markov source depicted in Figure 1, if $+1$ was the last symbol generated by the source, we are in the left state, labeled $+$, and p is the probability of generating another $+$. This is indicated by the curved arrow that starts and ends at the left state.

With probability $1 - p$ the source will emit the symbol -1 and thus jump to the right state, labeled $-$. In this state, the source will emit another -1 with probability q and thus remain in the right state. With probability $1 - q$, the source will emit a $+1$ and jump back to the left state.

For $p = q$, the entropy H_M of such a Markov source is

$$H_M = -p \log_2 p - (1 - p) \log_2 (1 - p) \qquad \text{bits per output symbol}$$

which happens to be the same as the entropy $H(p)$ of a *memoryless* binary source with probabilities p and $1 - p$ for the two possible outputs. This agreement is easily verified by modeling the first-order Markov source with $p = q$ as a zero-order (memoryless) source "kicking" a polarity reversal switch (i.e., changing $+$ to $-$ or $-$ to $+$) with probability $1 - p$ and *not* kicking the switch with probability p. The outputs from these two sources can be reversibly transformed into each other (except for an overall sign change) and therefore must have the same entropy.

Self-Similar and Non-Self-Similar Markov Processes

For $p = q$, the Markov source depicted in Figure 1, emits $+1$ or -1 with equal probability. For $p = \frac{1}{2}$, successive outputs are independent: the machine has turned into a memoryless honest-coin flipper.[1] The output sequence is an example of a statistically self-similar process: strike out every other symbol, and the decimated sequence is statistically indistinguishable from the original sequence, because we again have independent $+1$s and -1s, each occurring with probability $\frac{1}{2}$.

For $p \neq \frac{1}{2}$, however, self-similarity no longer holds. Adjacent samples are correlated, and when we skip symbols the remaining symbols will have an absolutely smaller correlation.

In fact, another look at Figure 1 reveals that, for $p = q > \frac{1}{2}$, a like symbol is more likely than an unlike one to follow a given symbol. For $p = q < \frac{1}{2}$, the situation is reversed: symbols prefer to alternate.

---

1. Not to be confused with the forgetful but honest coin-flipper.

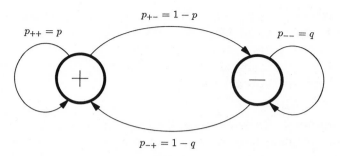

Figure 1   First-order Markov source with two states ($+$ and $-$) and four transition probabilities.

The correlation C_m between two symbols s_n and s_{n+m} from the source in Figure 1 with $p = q$ is given by

$$C_m = \lim_{N \to \infty} \frac{1}{N} \sum_{n=1}^{N} s_n s_{n+m} \tag{1}$$

Since our Markov source is stationary and ergodic for $0 \neq p \neq 1$, we can replace the "time" average in equation 1 by an ensemble expectation value, symbolized by angular brackets:

$$C_m = \langle s_n s_{n+m} \rangle$$

For C_1 we obtain, by averaging over the four distinct possibilities $(++, +-, -+, --)$,

$$C_1 = \tfrac{1}{2}[p - (1 - p) - (1 - p) + p] = 2p - 1$$

For $p = \tfrac{1}{2}$, $C_1 = 0$, as expected. Note that $C_1 < 0$ for $p < \tfrac{1}{2}$.

Here is a typical "random" sequence, generated by this inveterate stochastic generator while writing this (without mechanical or electronic assistance):

$$+ + - + - - - + - + - - - + - + + + + -$$

By coincidence, this sequence has equal numbers of $+$ and $-$. The sample correlation is $-\tfrac{3}{19}$, giving an estimate for the probability parameter $p = \tfrac{8}{19} \approx 0.42$. Considering that my intent was to generate a sequence with $p = \tfrac{1}{2}$, I fell about 16 percent short— a typical human failing. Most *human* random number generators can be characterized as Markov sources with $p < \tfrac{1}{2}$ when they are trying to generate independent binary events. People have the greatest trouble being really random; they almost always alternate too much (a common shortcoming that was judiciously exploited in Claude Shannon's "outguessing machine" [Sha 53], described in pages 149–150).

The Scaling of Markov Outputs

Since, in a first-order Markov process, the present is fully accounted for by the immediately preceding past, one obtains for the correlation coefficients

$$C_m = C_1^m = (2p - 1)^m$$

or, introducing a new parameter β defined by

$$e^{-\beta} = 2p - 1 \qquad \text{for} \qquad p \geq \frac{1}{2} \tag{2}$$

we have

$$C_m = e^{-\beta m}$$

Let us delete every other output symbol from our Markov source. The result can be viewed as the output of another Markov source with a different parameter p (for $p \neq \frac{1}{2}$). The correlation, $C_m^{(2)}$, of the decimated process is the square of that of the original process:

$$C_m^{(2)} = C_{2m} = C_m^2 = e^{-2\beta m}$$

Thus, we see that our parameter β has doubled, which means (see equation 2) that the new transition probability has changed from p to $p^{(2)}$ given by $2p^{(2)} - 1 = (2p - 1)^2$, or

$$p^{(2)} = 2p^2 - 2p + 1$$

The aforementioned value $p = \frac{8}{19} \approx 0.42$ then changes to $p^{(2)} \approx 0.51$. If we again take every other sample (every fourth term of the original sequence), we get $C_m^{(4)} = C_{4m} = C_m^4$ and $p^{(4)} = 0.5003$ for $p = 0.42$. As we leave out more and more intermediate samples, $p^{(2^n)}$, for $n \to \infty$, approaches $\frac{1}{2}$, the value for independent samples, from above.

Thus, while the output of our Markov source is not self-similar (except for $p = 0.5$ or 1 for nonnegative C_1), scaling the *index* of the output sequence s_k by an integer r, to yield the decimated sequence s_{rk}, is equivalent to taking the output of another Markov source with a rescaled parameter $\beta^{(r)}$, where $\beta^{(r)} = r\beta$. Thus, the parameter β scales exactly as the index.

The physical significance of $1/\beta^{(r)}$ is a *correlation length*, which goes to zero as r becomes larger and larger, reflecting again the fact that "skipping" samples makes the correlation smaller. In other contexts, the parameter β can also be identified with a temperature (as we will see shortly).

Together with periodic symmetries, as manifest in spatial rotations and other periodic phenomena, the scaling symmetry we just encountered is now one of the most important symmetries in physics and other fields. In fact, the ingenious Maurits Escher (1898–1972) has combined these two fundamental symmetries in several of his graphical representations.

In physics, rescaling has led to the by now ubiquitous *renormalization theories*. In a typical application, one might want to derive, from fundamental principles, the *critical exponent* α of the specific heat $c(T)$, say, near a critical temperature T_c. Measurements may suggest a simple power law like

$$c(T) - c(T_c) \approx |T - T_c|^{-\alpha}$$

In such problems it has been found again and again that the exponent α does not depend on the specific situation, but may be the same for very different physical systems like water, helium, xenon, or any other fluid near its liquid-gas critical point.[2] These are then said to fall into the same *universality class*, which typically depends on only a few pure *numbers*: the dimensionality of the space in which the phenomenon takes place and the number of degrees of freedom of the order parameter.

Of course, most physical systems are so complicated that one has to rely on simple models of reality. For example, for spin systems, an easy model is the one named after the German-born physicist Ising (originally, and appropriately it may seem, pronounced "easing"). In the Ising model, spins have only two possible values ("up" or "down"), and usually only adjacent spins are assumed to interact ("nearest-neighbor coupling").

The first-order Markov source that we studied in this chapter corresponds to the one-dimensional Ising model. This Ising model has two "critical temperatures": $T_c = 0$, in which all spins are aligned (corresponding to the fully correlated case in the Markov model with $\beta = 0$), and $T_c = \infty$, in which the spins are totally disordered (corresponding to the case $\beta = \infty$).

For the three-dimensional (3D) Ising model, computer simulations give, for the average spontaneous magnetization,

$$\overline{M} \sim (T_c - T)^\beta \quad \text{for} \quad T < T_c$$

---

2.   Under carefully controlled conditions, one can also push liquids beyond the critical point, resulting in "supercritical" liquids, which have numerous useful applications. Supercritical water, for example, can be used to extract caffeine from coffee beans without altering the taste, unlike chemical solvents.

with $\beta = 0.325$, the same exponent found for all other 3D systems with only one degree of freedom of the order parameter (spin up or down, in the Ising model).

Renormalization and Hierarchical Lattices

For renormalization to be applicable to an atomic lattice, the lattice must scale in the sense that the Koch flake and other fractals scale. To construct lattices having this property, one begins with an initiator—for example, just two spin sites ($k = 1$)—and a generator ($k = 2$); see Figure 2A. The next iteration yields the "lattice" shown for $k = 3$. Such self-similar lattices are described as *hierarchical* in this context [DDI 83, PR 86a, b].

For an antiferromagnetic lattice, one has to distinguish the two spins at the endpoints of the initiator (open and filled circles in Figure 2B). The generator

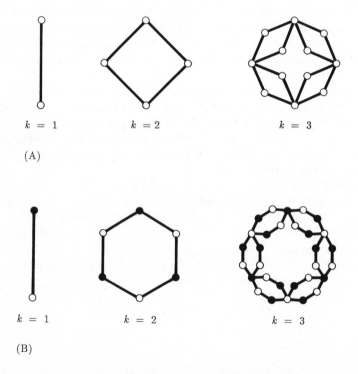

$k = 1$ $k = 2$ $k = 3$

(A)

$k = 1$ $k = 2$ $k = 3$

(B)

Figure 2   (A) Initiator, generator, and next generation of hierarchical lattice for ferromagnetic spin interactions. (B) Initiator, generator, and next generation of hierarchical lattice for antiferromagnetic interactions [PR 86a].

for the antiferromagnetic hierarchical lattice (a hexagon) is necessarily more complicated than the generator for the ferromagnetic lattice (a rhombus).

The fact that such hierarchical lattices do not exist in nature has not prevented physicists from playing endless computer games with them; they are great fun! And while the playing continues, everybody is waiting for even faster supercomputers and parallel processors to be able to study more realistic models. (In their spare time, these number crunchers can then factor giant integers, like the 100-digit monster that was reported on the front page of the *New York Times* on October 12, 1988, to have been cracked to yield its two prime factors, 41 and 60 digits long.)

Given that hierarchical lattices are defined recursively and exhibit self-similarity, it is not surprising that they can be characterized by a fractal dimension. However, the hoped-for universality has not materialized: hierarchical lattices with identical connectivities and fractal dimensions have been constructed whose phase transitions have different critical exponents [Hu 85].

In 1952, in one of the more daring attacks on phase transitions, Yang and Lee introduced complex numbers to represent such physical variables as temperature and magnetic field strength [YL 52]. Later it was found that the Julia sets of the renormalization transformation of hierarchical models are identical with the sets of complex zeros that Yang and Lee had worked with [PPR 85]. Similar Julia sets were obtained for the zeros of the partition function of Ising models on self-similar fractal lattices [SK 87]. These fractal Julia sets, like those of the quadratic map (Chapter 12), exhibit visually appealing self-similarities, which are engendered by the recursive construction of the underlying lattices. They are celebrated in Peitgen and Richter's *The Beauty of Fractals* [PR 86a].

The Percolation Threshold of the Bethe Lattice

Another type of hierarchical lattice is the Bethe lattice (see Figure 3), known in graph theory as a Cayley tree. In a Cayley tree each node has the same number z of branches or bonds. Thus, the size of the neighborhood grows *exponentially* with "diameter," as opposed to a power-law growth for physical lattices—fractal or nonfractal. It is therefore not surprising that in some respects the Bethe lattice behaves as if its number of dimensions were infinite. But the infinite Bethe lattice, being hierarchical, permits calculating both the percolation threshold and the probability P that a given lattice site is connected to infinity, by a beautifully simple similarity argument.

Starting at an arbitrary site and proceeding to one of its z neighbors, we find $z - 1$ new bonds or branches emanating from the neighbor (see Figure 4). Each of these $z - 1$ branches leads to a neighbor, which is occupied with probability p. Thus, on average there are $(z - 1)p$ new occupied neighbors to which the path can be continued. If this number is smaller than 1, the probability of

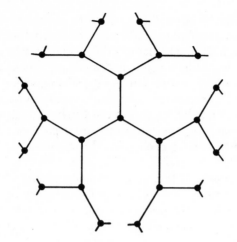

Figure 3 Caley tree, called Bethe lattice by physicists. Here each node has exactly three bonds [Sta 85].

finding a connected path of a given length decreases exponentially with length. On the other hand, if $(z - 1)p$ exceeds 1, there is a positive probability that an infinite path exists. Thus, the percolation threshold p_c (for either sites or bonds) is given by

$$p_c = \frac{1}{z - 1} \tag{3}$$

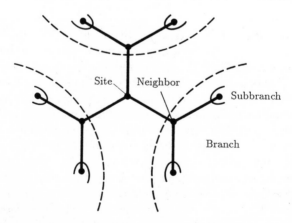

Figure 4 Hierarchical neighborhoods: branches and subbranches in a Bethe lattice [Sta 85].

which for $z = 3$ equals $\frac{1}{2}$, as for site percolation on a triangular lattice or bond percolation on a square lattice [Sta 85].

However, the probability P that a *given* lattice site is connected to infinity does not equal 1. After all, there is a finite probability $(1 - p)^z$ that its z neighbors are unoccupied. What is the probability P that a given site does belong to an infinite cluster for $p > p_c$? (For $p < p_c$, P is obviously equal to zero.)

Let Q be the probability that a given site is *not* connected to infinity through *one* fixed branch originating from this site. The probability that *all* $z - 1$ subbranches from a neighbor site are not connected to infinity equals Q^{z-1}. (Because of the statistical independence of the occupation probabilities, the probabilities Q are simply multiplied.) Thus, pQ^{z-1} is the probability that the neighbor is occupied but not connected to infinity. With probability $1 - p$ the neighbor is not even occupied, in which case it provides no link to infinity even if it is well connected. Thus, we find the fundamental relation

$$Q = 1 - p + pQ^{z-1} \tag{4}$$

which for $z = 3$ has two solutions:

$$Q = 1 \qquad \text{and} \qquad Q = \frac{1 - p}{p}$$

The probability $p - P$ that a given site is occupied but not connected to infinity equals pQ^z. Thus,

$$P = p(1 - Q^z) \tag{5}$$

or, for $z = 3$,

$$P = p\left(1 - \left(\frac{1 - p}{p}\right)^3\right)$$

At the percolation threshold $p = p_c = \frac{1}{2}$, this relation gives $P = 0$, demonstrating that although an infinite cluster exists, it is infinitely dilute.

The ratio P/p is plotted as the solid line in Figure 5, together with P/p for the triangular lattice (dashed line). Note the steep rate of increase of P/p for the triangular lattice. For example, for $p = 0.6$, the probability P that a given Bethe lattice site is a member of an infinite cluster equals 0.422, while for the triangular lattice the probability P that an *occupied* site is a member of an infinite cluster is practically 1.

The solution $Q = 1$ of equation 4, which, with equation 5, gives $P = 0$, obviously corresponds to $p < p_c$. Indeed, for $z = 2$, for which equation 3 gives $p_c = 1$, the only solution of equation 4 for $p < 1$ is $Q = 1$, that is, $P = 0$.

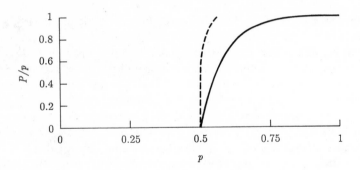

Figure 5   Strength P of the infinite network plotted against concentration p in the Bethe lattice (solid line), and in the triangular lattice (dashed line) [Sta 85].

For P near but above $p_c = \frac{1}{2}$, P increases with p according to the relation

$$P = 6(p - p_c) \qquad p > p_c \tag{6}$$

Thus, the critical exponent β of P equals 1.

Below the percolation threshold, the *mean cluster size* S can be calculated in a similar manner, giving

$$S = p\,\frac{1 + p}{1 - 2p}$$

[Sta 85], or, for p near but below $p_c = \frac{1}{2}$,

$$S = \frac{3}{8}(p_c - p)^{-1} \qquad p < p_c \tag{7}$$

Hence, the critical exponent for the cluster size is equal to -1.

Equations 6 and 7 reflect the behavior, for the Bethe lattice with $z = 3$, of two important "order parameters," P and S, near a critical point, the percolation threshold $p = p_c$. This behavior is characterized by two simple power laws with exponents 1 and -1, respectively. Such behavior is now often described as "algebraic," as opposed to logarithmic, exponential, or other transcendental behavior.

Another, early instance in which the Bethe lattice has resulted in an *exactly solvable model* is Anderson localization, an important phenomenon in disordered systems [And 58]. For its discovery, Philip Anderson was awarded the 1977 Nobel Prize in physics. Disordered systems are now a central theme and the subject of intense study in several fields of physics, such as spin glasses and neural networks.

Figure 6   Random resistor network with one conducting path (arrows) between two copper bars [Sta 85].

There are other lattices for which different exponents may be found although the behavior is still algebraic. For example, on the square lattice, P varies as $(p - p_c)^\beta$ with a critical exponent $\beta = \frac{5}{36}$. The very value of this exponent is an indication that its theoretical derivation is anything but trivial.

An important application of percolation theory [Sta 85, Kes 87, Gri 89] is the conductance Σ of random resistor networks (see Figure 6). Of course, for $p < p_c$, the conductance is zero. But even for $p > p_c$ it grows rather slowly with p, compared with the growth of the probability P that a given site is a member of an infinite cluster. The reason is that most sites in an infinite cluster near the percolation threshold belong not to the "backbone" but to dangling dead ends that do not contribute to the conductance.

A Simple Renormalization

The fundamental requirement for renormalization to work is self-similarity. Since many critical phenomena in physics show self-similar behavior near the critical point (the percolation threshold, or the Curie temperature, for instance), these phenomena are therefore amenable to a renormalization-theoretic treatment, yielding the critical exponents for the correlation length ξ and other important parameters.

Following Stauffer [Sta 85], we shall illustrate the renormalization method for a case for which an exact answer is known: the triangular lattice (see Figure 7). For this lattice, the *bond* percolation threshold p_c equals precisely $\frac{1}{2}$ and the correlation length exponent v is believed to be $\frac{4}{3}$. Let us see whether we can derive p_c and v by a space renormalization of the lattice.

For this purpose, we replace three adjacent lattice sites in the triangular lattice by a "supersite" (the open circles in Figure 7). Suppose the occupation probability of the original sites is p. What is the corresponding probability, p', of a supersite? We consider a supersite occupied if its original sites form a

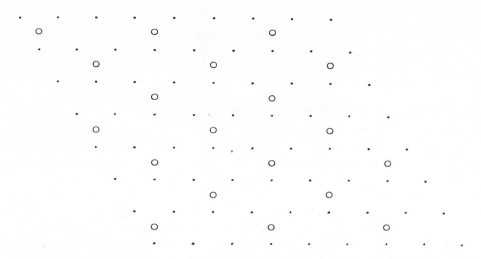

Figure 7   Space renormalization of the triangular lattice."Supersites" (open circles) each replace three adjacent sites and again form a triangular lattice [Sta 85].

"spanning cluster," that is, if *at least* two out of its three sites are occupied. The probability that all three sites are occupied is p^3, and the probability that exactly two out of three sites are occupied equals $\binom{3}{2} p^2(1 - p)$. Thus,

$$p' = p^3 + 3p^2(1 - p) \tag{8}$$

At a critical point $p = p_c$, we should have $p' = p$. Hence, with equation 8, we find three critical points $p_c = 0, \frac{1}{2}$, and 1, of which only $p_c = \frac{1}{2}$ is nontrivial. This renormalization result corresponds precisely to the known site percolation threshold for the triangular lattice. Renormalization seems to work! But will we be as lucky with the correlation-length exponent v?

The correlation length ξ near a critical point is given by

$$\xi = c|p - p_c|^{-v} \tag{9}$$

where c is a constant. In the renormalized lattice we have

$$\xi' = bc|p' - p_c|^{-v} \tag{10}$$

where b is the length scaling factor between lattice and superlattice. If we set $\xi = \xi'$, equations (9) and (10) give

$$v = \frac{\log b}{\log \left[(p' - p_c)/(p - p_c) \right]} \tag{11}$$

Expanding equation 8 about the fixed point $p = p_c = \frac{1}{2}$ gives

$$p' = p_c + \frac{3}{2}(p - p_c) + \cdots$$

or

$$\frac{p' - p_c}{p - p_c} = \frac{3}{2}$$

With $b = 3^{1/2}$ (see Figure 7), equation (11) yields, to first order in $(p - p_c)$,

$$v = \frac{\log 3^{1/2}}{\log (3/2)} = 1.355$$

which is reassuringly close to the exact value $v = \frac{4}{3}$.

Another powerful approach to renormalization is *conformal mapping* [Car 85]. Like self-similar scaling, conformal mapping preserves angles. Its usefulness results from the conformal invariance—real or assumed—of the system under study. Mirroring a given space at a fixed sphere is a well-known example of a conformal mapping. In the plane, any analytic function defines a local conformal mapping at points where its derivative does not vanish. As an instance of conformal invariance in physics, we might mention Maxwell's famous equations, which were revealed as such in 1909—several decades after their conception.

Here we reluctantly leave renormalization theories and phase transitions to transit to the self-similarities engendered by *cellular automata*.

17

Cellular Automata

> *Truth is much too complicated to allow anything but approximations.*
>
> —JOHN VON NEUMANN

Cellular automata were originally conceived by Konrad Zuse and Stanislaw Ulam and put into practice by John von Neumann to mimic the behavior of complex, spatially extended structures [TM 87]. As early as the early 1940s, Zuse thought of "computing spaces," as he suggestively called them, as discrete models of physical systems. Ulam's proposal came in the late 1940s, shortly after his invention, with Nicholas Metropolis, of the Monte Carlo method. (The astonishingly broad scope of Ulam's mind can be sampled in the selection from his works titled *Sets, Numbers, and Universes* [Ula 74].) An anthology surveying the present state of cellular automata was edited by Stephen Wolfram [Wol 86].

A one-dimensional cellular automaton consists of a row of *cells*, each cell containing some initial numbers, and a set of *rules* specifying how these numbers are to be changed at every clock time. Suppose in the initial state of the automaton all cells are filled with 0s, except a single cell which is occupied by a 1:

$$\ldots 01000000 \ldots$$

And suppose the rule states that the number in each cell is to be replaced by the sum of itself and its left neighbor. Thus, after one clock time, the state of the automaton will be as follows:

$$\ldots 01100000 \ldots$$

Another clock time later the state will be

$$\ldots 01210000 \ldots$$

followed by

$$\ldots 01331000 \ldots$$

$$\ldots 01464100 \ldots$$

and so on. Such cellular automata are in fact computers, and cellular computers are being put to increasing use in calculating intricate functions because they are naturally amenable to fast parallel processing. In the example just given, the cellular computer calculates the binomial coefficients that appear in the expansion of the powers of binomials such as $(a + b)^4$, which equals $a^4 + 4a^3b + 6a^2b^2 + 4ab^3 + b^4$.

Cellular automata come in one, two, or many dimensions. To calculate a two-dimensional fluid flow, one uses cellular automata that are two-dimensional *arrays* of cells, each cell filled with a number (representing fluid density, for example) that changes at clock times in accordance with fixed rules acting on a neighborhood of cells. These rules embody local interactions between neighboring cells, reflecting the dynamics of the system under study.

Instead of forming a *square* lattice, the cells can form a hexagonal pattern, for example; and the "numbers" in each cell can in fact be vectors, representing the velocity of a fluid or gas at each lattice point (see Figure 1). Such cellular

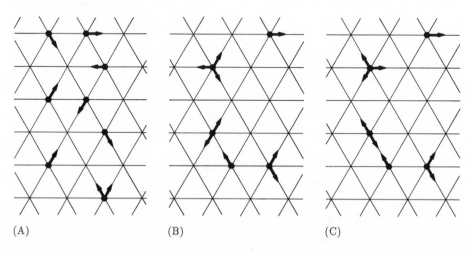

(A) (B) (C)

Figure 1 Hydrodynamic flow modeled by a cellular automaton ("lattice gas"). Parts A, B, and C show successive stages in the history of the gas. The arrows show the directions of the particle velocities.

Figure 2   Flow behind a moving cylinder simulated by a lattice gas [SW 86].

automata, called *lattice gas* models by physicists, have been used to great advantage to simulate otherwise intractable flow problems [SW 86, MBVB 89]. Figure 2 shows the flow behind a cylinder moving from right to left through a viscous gas, exhibiting the well-known vortex shedding behind the obstacle. Such fluid-flow phenomena are still studied by physical experiments in wind tunnels and ship-model basins. But they are now increasingly being analyzed by computer simulations based on cellular automata.

Since typical cellular automata employ repetitive application of fixed rules, we should expect to find self-similarities—as we did with so many other iterative procedures. And indeed, many cellular automata do produce self-similar patterns, often of considerable visual appeal.

The Game of Life

The best-known cellular automaton is probably John Horton Conway's game of "Life" [Gar 70]. "Life" describes the growth and decline of a population of cells according to rather simple rules—rules that nevertheless lead to a rich zoo of creatures with truly astounding behavior [BCG 82].

In "Life" as conceived by Conway, each cell is either dead (0) or alive (1) and changes its state according to the states in its immediate neighborhood including its own state. Specifically, at each clock time ("tick"), a cell that is alive will stay alive when it is surrounded by precisely two or three live cells among its eight neighbors on a square lattice. If more than three neighbors are alive, the cell will feel overcrowded and "suffocate" to death. If fewer than two neighbors

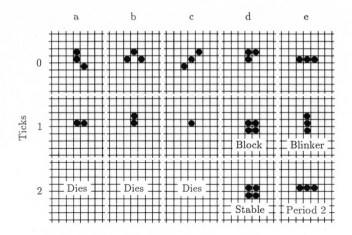

Figure 3   Conway's "Life": the fates of five triplets [Gar 83].

are alive, the cell will die from loneliness. On the other hand, a dead cell will come to life when surrounded by exactly three live neighbors (two parents and a midwife, so to speak). Figure 3 illustrates the fates of five different triplets. The plethora of patterns generated by these simple rules is beyond belief. Figures 4 to 8 show a sparse sampling of stationary, periodic, disappearing, and surviving "organisms."

Conway's set of rules, or *law*, is but one of many imaginable. For binary-valued cells and a neighborhood of eight cells acting on a center cell, there are $2^{2^9} \approx 10^{154}$ different "life"-like laws, of which, it seems, only one, the one decreed by Conway, really comes to life.

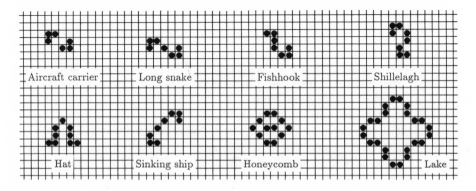

Figure 4   Six still "Life" forms [Gar 83].

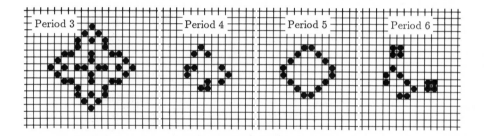

Figure 5   Four periodic patterns of "Life" [Gar 83].

Cellular Growth and Decay

Consider a two-dimensional cellular automaton in which a cell again has two possible states, 0 and 1, as in the game of "Life," but has only *four* neighbors, East, West, South, and North, acting on it. The present state of a cell *C* and its neighborhood EWSN is given by a 5-bit string, for example, EWSNC = 11000. The next state of *C*, say *C* = 1, is given by the prevailing rule 11000 → 11001 (see Figure 9). A complete set of rules, called a *law*, is given by a table of the 32 possible states and the subsequent values of the center cell *C* (see Figure 10). For binary-valued cells and four acting neighbors there are $2^{32} \approx 4$ billion possible different laws.

Figure 11 illustrates the variety of patterns obtained from the fixed law of Figure 10 called HGLASS for different initial conditions [TM 87].

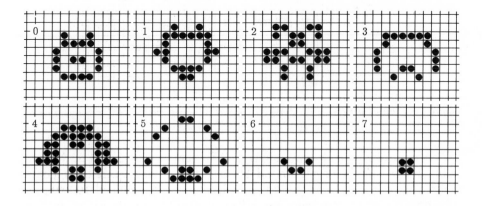

Figure 6   The Cheshire cat (0) leaves a grin (6) that turns into a permanent paw print [Gar 83].

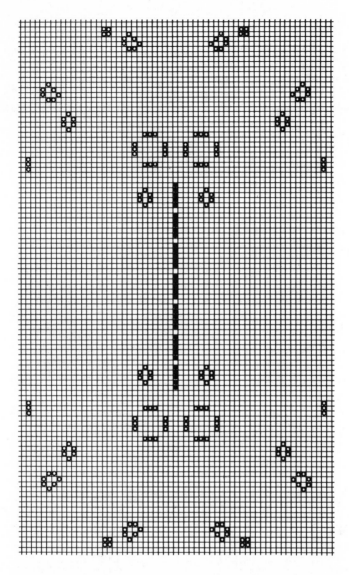

Figure 7   Initial pattern (solid dots) and final state (open circles) of 7 × 5 bits [Gar 83].

A particularly simple law assigns to C the sum modulo 2, that is, the *parity*, of the five cells of the neighborhood. Starting with a small square of 1s, floating in a sea of 0s, the patterns that have evolved after some 50 and 100 steps are shown in Figure 12. Are there any self-similarities? Indeed there are. In fact it can be shown that any initial pattern on a uniform background reproduces itself and surrounds itself with four identical copies after a certain number of steps.

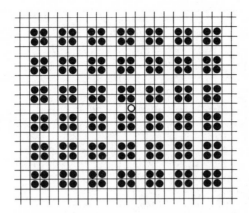

Figure 8 Killing "Life": a single "virus" in the position shown (open circle) destroys the entire pattern. In other positions, the virus is eliminated and the pattern repairs itself and survives intact [Gar 83].

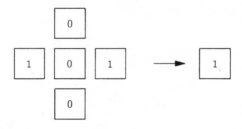

Figure 9 The center cell is turned on (switches from 0 to 1) by the rule (East, West, South, North, Center) = 11000 → 11001.

EWSNC	C_{new}	EWSNC	C_{new}	EWSNC	C_{new}	EWSNC	C_{new}
00000	0	01000	0	10000	0	11000	0
00001	1	01001	0	10001	0	11001	1
00010	1	01010	0	10010	0	11010	0
00011	1	01011	1	10011	0	11011	0
00100	0	01100	0	10100	0	11100	0
00101	0	01101	0	10101	1	11101	1
00110	0	01110	0	10110	0	11110	1
00111	0	01111	0	10111	0	11111	1

Figure 10 Table of rules HGLASS, one of 4 billion possible sets of rules.

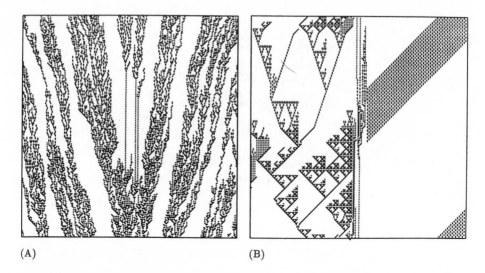

(A)   (B)

Figure 11   (A) An HGLASS pattern evolving from random seed. (B) HGLASS pattern resulting from a simple seed [TM 87].

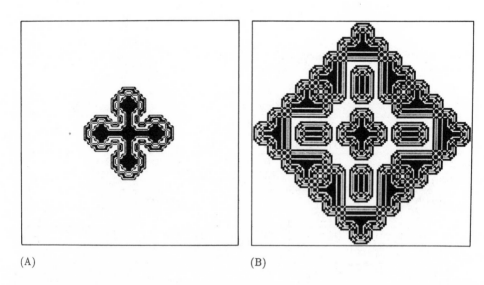

(A)   (B)

Figure 12   (A) Pattern produced by parity rules from a 32 × 32 square pattern after 50 steps. (B) Parity pattern after 100 steps [TM 87].

And after the same number of steps there will be 25 copies of the original, and so forth *ad infinitum*. Because summation modulo 2 is a linear operation, different patterns can penetrate each other without affecting the future growth. Specifically, any pattern can be obtained by the summation modulo 2 of patterns generated by a single, isolated point.

Another simple law turns a cell on (1) if exactly one of its eight neighbors is alive (1); otherwise it remains unchanged. The resulting growth is a self-similar fractal (Figure 13), whose Hausdorff dimension the reader may wish to calculate.

In still another law, each cell adjusts to the *majority* in its neighborhood: if four or fewer of the neighborhood of nine (including itself) are off, then the center cell will also turn (or stay) off. Otherwise it will turn or stay on. The resulting patterns resemble Ising spin systems at low temperature and are reminiscent of percolation; see Figure 14A for a pattern emerging from an initial configuration of random 1s occupying half the cells.

How sensitive the patterns are to slight amendments in the law is illustrated in Figure 14B, in which either five or fewer than four neighbors in the off state will turn the center cell off. This law, drafted by G. Vichniac, simulates annealing, as evidenced by the consolidation of domains [Vic 86]. These patterns are spatially homogeneous but not self-similar.

To obtain self-similar spin domains, the initial random configuration has to have a *critical "energy"* [Vic 84]. The energy pattern has a *broken symmetry* with magnetic domains on all size scales; see Figure 15.

While all laws passed so far have been of a "strictly enforced" nature, that is deterministic, many cellular automata are subjected to *random* rules to emulate

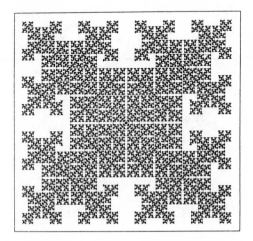

Figure 13   Self-similar fractal produced by the one-out-of-eight rule [TM 87].

(A)   (B)

Figure 14   (A) Pattern evolving from 50 percent random 1s by "majority-voting" rule of neighbors. (B) Result of "annealed-majority" rule [Vic 86].

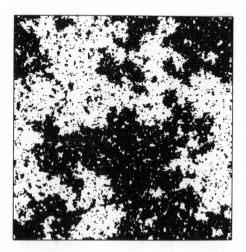

Figure 15   Equilibrium configuration of Ising spins at the critical temperature shows magnetic domains on all size scales [Vic 84].

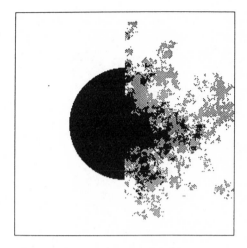

Figure 16   Exploding disk obtained from "copy random neighbor" rule [TM 87].

diffusion and other stochastic processes. For example, a lenient law may simply say, "Copy from a random neighbor." With one of four neighbors chosen with equal probability, an initial disk will explode as shown in the right half of Figure 16. Diffusion with drift can be simulated by "particles" (1s) that move with probability $\frac{3}{8}$ either east or south and with probability $\frac{1}{8}$ west or north (Figure 17).

Figure 17   Diffusion with and without drift: a pretty programming error [Schr 69].

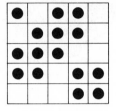

Figure 18   Odd-parity covers for a 4 × 4 and a 5 × 5 chessboard [Sut 89].

Such cellular automata can be bred into plausible models of genetic drift. Color Plate 9 shows the spatial intermingling for sixteen competing species of genes [TM 87].

A classical problem that can be phrased in the language of cellular automata is the "all-1s" problem [Sut 89]. Each square (cell) of an $n \times n$ chessboard is equipped with a light bulb and a switch that turns the bulbs in a given neighborhood on (or off, for those that are already on). Starting with a completely dark chessboard, which switches must be activated to light up all bulbs? It is clear that the number of activated buttons in the neighborhood of each square must be odd. This is called an *odd-parity cover*. If the neighborhood of a square consists of the square itself and its four edge-adjacent neighbors, odd-parity covers for a 4 × 4 and a 5 × 5 board are as shown in Figure 18.

What is the solution for an 8 × 8 board? Can the reader design the rules for a cellular automaton that will converge on the proper set of switches, that is, an odd-parity cover for a given neighborhood? Of course, if the odd-parity cover is a "Garden of Eden," it can never be reached. (A Garden-of-Eden pattern is defined as one that has no predecessor; once lost, it can never be regained.)

Biological Pattern Formation

Another field where cellular automata have proved their mettle for modeling is pattern formation in plants and animals. The formation of stripes in zebras and numerous other patterns in countless forms of life has been modeled with cellular automata by H. Meinhardt, A. Gierer, and others using combinations of local and long-range autocatalytic and inhibitory interactions [Mei 82].

A simple cellular automaton imitates the design on the shell of the snail *Olivia porphyria* (see Figure 19), characterized by diagonal lines that annihilate each other when they touch. Another mechanism causes a single line to bifurcate to keep the average pigmentation near a given level [MK 87].

Numerous other biological structures, including the formation of arms and legs, have been modeled by cellular automata employing simple self-reinforcing and antagonistic reactions. The variety of shapes thus engendered is truly astonishing.

Self-Similarity from a Cellular Automaton

Self-similarity arises in many fields in many forms. A set of Russian dolls, all looking alike but each a little smaller than its parent, is perhaps the most widely known example of discrete, if limited, self-similarity. Self-similarity can even be distilled from such a discrete and artless entity as the integers 0, 1, 2, 3, 4, 5, 6, 7, ... Let us write successive integers, starting with 0, in the binary number system (apparently invented by Leibniz while waiting to see the Pope in the Vatican with a proposal to reunify the Christian churches):[1]

$$0, 1, 10, 11, 100, 101, 110, 111, \ldots$$

The sums of the digits for each number form the sequence

$$B(t) = 0, 1, 1, 2, 1, 2, 2, 3, \ldots \qquad t = 0, 1, 2, \ldots$$

which can also be obtained *iteratively* as follows. To obtain the subsequence of length 2^{n+1} from the subsequence of length 2^n, repeat the latter with 1 added to each term. Thus, the initial subsequence of length $2^0 = 1$ (i.e., 0) generates the subsequence of length $2^1 = 2$ (i.e., 01) by appending to the initial 0 the number $0 + 1 = 1$. In this manner, successive generations of subsequences of length 2^n are generated:

$$0$$
$$01$$
$$0112$$
$$01121223$$
$$0112122312232334$$
$$\vdots$$

---

1.   The fact that this system uses only two digits, 0 and 1, is, of course, the reason why computers are so fond of it: a digit is simply and unambiguously represented by the two states of a switch, open or closed—nothing in between.

(A)

(B)

Figure 19 (A) Detail of the shell of the snail *Olivia porphyria*. (B) Wavelike design generated by a cellular automaton with local and long-range autocatalytic interactions [MK 87].

This generative rule is, of course, a direct consequence of how binary numbers are defined: for $k < 2^n$ the two integers k and $k + 2^n$ differ by precisely a single 1 in their binary notations.

It is interesting and important to note that our iterative rule for generating subsequences is a *fast algorithm*: each iteration *doubles* the length of the subsequences. Their lengths therefore grow *exponentially* with the number of iterations. (By contrast, a *linear* recursion, such as that for the sequence of Fibonacci numbers, $F_{n+2} = F_{n+1} + F_n$, adds only one additional term with each iteration.)

The infinite sequence $B(t)$ obtained in this manner is self-similar in the following sense: it reproduces itself when only even-indexed terms are retained, as indicated in the following by underlining:

$$B(t) = \underline{0},\ 1,\ \underline{1},\ 2,\ \underline{1},\ 2,\ \underline{2},\ 3,\ \underline{1},\ 2,\ \underline{2},\ 3,\ \underline{2},\ 3,\ \underline{3},\ 4,\ \ldots$$

Thus, $B(2t) = B(t)$.

This self-similarity is a near-trivial consequence of the fact that, in the binary system, multiplication by 2 results in a mere left shift of the digits, which, of course, does not change the sum of the digits.

The sequence $B(t)$ can be converted into a sequence that is self-similar also in the *magnitude* of its terms. In fact, the sequence

$$C(t) = 2^{B(t)} = 1,\ 2,\ 2,\ 4,\ 2,\ 4,\ 4,\ 8,\ \ldots$$

has the same similarity factor of 2 not only in its index t but also in its magnitude. The second half of each subsequence of length 2^{n+1} equals twice the first half:

$$C(t + 2^n) = 2C(t) \qquad 0 \leq t < 2^n \tag{1}$$

Figure 20 illustrates the sequence $C(t)$ and its self-similarity. Alternatively, $C(t)$ can be obtained from the product $(1 + b_1)(1 + b_2)(1 + b_3) \ldots$, where the b_k are the bits of the binary expansion of t.

Interestingly, $C(t)$ can also be generated by a cellular automaton, and this is important for what follows. Let us ask how many of the binomial coefficients $\binom{t}{n}$ for a given t are odd as n runs from 0 to t. The answer (I leave the simple inductive proof to the reader) is $1, 2, 2, 4, 2, 4, 4, 8, \ldots = C(t)$. And the binomial coefficients themselves are generated by one of the simplest cellular automata. (See the introduction to this chapter.)

Note that $C(t)$ summed to $t = 2^m - 1$ is equal to 3^m. This follows directly from $C(0) = 1$ and equation 1.

Figure 20   Self-similar sequence obtained from the binary number system.

A Catalytic Converter as a Cellular Automaton

Sequences resembling $C(t)$, where t is interpreted as discrete time, have been observed in certain chemical reactions—for example, in catalytic oxidation processes (see Figure ·21). Now what on earth could the relation be between a chemical reaction and the sequence $C(t)$, that is, 2 raised to the sums of digits in the binary representations of successive integers? A simple explanation was found by Andreas Dress, who modeled such catalytic processes by one-dimensional cellular automata [DGJPS 85].

In a one-dimensional cellular automaton, each time epoch is characterized by a sequence of symbols or numbers. And as we learned in the introduction to this chapter, the sequence at time t, $g_t(n)$, is generated by some law from the sequence at time $t - 1$. For example,

$$g_t(n) = g_{t-1}(n) + g_{t-1}(n - 1)$$

which, with the initial generation $g_0(0) = 1$ and $g_0(n) = 0$ elsewhere, generates the binomial coefficients $\binom{t}{n}$ as arranged in Pascal's triangle. In Pascal's triangle

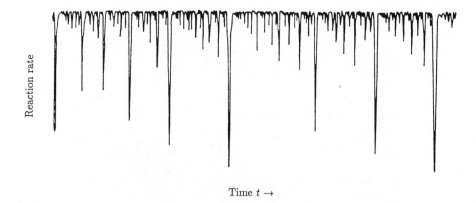

Time $t \rightarrow$

Figure 21 Chemical reaction rate as a function of time in a catalytic oxidation process [DGJPS 85].

each number is the sum of the two numbers directly above it. Now let us take Pascal's triangle modulo 2. That is, if the binomial coefficient $\binom{t}{n}$ is odd, then the number is replaced by 1; if it is even, it is replaced by 0. The resulting Pascal's triangle modulo 2 is illustrated in Figure 22. Black cells correspond to 1s and white cells to 0s.

In terms of the chemical reaction modeled by Pascal's triangle modulo 2, Dress assumed that a "molecule," represented by a cell at position n, becomes "infected" (e.g., oxidized) at time t if precisely one of its neighbors at positions n and $n - 1$ was infected (black) at time $t - 1$.

But by construction, the *number* of black squares (1s) at time t equals $C(t)$. Thus, $C(t)$ describes the chemical reaction rate in the specified catalytic converter, a fact originally suggested by the approximate self-similarity of the reaction as seen in Figure 21.

Pascal's Triangle Modulo N

The recursive generation of Pascal's triangle of binomial coefficients $\binom{t}{n}$ from a single 1 is a paradigm of a cellular automaton. While the numbers $\binom{t}{n}$ themselves become larger and larger with increasing t and $0 < n < t$, their divisibility properties form self-similar patterns. In fact, the even coefficients occupy triangles much like the holes in a Sierpinski gasket (see pages 17–18). The appearance of these triangles follows easily from the fact that the $\binom{t}{n}$ for $t = 2^m$ are all even

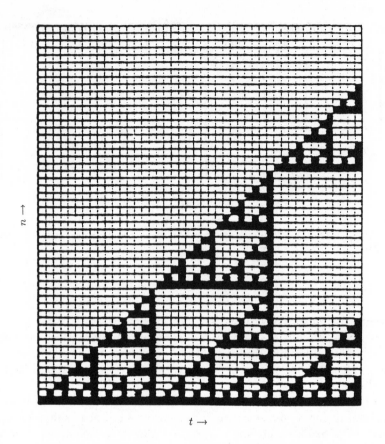

$n \rightarrow$

$t \rightarrow$

Figure 22   Pascal's triangle modulo 2: a discrete version of the Sierpinski gasket.

for $0 < n < t$. The two 1s for $n = 0$ and $n = t$ then progressively "eat up" the even coefficients as t is increased, until for $t = 2^{m+1} - 1$ all coefficients are odd.

Similar mechanisms produce self-similar patterns from Pascal's triangle if the binomial coefficients are taken modulo any other prime [Wol 84]. The PC-equipped reader is invited to generate, in cellular-automaton fashion, Pascal's triangle modulo arbitrary prime numbers, powers of primes, and general composite numbers and to observe the resulting self-similarities in one or many colors. What are the Hausdorff dimensions D of the limiting patterns? The value of D for Pascal's triangle modulo 2, $D = \log 3/\log 2$, can be inferred from the fact that the total number of odd coefficients goes up by a factor of 3 every time the number of lines is doubled, beginning with the first line consisting of a single 1—just as the covered area of the infinite Sierpinski gasket triples for every doubling of the linear dimensions of the cover.

Bak's Self-Organized Critical Sandpiles

As we have seen in several chapters of this book, many natural phenomena—from flicker noise to the flow of the river Nile—have self-similar power spectra with an $f^{-\beta}$ frequency dependence. Such processes are called $1/f$ noise (even if the spectral exponent β is not exactly equal to 1). Such power-law spectra signal the absence of characteristic time scales; there are no such typical times as the half-life in radioactive decay, for example.

The absence of characteristic scales is also evident in the *spatial* aspects of numerous natural events; no characteristic lengths prevail, in contrast to nuclear forces or the mean free path of molecules in a gas.

To account for the ubiquity of such self-similar structures, Per Bak, Chao Tang, and Kurt Wiesenfeld have recently introduced the concept of *self-organized criticality* [BTW 87]. In their paper of that title and subsequent publications [TB 88, BTW 88], Bak and his collaborators argue persuasively that spatially extended dynamic systems evolve spontaneously into barely stable structures of critical states and that this self-organized criticality is the common underlying mechanism for many self-similar and fractal phenomena.

To make their proposal concrete, Bak and his coworkers have constructed several models, including a simple two-dimensional cellular automaton mimicking the flow of sand in a sandpile. If the slope becomes too large at some point (x, y) in the pile, sand flows to reduce the gravitational force z at the expense of the forces on the four neighboring points $(x \pm 1, y)$ and $(x, y \pm 1)$. Thus, if the (integer) variable z exceeds a critical value z_c, it is updated (synchronously) as follows:

$$z(x, y) \rightarrow z(x, y) - 4$$

$$z(x \pm 1, y) \rightarrow z(x \pm 1, y) + 1$$

$$z(x, y \pm 1) \rightarrow z(x, y \pm 1) + 1$$

The automaton is started with random initial conditions $z \gg z_c$; the boundary conditions are $z = 0$. Once all z are smaller than z_c, the evolution of the system stops; it has reached a *minimal stable state*—minimal, because the addition of a single grain of sand may set off an avalanche. In fact, backing Bak, the reader equipped with a personal computer may discover that sand avalanches on all length scales can be triggered by small local perturbations, that is, random additions of sand to a single site. The clusters of sites reached by this physical "domino effect" have power-law size distributions:

$$D(s) \sim s^{-\tau}$$

with $\tau \approx 1$ for cluster sizes s ranging up to 500 for a 50 × 50 array and $\tau \approx 1.35$ for a three-dimensional 20 × 20 × 20 array [BTW 87].

The *life-times* of these avalanches, too, follow power laws with exponents $\alpha \approx 0.43$ in two dimensions and $\alpha \approx 0.9$ in three, corresponding to power-spectrum exponents $2 - \alpha$ equal to 1.57 and 1.1, respectively.

Tang and Bak also found a power-law behavior of the flow, the correlation length, the largest cluster size, and other parameters if the average value of z is kept away from its critical value by an "external field" [TB 88]. While the critical exponents they found for these quantities may depend on the details of the system under consideration, they expect the power-law scaling as such to be more universal.

If so, self-organized criticality may lend itself as a generic model for a great variety of scale-invariant phenomena, from glassy systems, magnetic domains, water flow, and turbulence to traffic jams, economic interactions, and earthquakes [BW 90].

And do not the political upheavals in eastern Europe in 1989 also flow from long maintained minimally stable states?

The Hausdorff Dimension for Unequal Remainders

The following proof was suggested by H. W. Strube.

Let $I_k \subset [0, 1]$ be the intervals of the generator of a Cantor set F and $s_k = |I_k|$ their lengths. (In the original "ternary" Cantor set $I_1 = [0, \frac{1}{3}]$, $I_2 = [\frac{2}{3}, 1]$, and $s_1 = s_2 = \frac{1}{3}$.) Define those parts of F that lie in I_k as F_k:

$$F_k := F \cap I_k \tag{1}$$

Let $N(r)$ and $N_k(r)$ be the smallest numbers of intervals of radius r that cover F and F_k, respectively. Further, let L be the length of the *smallest gap* between the I_k. For $2r < L$, we then have

$$N(r) = \sum_k N_k(r) \tag{2}$$

or

$$\sum_k \frac{N_k(r)}{N(r)} = 1 \tag{3}$$

Since F_k is simply a scaled version of F, with the scaling factor s_k, we have $N_k(s_k r) = N(r)$, or

$$N_k(r) = N\left(\frac{r}{s_k}\right) \tag{4}$$

Thus, for $r \to 0$, since the Hausdorff dimension D is also a similarity dimension (that is, $N(r) \sim r^{-D}$),

$$\frac{N_k(r)}{N(r)} \to s_k^D \tag{5}$$

holds. Introducing this relation into equation 3, one obtains the generalized result

$$\sum_k s_k^D = 1 \tag{6}$$

for the Hausdorff dimension D of a Cantor set whose generator consists of intervals of arbitrary lengths s_k.

Noble and Near Noble Numbers

Noble numbers v are defined as irrational numbers whose continued fractions end in all 1s. For $0 < v < 1$, we have

$$v := [a_1, a_2, \ldots, a_n, \overline{1}] \tag{1}$$

where the bar over the 1 indicates an infinite sequence of 1s.

With the help of the golden mean

$$\gamma := [\overline{1}] = 0.618 \ldots \tag{2}$$

the noble numbers can be written as "equivalent numbers" [HW 84]:

$$v = \frac{A_n + \gamma A_{n-1}}{B_n + \gamma B_{n-1}} \tag{3}$$

where A_k and B_k are the numerator and denominator, respectively, of the kth approximating fraction ("convergent") of $[a_1, a_2, \ldots, a_n]$.

For example, a simple noble number, as defined in equation 1, for $n = 2$, with $a_1 = 1$ and $a_2 = 2$, is $v = [1, 2, \overline{1}] = 0.7236 \ldots$. Since the approximating fractions of the nonperiodic "leader" [1, 2] are equal to $A_1/B_1 = 1/1$ and $A_2/B_2 = 2/3$, v, according to equation 3, can also be written as

$$v = \frac{2 + \gamma}{3 + \gamma} \tag{4}$$

In general, because $\gamma = (\sqrt{5} - 1)/2$, the noble numbers form a subset of the field $\mathbb{Q}(\sqrt{5})$.

I have defined *near noble* numbers as all those real numbers $\tilde{v}$, $0 < \tilde{v} < 1$, whose continued fraction expansion is periodic, with period length P, the period comprising $(P - 1)$ 1s followed by an integer $n > 1$:

$$\tilde{v} = [\overline{1, 1, \ldots, 1, n}] \qquad \text{period length } P \tag{5}$$

where the horizontal bar indicates periodic repetition. A simple near noble number is $[\overline{1, 2}] = \sqrt{3} - 1$.

To see which $\tilde{v}$ have the continued fraction expansion represented in equation 5, we express its periodicity in the following form:

$$\tilde{v} = \left[1, 1, \ldots, 1, n, \frac{1}{\tilde{v}} \right] \tag{6}$$

where the last term on the right, $1/\tilde{v}$, although not an integer, is treated like any other term in a continued fraction expansion. For example,

$$\tilde{v} = [\overline{1,\,2}] = \left[1,\,2,\,\frac{1}{\tilde{v}}\right] = \frac{1}{1 + 1/(2 + \tilde{v})} \tag{7}$$

This is a quadratic equation for $\tilde{v}$ with the positive solution $\tilde{v} = \sqrt{3} - 1$.
  Calculating the value of $\tilde{v}$ from equation 6, we obtain

1 1 1 ... 1	n	$1/\tilde{v}$
1 0 1 1 2 ... $F_{P-2}\ F_{P-1}$	$nF_{P-1} + F_{P-2}$	$(nF_{P-1} + F_{P-2})\tilde{v}^{-1}\,F_{P-1}$
0 1 1 2 3 ... $F_{P-1}\ F_P$	$nF_P\ \ + F_{P-1}$	$(nF_P\ \ + F_{P-1})\tilde{v}^{-1}\,F_P$

where the F_n are the Fibonacci numbers [Schr 90]. From this, we obtain the following quadratic equation for $\tilde{v}$:

$$\tilde{v} = \frac{nF_{P-1} + F_{P-2} + \tilde{v}F_{P-1}}{nF_P + F_{P-1} + \tilde{v}F_P} \tag{8}$$

which has the solution

$$\tilde{v} = \frac{n}{2}\left(\sqrt{1 + 4\,\frac{nF_{P-1} + F_{P-2}}{n^2 F_P}} - 1\right) \tag{9}$$

For the special case $n = 2$, one obtains

$$\tilde{v} = \sqrt{\frac{F_P + 2F_{P-1} + F_{P-2}}{F_P}} - 1 \tag{10}$$

Applying   the   recursions $F_P + F_{P-1} = F_{P+1}$, $F_{P-1} + F_{P-2} = F_P$,   and $F_P + F_{P+1} = F_{P+2}$, the final result is

$$\tilde{v} = \sqrt{\frac{F_{P+2}}{F_P}} - 1 \tag{11}$$

  For $P = 3$, for example, equation 11 yields $\tilde{v} = \sqrt{\frac{5}{2}} - 1$, which indeed has a continued fraction expansion of period length 3, the period terminating in a 2: $\sqrt{\frac{5}{2}} - 1 = [\overline{1,\,1,\,2}]$.
  Asymptotically, for $P \to \infty$ any fixed n, our nearest noble numbers will approach the golden mean $\gamma = (\sqrt{5} - 1)_2 \approx 0.61803$. For example, for $n=2$

and $P = 10$, $\tilde{v} \approx 0.61808$. According to equation 10 or 11, successive values of $\tilde{v}$ for $P = 1, 2, 3, \ldots$ constitute, in fact, an approach to the golden mean γ through (quadratic) irrational numbers.

A possible sequence for *cubic* irrational numbers is

$$x_n = \left(\frac{F_n}{F_{n+k}} \right)^{1/k} \tag{12}$$

with $k = 3$, and $k > 3$ for quartic and higher-degree irrational numbers.

References

Single-author publications are identified by the first letters of the author's name followed by the year of publication. Multiple-author works are identified by the first letter of each author's name. Identical author codes are listed in the order of appearance in the text and distinguished by lowercase letters following the year.

[AAMA 82]   W. Alvarez, F. Asaro, H. V. Michel, and L. W. Alvarez: Iridium anomaly approximately synchronous with terminal eocene extinctions. *Science* **216**, 886–888.

[Agu 76]   M. Agu: A nonstationary time-series having $1/f$-type power spectrum. *J. Phys. Soc. Japan* **40**, 1510–1511.

[AH 48]   R. A. Alpher and R. Herman: Evolution of the universe. *Nature* **162**, 774–775.

[AL 85]   P. Alstrøm and M. T. Levinsen: Fractal structure of the complete devil's staircase in dissipative systems described by a driven damped-pendulum equation with a distorted potential. *Phys. Rev.* **B32**, 1503–1511.

[Alv 87]   L. W. Alvarez: *Adventures of a Physicist* (Basic Books, New York).

[And 58]   P. W. Anderson: Absence of diffusion in certain random lattices. *Phys. Rev.* **109**, 1492–1505.

[Arn 89]   V. I. Arnold: *Mathematical Methods of Classical Mechanics* (Springer, Berlin).

[AS 81]   J. E. Avron and B. Simon: Almost periodic Hill's equation and the rings of Saturn. *Phys. Rev. Lett.* **46**, 1166–1168.

[ASS 86]   G. Albinet, G. Searby, and D. Stauffer: Fire propagation in a 2-D random medium. *J. Phys.* (Paris) **47**, 1–7.

[ASSW 66]   B. S. Atal, M. R. Schroeder, G. M. Sessler, and J. E. West: Evaluation of acoustic properties of enclosures by means of digital computers. *J. Acoust. Soc. Am.* **40**, 428–433.

[AZLPAGNCFFMS-SBCGHHHKSW 89]   M. E. Ander, M. A. Zumberge, T. Lautzenhiser, R. L. Parker, C. L. V. Aiken, M. R, Gorman, M. M. Nieto, A. P. R. Cooper, J. F. Ferguson, E. Fisher, G. A. McMechan, G. Sasagawa, J. M. Stevenson, G. Backus, A. D. Chave, J. Greer, P. Hammer, B. L. Hansen, J. A. Hildebrand, J. R. Kelty, C. Sidles, and J. Wirtz: Test of Newton's inverse-square law in the Greenland ice cap. *Phys. Rev. Lett.* **62**, 985–988.

[Bar 1509]   A. Barclay: *The Ship of Fools*. (Translation from the original Alsatian German of *Das Narrenschiff*, 1494, by Sebastian Brant; an exposition of abuses within the church and precursor of the Protestant Reformation.)

[Bar 88]   M. Barnsley: *Fractals Everywhere* (Academic Press, San Diego).

[Bas 90]   T. A. Bass: *The Newtonian Casino* (Longman, New York).

[BB 79]   J. von Boehm and P. Bak: Devil's stairs and the commensurate-incommensurate transition in CeSb. *Phys. Rev. Lett.* **42**, 122–125.

[BB 82]   P. Bak and R. Bruinsma: One-dimensional Ising model and the complete devil's staircase. *Phys. Rev. Lett.* **49**, 249–251.

[BB 83]   R. Bruinsma and P. Bak: Self-similarity and fractal dimension of the devil's staircase in the one-dimensional Ising model. *Phys. Rev.* **B27**, 5824–5825.

[BB 87]   J. M. Borwein and P. B. Borwein: *Pi and the AGM—A Study in Analytic Number Theory and Computational Complexity* (Wiley, New York).

[BBB 89]   J. M. Borwein, P. B. Borwein, and D. H. Bailey: Ramanujan, modular equations, and approximations to pi, or how to compute one billion digits of pi. *Am. Math. Monthly* **96**, 201–219.

[BBFT 89]   P. G. Bizetti, A. M. Bizetti-Sona, T. Fazzini, and N. Taccetti: Search for a composition-dependent fifth force. *Phys. Rev. Lett.* **62**, 2901–2904.

[BBJ 84]   T. Bohr, P. Bak, and M. H. Jensen: Transition to chaos by interaction of resonances in dissipative systems. II. Josephson junctions, charge-density, and standard maps. *Phys. Rev.* **A30**, 1970–1981.

[BCG 82]   E. R. Berlekamp, J. H. Conway, and R. K. Guy: *Winning Ways for Your Mathematical Plays*, vol. 2 (Academic Press, New York), chapter 25.

[Ber 62]   L. L. Beranek: *Music, Acoustics and Architecture* (Wiley, New York).

[Ber 79]   M. V. Berry: Diffractals. *J. Phys.* **A12**, 781–797.

[Bil 83]   P. Billingsley: The singular function of bold play. *Am. Sci.* **71**, 392–397.

[BPW 87]   J. Bahcall, T. Piran, and S. Weinberg (eds.): *Dark Matter in the Universe* (World Scientific, Singapore).

[BS 84]   H. D. Bale and P. W. Smith: Small-angle x-ray scattering investigation of submicroscopic porosity with fractal properties. *Phys. Rev. Lett.* **53**, 596–599.

[BTW 87]   P. Bak, C. Tang, and K. Wiesenfeld: Self-organized criticality: An explanation of $1/f$ noise. *Phys. Rev. Lett.* **59**, 381–384.

[BTW 88]   P. Bak, C. Tang, and K. Wiesenfeld: Self-organized criticality. *Phys. Rev.* **A38**, 364–374.

[Bur 87]   R. W. Burchfield (ed.): *Oxford English Dictionary* (Oxford University Press, Oxford).

[BW 90]   K. L. Babcock and R. M. Westerverlt: Avalanche and self-organization in cellular magnetic-domain patterns. *Phys. Rev. Lett.* **64**, 2168–2171.

[Cam 86]   X. Campi: Multifragmentation: Nuclei break up like percolation. *J. Phys.* **A19**, L917–921. See also *Nuclear Phys.* **A459**, 692.

[Car 85]   J. L. Cardy: Conformal invariance and the Yang-Lee edge singularity in two dimensions. *Phys. Rev. Lett.* **54**, 1354–1356.

[CE 80]   P. C. Collet and J.-P. Eckmann: *Iterated Maps of the Interval as Dynamical Systems* (Birkhäuser Boston, Cambridge, Mass.).

[CJ 87]   D. R. Chiavalo and J. Jalife: Nonlinear dynamics of cardiac excitation and impulse propagation. *Nature* **330**, 749–752.

[CJKP 85]   P. Cvitanović, M. H. Jensen, L. P. Kadanoff, and I. Procaccia: Renormalization, unstable manifolds, and the fractal structure of mode locking. *Phys. Rev. Lett.* **55**, 343–346.

[CLK 88]   H. Chen, D. X. Li, and K. H. Kuo: New type of two-dimensional quasicrystal with twelvefold rotational symmetry. *Phys. Rev. Lett.* **60**, 1645–1648.

[CM 90]   R. G. Corzine and J. A. Mosko: *Four-Arm Spiral Antennas* (Artech House, Norwood, Mass.).

[CMP 87]   D. D. Coon, S. N. Maa, and A. G. U. Perera: Farey fraction frequency modulation in the neuron-like output of silicon *p-i-n* diodes at 4.2 K. *Phys. Rev. Lett.* **58**, 1139–1142.

[COM 66]   J. H. Comroe: The lung. *Scientific American* **214**, 56–68 (February 1966).

[Cri 81]   F. Crick: *Life Itself* (Simon & Schuster, New York).

[CS 88]   D. S. Clark and O. Shisha: Invulnerable queens on an infinite chessboard. *3rd Int. Conf. Combinatorial Math.*, Annals of the New York Academy of Sciences.

[D'An 90]   P. D'Antonio: A new 1- or 2-dimensional fractal sound diffusor. *J. Acoust. Soc. Am.*, Suppl. 1, **87**, S10.

[DDI 83]   B. Derrida, L. De Seze, and C. Itzykson: Fractal structure of zeroes in hierarchical models. *J. Statist. Phys.* **33**, 559–569.

[deB 81]   N. G. de Bruijn: Sequences of zeros and ones generated by special production rules. *Kon. Ned. Akad. Wetensch. Proc. Ser.* **A84** (*Indaginationes Mathematicae* **43**) 27–37. See also N. G. de Bruijn: Algebraic theory of Penrose's nonperiodic tilings of the plane. *Kon. Ned. Akad. Wetensch.* **84**, 39–66.

[deW 51]   H. J. de Wijs: Statistics of ore distribution. *Geologie en Mijnbouw* (Amsterdam) **13**, 365–375. See also **15**, 12–24 (1953).

[DGJPS 85]   A. W. M. Dress, M. Gerhardt, N. I. Jaeger, P. J. Plath, and H. Schuster: Some proposals concerning the mathematical modelling of oscillating heterogeneous catalytic reactions on metal surfaces. In L. Rensing and N. I. Jaeger (eds.): *Temporal Order* (Springer, Berlin), pp. 67–74.

[DGP 78]   B. Derrida, A. Gervois, and Y. Pomeau: Iteration of endomorphisms on the real axis and representation of numbers. *Ann. Inst. Henri Poincaré* **XXIX**, 305–356.

[DH 82]   A. Douady and J. H. Hubbard: Iterations des polynomes quadratique complexes. *Compt. Rend. Acad. Sci. Paris* **294**, 123–126.

[DH 85]   A. Douady and J. H. Hubbard: On the dynamics of polynomial-like mappings. *Ann. Sci. École Normale Sup. 4e Série*, **18**, 287–343.

[Dia 77]   P. Diaconis: The distribution of leading digit and uniform distributions mod 1. *Ann. Prob.* **5**, 72–81.

[DK 85]   M. Duneau and A. Katz: Quasiperiodic patterns. *Phys. Rev. Lett.* **54**, 2688–2691.

[DK 88]   C. Davies and D. Knuth: Number representations and dragon curves. *J. Recreational Math.* **3**, 66–81 and 133–149.

[DMP 82]   F. M. Dekking, M. Mendès France, and A. van der Poorten: Folds! *Math. Intelligencer* **4**, 130–138, 173–181, and 190–195.

[Dou 86]   A. Doudy: Julia sets and the Mandelbrot set. In H.-O. Peitgen and P. H. Richter: *The Beauty of Fractals* (Springer, Berlin). pp. 161–173.

[Du 84]   B.-S. Du: A chaotic function whose nonwandering set is the Cantor ternary set. *Proc. Am. Math. Soc.* **92**, 277–278.

[EG 88]   R. B. Eggleton and R. K. Guy: Catalan strikes again! *Mathematics Magazine* **61**, 211–219.

[Ein 05]   A. Einstein: Über die von der molekularkinetischen Theorie der Wärme geforderte Bewegung von in ruhenden Flüssigkeiten suspendierten Teilchen. *Annalen der Physik* **17**, 549–560.

[Els 85]   V. Elsner: Indexing problems in quasi-crystal diffraction. *Phys. Rev.* **B32**, 4892–4898.

[Erd 54]   A. Erdély: *Tables of Integral Transforms* (McGraw-Hill, New York).

[Esc 71]   M. C. Escher: *The World of M. C. Escher* (H. N. Abrams, New York).

[EU 86]   R. Eykholt and D. K. Umberger: Characterization of fat fractals in nonlinear dynamical systems. *Phys. Rev. Lett.* **57**, 2333–2336.

[Fal 87]   K. J. Falconer: Digital sun dials, paradoxical sets, and Vitushkin's conjecture. *Math Intelligencer* **9**, 24–27.

[Far 82]   J. D. Farmer: Dimension, fractal measure and chaotic dynamics. In H. Haken (ed.): *Evolution of Order and Chaos* (Springer, Berlin/New York).

[Far 85]   J. D. Farmer: Sensitive dependence on parameters in nonlinear dynamics. *Phys. Rev. Lett.* **55**, 351–354.

[Fed 88]   J. Feder: *Fractals* (Plenum, New York).

[Fei 79]   M. J. Feigenbaum: The universal metric properties of nonlinear transformations. *J. Statis. Phys.* **21**, 669–706.

[Fei 83]   M. J. Feigenbaum: Universal behavior in nonlinear systems. *Physica* **7D**, 16–39.

[Fel 68]   W. Feller: *An Introduction to Probability Theory and Its Applications* (Wiley, New York).

[FHG 85]   A. P. Fein, M. S. Heutmaker, and J. P. Gollub: Physical scaling at the transition from quasiperiodicity to chaos in a hydrodynamic system. *Phys. Scr.* **T9** 79–84.

[Fib 1202]   Leonardo Fibonacci: *Liber Abaci* (Pisa).

[Fis 90]   D. E. Fisher: *Fire and Ice: The Greenhouse Effect, Ozone Depletion, and Nuclear Winter* (Harper & Row, New York).

[FL 84]   F. Family and D. P. Landau (eds.): *Kinetics of Aggregation and Gelation* (North Holland, Amsterdam).

[Fri 1240]   Friedrich II.: *De Arte Venandi cum Avibus.* (The Latin original of *On the Art of Hunting with Birds,* with the emperor's own drawings, was stolen during the siege of Parma in 1248 and subsequently lost. A copy made by his son Manfred was republished by Schneider, Leipzig, 1789.)

[Gar 70]   M. Gardner: The fantastic combinations of John Conway's new solitaire game of life. *Scientific American* **223**, 120–123 (Mathematical Games, April 1970).

[Gar 77]   M. Gardner: Extraordinary nonperiodic tiling that enriches the theory of tiles. *Scientific American* **236**, 110–121 (Mathematical Games, January 1977).

[Gar 78]   M. Gardner: White and brown music, fractal curves and one-over-f noise. *Scientific American* **238**, 16–32 (Mathematical Games, April 1978).

[Gar 83]   M. Gardner: *Wheels, Life and Other Mathematical Amusements* (W. H. Freeman, New York).

[Gar 89]   M. Gardner: *Penrose Tiles to Trapdoor Ciphers* (W. H. Freeman, New York).

[Gil 58]   E. N. Gilbert: Gray codes and paths on the n-cube. *Bell Syst. Tech. J.* **37**, 815–826.

[Gil 84]   W. J. Gilbert: A cube-filling Hilbert curve. *Math. Intelligencer* **6**, 78.

[Gle 87]   J. Gleick: *Chaos: Making a New Science* (Viking, New York).

[GLL 78]   R. L. Graham, S. Lin, and C.-S. Lin: Spectra of numbers. *Mathematics Magazine* **51**, 174–176.

[GP 83]   P. Grassberger and I. Procaccia: Characterization of strange attractors. *Phys. Rev. Lett.* **50**, 346–349.

[Gra 81]   P. Grassberger: On the Hausdorff dimension of fractal attractors. *J. Statist. Phys.* **26**, 173–179.

[Gra 83]   P. Grassberger: Generalized dimensions of strange attractors. *Phys. Lett.* **97A**, 227–230.

[Gra 85]   P. Grassberger: On the spreading of two-dimensional percolation. *J. Phys.* **A18**, L215–219.

[Gri 89]   G. Grimmett: *Percolation* (Springer, New York).

[Gro 82]   S. Grossmann: Diversity and universality. Spectral structure of discrete time evolution. In H. Haken (ed.): *Evolution of Order and Chaos* (Springer, Berlin).

[GS 87]   B. Grünbaum and G. C. Shephard: *Tilings and Patterns* (W. H. Freeman, New York).

[GT 77]   S. Grossmann and S. Thomae: Invariant distributions and stationary correlations of one-dimensional discrete processes. *Zeitschrift für Naturforschung* **32a**, 1353–1363.

[GW 87]   E. G. Gwinn and R. M. Westervelt: Scaling structure of attractors at the transition from quasiperiodicity to chaos in electronic transport in Ge. *Phys. Rev. Lett.* **59**, 157–160.

[GZR 87]   T. Geisel, A. Zacherl, and G. Radons: Generic 1/f noise in chaotic Hamiltonian systems. *Phys. Rev. Lett.* **59**, 2503–2506.

[Hak 78]   H. Haken: *Synergetics* (Springer, Berlin/New York).

[Hal 28]   J. B. S. Haldane: *On Being the Right Size* (Oxford University Press, London).

[Hal 74]   P. R. Halmos: *Naive Set Theory* (Springer, New York).

[Har 77]   S. Harris: *What's So Funny about Science* (Kaufmann, Los Altos, Calif.).

[Haw 88]   S. Hawking: *A Brief History of Time* (Bantam, New York).

[HBM 39]   F. V. Hunt, L. L. Beranek, and D. Y. Maa: *Analysis of sound decay in rectangular rooms. J. Acoust. Soc. Am.* **11**, 80–94.

[HBS 65]   H. E. Hurst, R. P. Black, and Y. M. Simaika: *Long Term Storage: An Experimental Study* (Constable, London).

[Hen 76]   M. Hénon: A two-dimensional map with a strange attractor. *Commun. Math. Phys.* **50**, 69–77.

[Hen 88]   M. Hénon: Chaotic scattering modelled by an inclined billiard. *Physica* **D33**, 132–156.

[HFJ 87]   E. Hinrichsen, J. Feder, and T. Jøssang: DLA growth from a line. Report Series, Cooperative Phenomena Project, Department of Physics, University of Oslo **87-11**, 1–21.

[HJKPS 86]   T. C. Halsey, M. H. Jensen, L. P. Kadanoff, I. Procaccia, and B. I. Shraiman: Fractal measures and their singularities: The characterization of strange sets. *Phys. Rev.* **A33**, 1141–1151.

[HLSTLW 86]   G. Hasinger, A. Langmeier, M. Sztajno, J. Trümper, W. H. G. Lewien, and N. E. White: Quasi-periodic oscillations in the X-ray flux of Cyg X-2. *Nature* **319**, 469–471.

[Hof 80]   D. R. Hofstadter: *Gödel, Escher, Bach: An Eternal Golden Braid* (Vintage, 1980).

[HP 83a]   H. G. E. Hentschel and I. Procaccia: The infinite number of generalized dimensions of fractals and strange attractors. *Physica* **8D**, 435–444.

[HP 83b]   H. G. E. Hentschel and I. Procaccia: Fractal nature of turbulence as manifested in turbulent diffusion. *Phys. Rev.* **A27**, 1266–1269.

[HP 84]   H. G. E. Hentschel and I. Procaccia: Relative diffusion in turbulent media: The fractal dimension of clouds. *Phys. Rev.* **A29**, 1461–1470.

[HS 79]   M. Harwit and N. J. A. Sloane: *Hadamard Transform Optics* (Academic Press, New York).

[Hu 85]   B. Hu: Problem of universality in phase transitions on hierarchical lattices. *Phys. Rev. Lett.* **55**, 2316–2319.

[Huy 1673]   C. Huygens: *Horologium Oscillatorium* (Muguet, Paris).

[HW 84]   G. H. Hardy and E. M. Wright: *An Introduction to the Theory of Numbers* (Clarendon, Oxford).

[INF 85]   T. Ichimasa, H.-U. Nissen, and Y. Fukano: New ordered state between crystalline and amorphous in Ni-Cr particles. *Phys. Rev. Lett.* **55**, 511–513.

[INF 88]   T. Ichimasa, H.-U. Nissen, and Y. Fukano: Electron microscopy of crystalloid structure in Ni-Cr small particles. *Phil. Mag.* **A58**, 835–863.

[Jak 81]   M. V. Jakobson: Absolutely continuous invariant measures for one-parameter families of one-dimensional maps. *Commun. Math. Phys.* **81**, 39–88.

[Jan 89]   A. Janner: Symmetries in higher-dimensional crystallography. *Phase Transitions* **16/17**, 87–101.

[JBB 83]   M. H. Jensen, P. Bak, and T. Bohr: Complete devil's staircase, fractal dimension, and universality of mode-locking structure in the circle map. *Phys. Rev. Lett.* **50**, 1637–1639.

[JBB 84]   M. H. Jensen, P. Bak, and T. Bohr: Transition to chaos by interaction of resonances in dissipative systems. *Phys. Rev.* **A30**, 1960–1969.

[JE 51]   E. Jahnke and F. Emde: *Tables of Higher Functions* (Teubner, Leipzig; also published by Dover, New York).

[JER 90]   C. Jekeli, D. H. Eckhardt, and A. J. Romaides: Tower gravity experiment: No evidence for non-Newtonian gravity. *Phys. Rev. Lett.* **64**, 1204–1206.

[JKLPS 85]   M. H. Jensen, L. P. Kadanoff, A. Libchaber, I. Procaccia, and J. Stavans: Global universality at the onset of chaos. *Phys. Rev. Lett.* **55**, 2798–2801.

[JM 85]   R. V. Jensen and C. R. Myer: Images of the critical points of nonlinear maps. *Phys. Rev.* **A32**, 1222–1224.

[Jon 88]   D. Jones: Abstract concrete. *Nature* **332**, 310.

[JSS 69]   B. Julesz, D. Slepian, and M. M. Sondhi: Correction of astigmatism by lens rotation and image processing. *J. Opt. Soc. Am.* **8**, 485.

[Kac 85]   M. Kac: *Enigmas of Chance* (Harper & Row, New York).

[KB 88]   B. Klein and I. Bivens: Proof without words. *Mathematics Magazine* **61**, 219.

[KBJ 83]   M. Kolb, R. Botet, and R. Jullien: Scaling of kinetically growing clusters. *Phys. Rev. Lett.* **51**, 1123–1126.

[KD 86]   A. Katz and M. Duneau: Quasiperiodic patterns and icosahedral symmetry. *J. Phys.* **47**, 181–196.

[Kel 56]   J. L. Kelly: A new interpretation of information rate. *Bell Syst. Tech. J.* **35**, 917–926.

[Kes 87]   H. Kesten: *Percolation Theory and Ergodic Theory of Infinite Particle Systems* (Springer, New York).

[Kim 81]   S. Kim: *Inversions* (Byte Books, McGraw-Hill, Peterborough, N. H.).

[Koh 1847]   R. Kohlrausch: Über das Dellman'sche Elektrometer. *Annalen der Physik und Chemie* (Poggendorf) **III-12**, 353–405.

[Koh 1854]   R. Kohlrausch: Theorie des elektrischen Rückstandes in der Leidener Flasche. *Annalen der Physik und Chemie* (Poggendorf) **IV-91**, 56–82 and 179–214.

[KS 87]   J. Koplowitz and A. P. Sundar Raj: A robust filtering algorithm for subpixel reconstruction of chain coded line drawings. *IEEE Trans. Pattern Analysis and Machine Analysis* **9**, 451–457.

[KW 89]   V. Klee and S. Wagon: *New and Old Unsolved Problems in Plane Geometry and Number Theory* (Mathematical Association of America, Washington, D. C.).

[Las 89]   J. Laskar: A numerical experiment on the chaotic behavior of the solar system. *Nature* **338**, 237–238.

[Law 88]   J. H. Lawton: More time means more variation. *Nature* **334**, 563.

[LC 81]   W. Lauterborn and E. Cramer: Subharmonic route to chaos observed in acoustics. *Phys. Rev. Lett.* **47**, 1445–1448.

[Lei 1714]   G. W. Leibniz: *Principia Philosophiae, More Geometrico Demonstrata*. Written in 1714 and generally known as *Monadologia* (*Monadology*), this great opus of Leibniz was first printed in 1720–1721. Several learned academies are still working on a complete edition of his works.

[Leo 62]   L. B. Leopold: Rivers. *Am. Scientist* **50**, 511–537.

[LF 88]   M. L. Lapidus and J. Fleckinger-Pellé: Tambour fractal: vers une résolution de la conjecture de Weyl-Berry pour less valeurs propres du laplacien. *Compt. Rend. Acad. Sci. Paris Math.* **306**, Sér. I, 171–175.

[LH 86]   W. Lauterborn and J. Holzfuss: Evidence for a low-dimensional strange attractor in acoustic turbulence. *Phys. Lett.* **A115**, 369–372.

[Lin 68]   A. Lindemeyer: Mathematical models of cellular interactions in development. *J. Theor. Biol.* **18**, 280–315.

[Liu 85]   S. H. Liu: Fractal model for the ac-response of a rough interface. *Phys. Rev. Lett.* **55**, 529–532.

[LL 76]   L. D. Landau and E. M. Lifschitz: *Mechanics* (Pergamon Press, Oxford), sec. 10.

[LM 80]   A. Libchaber and J. Maurer: Une experience de Rayleigh-Bénard de géometrie reduite; Multiplication, Accrochage et démultiplication de fréquences. *J. Phys. (Paris) Coll.* **41**, C3–51.

[LM 85]   S. Lovejoy and B. B. Mandelbrot: Fractal properties of rain, and a fractal model. *Tellus* **37A**, 209–232.

[Lor 80]   E. N. Lorenz: Noisy periodicity and reverse bifurcation. In R. H. G. Helleman: *Nonlinear Dynamics (Annals of the New York Academy of Sciences* **357**, 282–291).

[Lov 82]   S. Lovejoy: Area-perimeter relation for rain and cloud areas. *Science* **216**, 185–187.

[LP 88]   W. Lauterborn and U. Parlitz: Methods of chaos physics and their application to acoustics. *J. Acoust. Soc. Am.* **84**, 1975–1993.

[LS 84]   L. Levin and P. J. Steinhardt: Quasicrystals: A new class of ordered structures. *Phys. Rev. Lett.* **53**, 2477–2480.

[LY 75]   T. -Y. Li and J. A. Yorke: Period three implies chaos. *Am. Math. Monthly* **82**, 985–992.

[Mac 82]   A. Mackey: Crystallography and the Penrose pattern. *Physica* **114A**, 609–613.

[Mal 91]   C. Mallows: Conways challenging sequence. *Am. Math. Monthly* **98**, 5–20.

[Man 61]   B. B. Mandelbrot: On the theory of word frequencies and on related Markovian models of discourse. In R. Jacobson (ed.): *Structures of Language and Its Mathematical Aspects* (American Mathematical Society, New York).

[Man 63a]   B. B. Mandelbrot: The stable Paretian income distribution when the apparent exponent is near zero. *Int. Econ. Rev.* **4**, 111–115.

[Man 63b]   B. B. Mandelbrot: The variation of certain speculative stock prices. *J. Bus.* (Chicago) **36**, 394–419.

[Man 63c]   B. B. Mandelbrot: New methods in statistical economics. *J. Polit. Econ.* **71**. 421–440.

[Man 74]   B. B. Mandelbrot: Intermittent turbulence in self-similar cascades: Divergence of high moments and dimension of the carrier. *J. Fluid Mech.* **62**, 331–358.

[Man 80]   B. B. Mandelbrot: Fractal aspects of the interation $z \to \lambda z(1 - z)$ for complex λ and z. In R. H. G. Helleman (ed.): *Nonlinear Dynamics* (*Annals of the New York Academy of Sciences* **357**, 249–259).

[Man 83]   B. B. Mandelbrot: *The Fractal Geometry of Nature*, updated and augmented (W. H. Freeman, New York).

[Man 91]   B. B. Mandelbrot: *Fractals and Multifractals*, Selecta Vol. 1 (Springer, New York).

[May 88]   R. M. May: How many species are there on earth? *Science* **214**, 1441–1449.

[MBVB 89]   P. Manneville, N. Boccara, G. Y. Vichniac, and R. Bidaux (eds.): *Cellular Automata and Modeling of Complex Physical Systems* (Springer, Berlin).

[McG 71]   *McGraw-Hill Encyclopedia of Science and Technology* (New York).

[MCSW 86]   P. Meakin, A. Coniglio, H. E. Stanley, and T. A. Witten: Scaling properties for the surfaces of fractal and nonfractal objects: An infinite hierarchy of critical exponents. *Phys. Rev.* **A34**, 3325–3340.

[Mea 83]   P. Meakin: Formation of fractal clusters and networks by irreversible diffusion-limited aggregation. *Phys. Rev, Lett.* **51**, 1119–1122.

[Mea 87]   P. Meakin: Scaling properties for the growth probability measure and harmonic measure of fractals. *Phys. Rev.* **A35**, 2234–2245.

[Mei 82]   H. Meinhardt: *Models of Biological Pattern Formation* (Academic Press, London).

[Mek 90]   A. Z. Mekjian: Model of a fragmentation process and its power-law behavior. *Phys. Rev. Lett.* **64**, 2125–2128.

[Men 79]   K. Menger: *Selected Papers in Logic and Foundations, Didactics and Economics* (Reidel, Boston).

[MH 87]   J. E. Martin and A. J. Hurd: Scattering from fractals. *J. Appl. Crystallog.* **20**, 61–78.

[MK 87]   H. Meinhardt and M. Klinger: A model for pattern formation on the shells of molluscs. *J. Theor. Biol.* **126**, 63–89.

[MM 86]   S. Martin and W. Martienssen: Circle maps and mode locking in the driven electrical conductivity of barium sodium niobate crystals. *Phys. Rev. Lett.* **56**, 1522–1525.

[Moo 84]   F. C. Moon: Fractal boundary for chaos in a two-state mechanical oscillator. *Phys. Rev. Lett.* **53**, 962–964.

[Mor 21]   M. Morse: Recurrent geodesics on a surface of negative curvature. *Trans. Am. Math. Soc.* **22**, 84–100.

[MPP 84] B. B. Mandelbrot, D. Passoja, and A. Paullay: Fractal character of fracture surfaces of metal. *Nature* **308**, 721–722.

[MPRR 88] M. V. Mathews, J. R. Pierce, A. Reeves, and L. Roberts: Theoretical and experimental explorations of the Bohlen-Pierce scale. *J. Acoust. Soc. Am.* **84**, 1214–1222.

[MS 87] C. Meneveau and K. R. Sreenivasan: Simple multifractal cascade model for fully developed turbulence. *Phys. Rev. Lett.* **59**, 1424–1427.

[MSCW 85] P. Meakin, H. E. Stanley, A. Coniglio, and T. A. Witten: Surfaces, interfaces and screening of fractal structures. *Phys. Rev.* **A32**, 2364–2369.

[MSS 73] N. Metropolis, M. L. Stein, and P. R. Stein: On finite limit sets for transformations of the unit interval. *J. Combinatorial Theory* **(A)15**, 25–44.

[MW 69] B. B. Mandelbrot and J. R. Wallis: Some long-run properties of geophysical records. *Water Resources Research* **5**, 321–340.

[NPW 84] L. Niemeyer, L. Pietronero, and H. J. Wiesmann: Fractal dimension of dielectric breakdown. *Phys. Rev. Lett.* **52**, 1033–1040.

[NPW 86] L. Niemeyer, L. Pietronero, and H. J. Wiesmann: Response to comments on paper on dielectric breakdown [NPW 84]. *Phys. Rev. Lett.* **57**, 649–650.

[NS 89] M. Nauenberg and H. J. Schellnhuber: Analytic evaluation of the multifractal properties of a Newtonian Julia set. *Phys. Rev. Lett.* **62**, 1807–1810.

[OT 86] G. Y. Onoda and J. Toner: Fractal dimensions of model particle packings having multiple generations of agglomerations. *Comm. Am. Ceramic Soc.* **69**, C-278 to C-279.

[Ott 89] J. M. Ottino: The mixing of fluids. *Scientific American* **260**, 40–49. (January 1989).

[PA 83] P. Pfeifer and D. Avnir: Chemistry in noninteger dimensions between two and three. *J. Chem. Phys.* **79**, 3558–3565. Erratum: **80**, 4573 (1984).

[Pag 81] W. Page: Proof without words: Geometric sums. *Mathematics Magazine* **54**, 201.

[Pai 82] A. Pais: *Subtle Is the Lord . . .* (Clarendon Press, Oxford).

[Par 1896] V. Pareto: *Oeuvres Complètes* (Droz, Geneva).

[Pau 85] L. Pauling: Apparent icosahedral symmetry is due to directed multiple twinning of cubic crystals. *Nature* **317**, 512–514.

[Pen 74] R. Penrose: The role of aesthetics in pure and applied mathematical research. *Bull. Inst. Math. & Its Appl.* **10**, 266–271. See also R. Penrose: Pentaplexity: A class of non-periodic tilings of the plane. *Math. Intelligencer* **2**, 32–37 (1979).

[PI 87] W. Purkert and H. J. Ilgauds: *Georg Cantor* (Birkhäuser, Basel/Boston).

[Pie 83] J. R. Pierce: *The Science of Musical Sound* (Scientific American Books, W. H. Freeman, New York).

[Pin 62] R. S. Pinkham: On the distribution of first significant digits. *Ann. Math. Stat.* **32**, 1223–1230.

[PPR 85] H. -O. Peitgen, M. Prüfer, and P. H. Richter: Phase transitions and Julia sets. In W. Ebeling and M. Peschel (eds.): *Lotka-Volterra Approach to Cooperation and Competition in Dynamical Systems* (Akademie-Verlag, Berlin).

[PR 84] H. -O. Peitgen and P. Richter: *Harmonie in Chaos und Kosmos: Bilder aus der Theorie der dynamischen Systeme* (Die Sparkasse in Bremen, Bremen).

[PR 86a] H. -O. Peitgen and P. H. Richter: *The Beauty of Fractals (Springer, Berlin).*

[PR 86b] H. -O. Peitgen and P. H. Richter: *Fraktale und die Theorie der Phasenübergänge. Phys. Blätter* **42**, 9–22.

[Pru 87] P. Prusinkiewicz: Applications of L-systems to computer imagery. In H. Ehrig, M. Nagl, A. Rosenfeld, and G. Rozenberg (eds.): *Graph Grammars and Their Application to Computer Science, 3rd Int. Workshop* (Springer, Berlin) pp. 534–548 (*Lecture Notes in Computer Science* **291**).

[PS 86] L. Pietronero and A. P. Siebesma: Self-similarity of fluctuations in random multiplicative processes. *Phys. Rev. Lett.* **57**, 1098–1101.

[PS 88] H. -O. Peitgen and D. Saupe: *The Science of Fractal Images* (Springer, New York).

[PTT 87] I. Procaccia, S. Thomae, and C. Tressor: First return maps as a unified renormalization scheme for dynamical systems. *Phys. Rev.* **A 35**, 1884–1900.

[Pur 77] E. M. Purcell: Life at low Reynolds number. *Am. J. Phys.* **45**, 3–11.

[Pye 85] L. Pyenson: *The Young Einstein* (Hilger, Bristol/Boston).

[Rai 76] R. A. Raimi: The first digit problem. *Am. Math. Monthly* **83**, 521–538.

[Ram 14] S. Ramanujan: Modular equations and approximations to π. *Quar. J. Math.* **45**, 350–372.

[Ren 55] A. Rényi: On a new axiomatic theory of probability. *Acta Mathematica Hungarica* **6**, 285–335.

[RI 57] H. Rouse and S. Ince: *History of Hydraulics.* (Iowa Inst. of Hydraulics; republished in 1963 by Dover Publ., New York).

[Ris 71] J. -C. Risset: Paradoxes de hauteur. *Proc. 7th Int. Cong. Acoustics,* Budapest, paper S10, p. 20.

[Ris 75]   J.-C. Risset: Jugements relatifs de hauteur. *Compt. Rend. Acad. Sci. Paris* **B281**, 289–292.

[RK 82]   A. Rosenfeld and C. E. Kim: How a digital computer can tell whether a line is straight. *Am. Math. Monthly* **89**, 230–235.

[Rös 86]   J. Röschke: Eine Analyse der nichtlinearen EEG-Dynamik. (Dissertation, Göttingen).

[RP 88]   A. Redfearn and S. L. Pimm: *Ecological Monogr.* **58**, 39–55.

[RS 87]   P. Richter and H.-J. Scholz: Der goldene Schnitt in der Natur. In B.-O. Küppers (ed.): *Ordnung aus dem Chaos* (Piper, München), pp. 175–214.

[RTV 86]   R. Rammal, G. Toulouse, and M. A. Virasoro: Ultrametricity for physicists. *Rev. Mod. Phys.* **58**, 765–788.

[SA 85]   M. R. Schroeder and B. S. Atal: Stochastic coding of speech signals at very low bit rates: The importance of speech prediction. *Speech Comm.* **4**, 155–162.

[Sag 89]   Y. Sagher: Counting the rationals. *Am. Math. Monthly* **96**, 823.

[SASW 66]   M. R. Schroeder, B. S. Atal, G. M. Sessler, and J. E. West: Acoustical measurements in Philharmonic Hall (New York). *J. Acoust. Soc. Am.* **40**, 434–440.

[SBGC 84]   D. Shechtman, I. Blech, D. Gratias, and J. W. Cahn: Metallic phase with long-range orientational order and no translational symmetry. *Phys. Rev. Lett.* **53**, 1951–1953.

[Schr 54]   M. Schröder: Die statistischen Parameter der Frequenzkurven von grossen Räumen. *Acustica (Beiheft 2)* **4**, 594–600. For an English translation, see M. R. Schroeder: Statistical parameters of the frequency response curves of large rooms. *J. Audio Eng. Soc.* **35**, 299–306 (1987).

[Schr 64]   M. R. Schroeder: Improvement of acoustic feedback stability. *J. Acoust. Soc. Am.* **36**, 1718–1724.

[Schr 67]   M. R. Schroeder: Determination of the geometry of the human vocal tract by acoustic measurements. *J. Acoust. Soc. Am.* **41**, 1002–1010.

[Schr 69]   M. R. Schroeder: Images from computers and microfilm plotters. *Comm. Assoc. Comp.–Mach.* **12**, 95–101. See also M. R. Schroeder: Images from computers. *IEEE Spectrum* **6**, 66–78 (March 1969).

[Schr 70]   M. R. Schroeder: Digital simulation of sound transmission in reverberated spaces. *J. Acoust. Soc. Am.* **47**, 424–431.

[Schr 73]   M. R. Schroeder: An integrable model for the basilar membrane. *J. Acoust. Soc. Am.* **53**, 429–433.

[Schr 86]   M. R. Schroeder: Auditory paradox based on fractal waveform. *J. Acoust. Soc. Am.* **79**, 186–189.

[Schr 87]   M. R. Schroeder: Statistical parameters of the frequency response curves of large rooms. *J. Audio Eng. Soc.* **53**, 299–306.

[Schr 90]   M. R. Schroeder: *Number Theory in Science and Communication, with Applications in Cryptography, Physics, Digital Information, Computing, and Self-Similarity,* 2nd enlarged ed. (Springer, Berlin/New York).

[Schu 84]   H. G. Schuster: *Deterministic Chaos* (Physik-Verlag, Weinheim).

[Ses 80]   G. M. Sessler: *Electrets* (Springer, Berlin).

[SGS 74]   M. R. Schroeder, D. Gottlob, and F. K. Siebrasse: Comparative study of European concert halls: Correlation of subjective preference with geometric and acoustic parameters. *J. Acoust. Soc. Am.* **56**, 1195–1201.

[Sha 51]   C. E. Shannon: Prediction and entropy of printed English. *Bell Syst. Tech. J.* **30**, 50–64.

[Sha 53]   C. E. Shannon: Computers and automata. *Proc. IRE* **41**, 1234–1241.

[Sha 64]   A. N. Sharkovski: Coexistence of cycles of a continuous map of a line into itself. *Ukrain. Math. Zeitschrift* **16**, 61–71.

[She 62]   R. N. Shepard: The analysis of proximities: Multidimensional scaling with unknown distance function. *Psychometrica* **27**, 125–140 and 219–246.

[She 64]   R. N. Shepard: Circularity in pitch judgment. *J. Acoust. Soc. Am.* **36**, 2346–2353.

[She 82]   S. J. Shenker: Scaling behavior in a map of a circle onto itself: Empirical results. *Physica* (Utrecht) **5D**, 405–411.

[SHJ 88]   J. L. C. Sanz, E. B. Hinkle, and A. K. Jain: *Radon and Projection Transform-Based Computer Vision* (Springer, Berlin/New York).

[Sin 78]   D. Singer: Stable orbits and bifurcations of maps of the interval *SIAM J. Appl. Math.* **35**, 260–267.

[SK 87]   W. von Saarloos and D. A. Kurtze: Location of zeros in the complex temperature plane: Absence of Lee-Young theorem. *J. Phys.* **A17**, 1301–1311.

[Sla 60]   P. Slater: The analysis of personal preferences. *Br. J. Stat. Psychol.* **8**, 119–135.

[Slo 73]   N. J. A. Sloane: *A Handbook of Integer Sequences* (Academic Press, New York).

[SM 88]   H. E. Stanley and P. Meakin: Multifractal phenomena in physics and chemistry. *Nature* **335**, 405–409.

[Sma 67]   S. Smale: Differentiable dynamical systems. *Bull. Am. Math. Soc.* **73**, 747–817.

[SMWC 84]   D. W. Schaefer, J. E. Martin, P. Wiltzius, and D. S. Cannell: Fractal geometry of collidal aggregates. *Phys. Rev. Lett.* **52**, 2371–2374.

[SO 85]   H. E. Stanley and N. Ostrowsky (eds.): *On Growth and Form: Fractal and Non-Fractal Patterns in Physics* (NATO ASI, Martinus Nijhoff, Dordrecht).

[ST 71]   N. Suwa and T. Takahashi: *Morphological and morphometrical analysis of circulation in hypertension and ischemic kidney* (Urban & Schwarzenberg, Munich).

[Sta 85]   D. Stauffer: *Introduction to Percolation Theory* (Taylor & Francis, London).

[Ste 69]   S. S. Stevens: On predicting exponent for cross-modality matches. *Percep. Psychophys.* **6**, 251–256.

[Sut 89]   K. Sutner: Linear cellular automata and the Garden-of-Eden. *Math. Intelligencer* **11**, 49–53.

[SW 86]   J. Salem and S. Wolfram: Thermodynamics and hydrodynamics of cellular automata. In S. Wolfram (ed.): *Theory and Application of Cellular Automata* (World Scientific, Singapore/Teaneck, N.J.), pp. 362–366.

[SWGDRCL 86] J. P. Stokes, D. A. Weitz, J. P. Gollub, A. Dougherty, M. O. Robbins, P. M. Chaikin, and H. M. Lindsay: Interfacial stability of immiscible displacement in a porous medium. *Phys. Rev. Lett.* **57**, 1718–1721.

[TB 88]   C. Tang and P. Bak: Critical exponents and scaling relations for self-organized critical phenomena. *Phys. Rev. Lett.* **60**, 2347–2350.

[Tho 61]   D'A. W. Thompson: *On Growth and Form* (Cambridge University Press, Cambridge).

[Thu 06]   A. Thue: Über die gegenseitige Lage gleicher Teile gewisser Zeichenreihen. *K. Nord. Vid. Skrifter I Math. Nat.* (Oslo) **7**, 1–22.

[TKFFHKMM 89] J. Thomas, P. Kasameyer, O. Fackler, D. Felske, R. Harris, J. Kammeraad, M. Millett, and M. Mugge: Testing the inverse-square law of gravity on a 465-m tower. *Phys. Rev. Lett.* **63**, 1902–1905.

[TM 87]   T. Toffoli and N. Margolus: *Cellular Automata Machines* (MIT Press, Cambridge, Mass.).

[Ula 74]   S. Ulam: *Sets, Numbers, and Universes* (MIT Press, Cambridge, Mass.).

[Ulr 72]   S. Ulrich: *Schätze deutscher Kunst* (Bertelsmann, Munich).

[VC 78]   R. V. Voss and J. Clark: 1/f noise in music: Music from 1/f noise. *J. Acoust. Soc. Am.* **63**, 258–263.

[Ver 1845] P.-F. Verhulst: Récherches mathématiques sur la loi d'accroissement de la population. *Nouv. Mém. de l'Acad. Roy. des Sciences et Belles-Lettres de Bruxelles* **XVIII.8**, 1–38.

[Vic 84]   G. Vichniac: Simulating physics with cellular automata. *Physica* **10D**, 96–115.

[Vic 86]   G. Vichniac: Cellular automata models of disorder and organization. In Bienenstock, F. Fogelman Soulié, and G. Weisbuch (eds.): *Disordered Systems and Biological Organisation* (Springer, Berlin), pp. 1–20.

[Vil 87]   A. Vilenkin: Cosmic strings. *Scientific American* **257**, 52–60 (December 1987).

[Vos 85]   R. F. Voss: Random fractal forgeries. In R. A. Earnshaw (ed.): *Fundamental Algorithms for Computer Graphics* (Springer, Berlin/New York).

[Vos 88]   R. F. Voss: Fractals in nature: From characterization to simulation. In H.-O. Peitgen and D. Saupe: *The Science of Fractal Images* (Springer, New York).

[Wag 85]   S. Wagon: Is π normal? *Math. Intelligencer* **7**, 65–67.

[WCK 87]   N. Wang, H. Chen, and K. H. Kuo: Two-dimensional quasicrystal with eightfold rotational symmetry. *Phys. Rev. Lett.* **59**, 1010–1013.

[Wei 90]   J. Weiner: *The Next One Hundred Years: Shaping the Fate of Our Living Earth* (Bantam, New York).

[Wey 81]   H. Weyl: *Symmetrie* (Birkhäuser, Basel).

[Wil 67]   T. A. Wilson: Design of the bronchial tree. *Nature* **213**, 668–669.

[Wis 87]   J. Wisdom: Chaotic dynamics in the solar system. (Urey Prize Lecture): *Icarus* **72**, 241.

[WO 84]   D. A. Weitz and M. Oliveria: Fractal structures formed by kinetic aggregation of aqueous gold colloids. *Phys. Rev. Lett.* **52**, 1433–1436.

[Wol 84]   S. Wolfram: Cellular automata as models of complexity. *Nature* **311**, 419–424.

[Wol 86]   S. Wolfram (ed.): *Theory and Application of Cellular Automata* (World Scientific, Singapore/Teaneck, N.J.).

[WPM 83]   J. Wisdom, S. Peale, and F. Mignard: The chaotic rotation of Hyperion. *Icarus* **58**, 137–152.

[WS 83]   T. A. Witten and L. M. Sander: Diffusion-limited aggregation: A kinetic critical phenomenon. *Phys. Rev. Lett.* **47**, 1400–1403. See also *Phys. Rev.* **B27**, 5686–5697.

[WWM 87]   M. Werman, A. Wu, and R. A. Melter: Recognition and characterization of digitized curves. *Pattern Recognition Lett.* **5**, 207–213.

[YL 52]   C. N. Yang and T. D. Lee: Statistical theory of equations of state and phase transitions. *Phys. Rev.* **87**, 404–419.

[ZD 85]   R. K. P. Zia and W. J. Dallas: A simple derivation of quasi-crystalline spectra. *J. Phys. A: Math. Gen.* **18**, L341–L345.

[Zim 78]   M. H. Zimmerman: Hydraulic architecture of some diffuse-porous trees. *Can. J. Botany* **56**, 2286–2295.

[Zip 49]   G. K. Zipf: *Human Behavior and the Principle of Least Effort* (Addison-Wesley, Cambridge, Mass.).

[ZLP 76]   G. K. Zweig, R. Lipes, and J. R. Pierce: The cochlear compromise. *J. Acoust. Soc. Am.* **59**, 975–982.

Author Index

The codes in brackets refer to the publications listed in the References.

Index

Writing Under Pressure

WRITING UNDER PRESSURE

The Quick Writing Process

SANFORD KAYE

OXFORD UNIVERSITY PRESS

New York   Oxford

Oxford University Press

Oxford   New York   Toronto
Delhi   Bombay   Calcutta   Madras   Karachi
Petaling Jaya   Singapore   Hong Kong   Tokyo
Nairobi   Dar es Salaam   Cape Town
Melbourne   Auckland
and associated companies in
Berlin   Ibadan

Copyright © 1989 by Sanford Kaye

First published in 1989 by Oxford University Press, Inc.,
200 Madison Avenue, New York, New York 10016

First issued as an Oxford University Press paperback, 1990

Oxford is a registered trademark of Oxford University Press

Library of Congress Cataloging-in-Publication Data
Kaye, Sanford.
Writing under pressure: the quick writing process / Sanford Kaye.
p.   cm.
1. Authorship.   I. Title.   PN151.K38   1988   808′.02—dc19   87-35922   CIP
ISBN 0-19-505223-4
ISBN 0-19-506661-8 (pbk.)

2 4 6 8 10 9 7 5 3 1
Printed in the United States of America

for my mother and father

Preface

This is an unconventional book. Most of the thousands of books about writing fall into a few categories. The larger group consists of handbooks which provide variations on an idealized version of the "mechanics" of writing. A smaller group consists of texts that present a personal pedagogy with strong, practical recommendations. A few eminently useful books offer a consistent point of view on matters of style or diction.

This book assumes that anyone can write well under pressure, and sets out to make that possible. The first half combines theory and practice: a case study of a writing assignment with a close deadline that demonstrates what I call the Quick Writing Process—QWP—in an organic way. The writer's work of preparing, planning, and generating material is governed by a timetable to ensure good writing without the familiar last minute scrambling and rearranging that throws essays out of balance and confuses the reader. In the producing stage, QWP employs two levels of editing: a rough first cut for coherence and consistency, and then a second reading for unity.

The second half of the book, applying this system, enriches the general theory by illustrating specific practices. The chapter on exam-writing provides the most direct application, and presents a good example of how to plan and then produce good writing under pressure for a reader who holds power (such as a grade or promotion) over you. The chapters on research writing develop the system for discovering and presenting meaning. The chapters on organizational writing take the system into the environment, analyzing some of the subtle pressures against writing well that we all face in our daily work. These applications, with the final chapter, illustrate how you can adapt your own, thorough version of QWP to any writing assignment.

The idea that anyone can write well is not a popular notion. There is so much mythology and mystery surrounding the writing process and product that writing a book for everyone who needs or wants to write requires a good deal of support, encouragement, and constructive criticism. I have been extremely lucky in this regard. My brother, Professor Jerome Kaye, guided me in the idea that everyone deserves the best education possible, and that this often means the freedom to learn on one's own. I hope this book provides its readers with some helpful theory and practice as they teach themselves to write. My sons, Anders and Anson Kaye, patiently but persistently encouraged me through their understanding and experience of the writing process, as I hope this book will sustain readers in their independent work. My wife, Susan Shepherd, elegant writer and profound reader, is responsible for whatever aspects of content and style make this book readable. Her advice is the kind that I have incorporated into the QWP model: talking through a topic or a writing block; getting constructive feedback on a rough draft; and completing the conscious act of writing by shaping it into a product for readers. The idea of putting twenty-five years of teaching and consulting experience into book form arose in discussions with my colleague, Sandra Sanneh, at the Kennedy School of Government, Harvard University. Incisive, inspiring comments, questions, and recommendations by Paul Schlotthauer, my editor at Oxford University Press, made preparing the manuscript for publication as stimulating as writing the first draft. This book envisions writing as an independent, lifelong skill; that vision is put into practice with the support of people who believe we all have a right to speak, and to be heard.

Finally, anyone who has taught for twenty-five years has had thousands of students. For a writing teacher, this does not mean people in a lecture hall, but writers—students and clients—seated around a table sharing their work. These people from different backgrounds, with a variety of needs, have taught me what I know about the teaching and learning of writing. *Writing Under Pressure* comes from all of them— their experiences, frustrations, and discoveries in the act of writing—to help you develop your own personal, adaptable system for good writing on time.

Hanson, Mass.   S.K.
January 1988

Contents

PART ONE

QWP: The Quick Writing Process

CHAPTER
1

Preparing

Drawing a Blank

Most people would rather talk than write. Pressed to write, they draw a blank, or feel either that they have nothing new to say or that whatever they have to say could be said better by someone else. Yet when you talk to hesitant writers about assignments at work or school, you find they have *too* much to say: they are swamped by their research, and feel helpless in trying to sort out their own arguments in a clear and forceful way.

The concept of writing with confidence or pleasure is even more mysterious, one reserved for "born writers." This debilitating illusion is reenforced by our cultural myth that "writing can't be taught." Yet most of us *have* to write; and we wish we could do it well, with less frustration and more control. Fortunately, common sense tells us anyone can learn to write well, with both energy and satisfaction.

After twenty-five years of teaching writing, I've found that most writers share some unpleasant experiences. We feel a sense of dread when we are asked to answer a question that is often not of great interest to us. We have the dull knowledge that we most often write not to explore or to understand, but to demonstrate to a teacher or a supervisor that we have covered certain material, or adopted a particular way of looking at an issue. We remember those writing assignments that require us to condense a book's worth of material or six months of research into two pages. We've discovered that what we have struggled to say is not always read carefully. It isn't any wonder, then, that people normally write without a sense of self-discovery; they write not to engage the reader but to get an onerous job done.

3

When I was very young, it never occurred to me that I had "nothing to say." My friends and I talked all the time about everything. We asked everyone we knew about the questions on our minds, the discoveries we made every day. We wanted to know everything; and we certainly didn't mind telling anyone what we'd learned. But during the early years of school, we realized that what excited us most had almost nothing to do with education. Our teachers presided over a strict world of words and numbers and expected responses, none of which seemed to have anything to do with what we needed to know. The best we could do, trusting the advice of an older sister or a respected uncle, was to create a second self, that of the student who learned the things in school that were supposed to make sense later on in life. We relegated our own private searches to after-school hours. In time, this habit of neglecting what we really cared about took over. School was serious and exhausting; our feelings and ideas were only kid stuff. We stopped sharing, stopped comparing our ideas and feelings. We stopped wondering out loud.

In high school, the time-killing, predictable exercises meant to discipline our writing skills confirmed this need for a second self. We could not ignore the red marks in the margins of our compositions. Although it was quite clear that what we had to say was of little interest to our readers, the way we said it did elicit feelings, usually anger, irritation, and a lot of exclamation points. Our anxiety about mechanical errors was linked with an even deeper concern about grades, or winning a scholarship to college. Although I did write a column for our school newspaper ("Kaye's Korner"), my compositions in class were increasingly mindless as they became increasingly perfect. The goal, as our honors English teacher told us, was to leave her with "no work to do." In this system the best papers received no response at all.

I did win a scholarship, but when my first writing assignment in college was to "define Thoreau's concept of conscience," I drew a complete blank. I read Thoreau's "Civil Disobedience" three times, and still had nothing to say. Why ask *me* for Thoreau's concept of conscience? Didn't he know himself? I asked my roommate, who had written a Thoreau paper as a sophomore in prep school. He lectured me for an hour about the relationship between the individual and the state.

He could have written fifty pages on the spot, and I couldn't think of two! I got my paper done, but it was a humiliating experience. I pieced together a series of quotes from Thoreau and threw in a few transitions: "Then Thoreau says," "Then he points out." There was nothing of my own thinking in the paper, and certainly no pleasure or pride in writing it. Turning it in was an act of survival, which left me dreading the next paper even more.

I wish I had known then that everybody goes through writing anxieties, and that I was just beginning to teach myself how to find out what I had to say, and then apply it to the question posed. Through the process and product, I could enlighten myself and the reader; and I could make the whole experience open-ended, so that each time I wrote I not only answered the reader's questions, but also learned something more about my own writing.

That first semester in college, I realized that although my roommate might be one of those "born writers," I was not. My job was to rebuild the bridge between my instincts, feelings, experiences, and ideas, and my words on the page. I had to heal the torn and atrophied muscle of communication so that writing became what talking had been earlier in my life: an extension of personality, a reaching out to others with the excitement of real inquiry, and the conviction that it all mattered. I didn't need checklists or rules for making writing into a game to be played with teachers. I had to transform my cynicism about writing into a lifelong skill for staying in touch with myself and with the world. QWP, the Quick Writing Process, is a way of doing this, of transforming all our pent-up resistance to writing into a process of writing efficiently, with confidence and pleasure.

QWP depends on two different kinds of planning. First, you prepare to write by making an assessment of what is involved in answering the question or fulfilling the requirement. Then you plan the time and structure to get the job done. The QWP timetable is practical, not theoretical. Although you will want to establish the timetable for your whole project *before* you set out to write, the technique for planning writing time will be clearer if demonstrated in the context of a case study position paper in Chapter 3. Here, I want to consider the writer's preparation.

The Reader's Questions

Scientists speak of elegant solutions, engineers of "quick and dirty" results. In both cases, the issue is a process for getting work done in the most efficient manner. This simplicity, cutting away the fat in both the thinking and doing processes, is the essence of QWP. The result, at least in the beginning of one's experience with QWP, will fall somewhere between elegant and quick and dirty, depending upon the preparation and the kind of deadline. Perhaps the best way of viewing the goal is through the operative description, "quick and clean": a process for producing results somewhat short of perfection, but in a far more comprehensive, and persuasive manner than the untrained writer's unbalanced, unfocused, frantic method. QWP practice and experience strengthens the writer's capacity to write at his or her best in the most spontaneous way, by consciously integrating audience awareness; a provisional thesis that generates a balanced argument-outline with a beginning, middle, end; appropriate examples, and a practical look at the "other side"; and ideas and sentences connected by the internal logic of the thesis. All these elements drive toward a forceful conclusion that puts the question into a new context and provides the reader with a clearer perspective.

Any writing job requires some adjustment of the relationship between writer and reader. If you are writing under pressure, it is particularly important to make this relationship as conscious as possible when planning, and as visible as possible in the writing. The immediate benefit of such conscious planning is that it counteracts the sense of "drawing a blank." You begin at once by clarifying the relationship between writer and reader: is the reader someone who will understand a shorthand transcription of the way your mind works, so that you do not need to define your terms or make explicit all the connections among your thoughts? If the reader cannot provide those invisible connections, or supply his or her own examples for your abstract assertions, you need evidence and examples to reenforce your essay, memo, or letter.

Once this relationship between writer and reader becomes a conscious part of the planning process, it yields more useful information. A teacher or superior at work imposes personal or even institutional restrictions on the writer. Important decisions must be made if, for example, your writing makes an argument that will counter the beliefs of your reader. How much of a risk can you take? Is your writing a demonstration of knowledge, or of loyalty? Are you expected to speak your mind or, whether stated or implied, to "keep a low profile"? How much persuasion is appropriate: is there a genuine need to win the reader over to your point of view, or must you convince your reader you have covered a certain amount of material, sorted it out in your own mind, and integrated it with other information and ideas?

Everyone who writes has encountered each of these situations at one time or another. Whatever the writing task, this relationship must be considered when deciding your approach and style. Neutral prose definitely will not be appropriate in a personal essay for a college application, or in an experiential sociology paper. A strong personal voice may be out of place in a scientific monograph or government agency memo. The use of the first person pronoun may be risky in professional writing. There is no defense for dead prose, but in some environments, there are severe drawbacks to writing that is inappropriately intimate. The resolution of such questions affects the tone, the stance, and the texture of your writing. Once you define this relationship, you can anticipate some of the questions your reader will ask, and include the answers as you plan your writing.

Discovering the most appropriate relationship to the reader should not be a time-consuming process, of course. It's akin to a photographer's decision about which lens to use. The more consciously you make that decision, the better you control what most untrained writers consider a mystery: the reader's feeling of intimacy or distance—of personal involvement—with your writing.

The Writer's Question

Unless you've been part of a well-run writing group in which each person is responsible for giving constructive feedback, you may not have a sharp, balanced view of when to answer readers' questions as they naturally arise, and when to raise questions for readers to ponder. Writing is generally overkill. Studies have shown how little people remember of what they hear and read, but inexperienced writers often insist that either they thought their readers "already knew" what they had left out, or they didn't want to "shout at" readers for fear of losing them. This question about how much to say is cleared up immediately if you realize that there is never a good reason to leave the reader in doubt about what you mean.

Readers may find your argument difficult to grasp or wrongheaded: "Is the writer mistaken, or can I learn something here?" But if it is all there, you've done your job. When your writing is unclear, a reader wonders: "What's missing here? Why aren't there any good examples?" One set of questions results from the reader's active engagement, the other from the reader's confusion.

Writers under pressure do well to anticipate and accommodate the reader's and writer's inevitable questions. Writing under pressure is necessarily more deductive than inductive, leaving less to chance or mystery than writing that can be tested against readers, written again, and then rewritten. For a writer under pressure, there has to be a safety net underlying the process, and that safety net is established by making an early commitment to what you're going to say. This may be uncomfortable at first; and as the writing process continues, you may see the need to alter your basic argument. But the main purpose of QWP is to allow you to do the best job you can, in the time available, and that means accepting some of the discomfort and anxiety of knowing that, with *more* time, you might do a better, fuller, more elegant job. Moreover, as you continue to use QWP, you move closer and closer to fuller and more elegant pieces of writing.

Your goal is a process which allows you to write efficiently and confidently for a reader whom you have placed in an appropriate relationship to your words, and whose questions you have anticipated within the planned structure of your essay, memo, or exam. As these initial preparatory steps help you avoid "drawing a blank," you may begin to feel that now you have *too* much to say. Here you need to make one final preparatory decision.

Representation

No amount of planning can make the reader hear the music in your mind, or see the abstract design you perceive as you look out over a cranberry bog coming to fruition in late September. Writing can convey only certain things well, and others not at all. Therefore, you want a writing process based on realistic expectations. Your essay on banning cigarettes in public places, or your memo on how to maximize profits while still maintaining goodwill, almost never convinces readers directly or gives them a sense you have the final answer. Writing a scene between a husband and wife, even if based on an amalgam of your own experiences and those of your friends and relatives, will not match the actual time, space, or feelings of such moments. Writers can evoke similar responses, but such attempts involve risk and experimentation, because words on a page create a particular kind of experience in the reader's mind. Writers depend on *representations,* distillations of experience, thought, and imagination. These representations have the capacity to enact, in the theater of the reader's mind, passionate thoughts, feelings, and images. Conscious, careful selection, whether the representation is a word, symbol, or underlying metaphor, is crucial to both power and clarity in prose.

Representation is an act of economy in writing. I am not talking here about being brief, or "boiling it down." That is a different kind of economy, and one that is often dangerous. I had a writing client in government whose instructions were to reduce everything to one page, no matter what the subject or how much work had gone into the report. Almost half of the allotted page concerned who the writer was and where he could be found in the Pentagon. Distilling ideas or feelings into the symbols that are words on a page is not a matter of falsely or foolishly simplifying meaning. This skillful selection of the best words, evidence, examples, and structure out of an infinite number of possibilities, and of building a bridge to readers, forms the basic act of good writing. As a writer, you will always have too much material, most of which you will have to leave out. What you want is to suggest, to prompt the reader's mind. You are always sorting, scanning with incredible speed the different whorls of thought and feeling and instinct in order to pick out the best for your neat, linear rows of words on the page, or the bright letters on the screen. Writing well, under pressure, pushes this process to an even higher intensity. Writers constantly look for ways, in both structure and content, to signal, to indicate, to say more of what they have to say than they can actually get down on the page. If your task is to write one page, you need to fill it with the knowledge of a hundred pages. If your job is to read twenty books and thirty articles and come up with a twelve-page paper in a week, you must *select* with the greatest concentration, leaving crucial evidence behind, striving always to find the most representative, the most characteristic examples of your argument: not only the facts, experiences, and opinions, but the words themselves.

Many people cannot come to terms with this need to select. Some do not write at all or fail to finish their writing on time, because they are so keenly aware that they can never tell the *whole* story. When writing under pressure, it is as important to understand this need to select, to represent, as it is to analyze the writer's relationship to readers, to anticipate the readers' questions, and to resolve the writer's questions. But as with all of these preparatory questions, a writer's appreciation of representation must be tucked away finally, to let the process of writing begin.

2

Planning and Generating

Pressures

Suppose I am under pressure to write about whether to have an English competency exam as a requirement for graduation at, say, Shakespeare College of the Communicable Arts, where I teach writing. This (thankfully) purely fictional assignment is due tomorrow morning, and I am expected to write "about three pages." The dean is the audience. He doesn't need background: he wants evidence for a point of view—a position paper.

This imaginary assignment has the familiar characteristics of many such writing tasks: there isn't enough time to do a good job; there is a strong likelihood that what I say will not *convince* the reader as he weighs my views against those of others closer to him; but there is the possibility of doing something useful. It is also a topic I would never willingly choose to spend time on, so I immediately resist the task. If we took the example of a student paper in a college course, we could substitute other limitations roughly in the same range: the required length might be anywhere from two to fifteen pages; the paper might be due not in twenty-four hours but in a week; the audience would be a professor who either wanted to compare the student paper with some ideal paper in his or her mind, or hoped to find some fresh thinking on an old topic. Or we might take a memo, due in a few hours, for a supervisor who is depending on the writer to survey options and put forward a recommendation in one easily readable page. Each of these cases presents particular problems of its own; but they all share the general characteristics of writing under pressure. And probably all are accompanied by a certain sense of dread.

Writers react to the pressures of time constraints in different ways. Even if a deadline is several months or a year away, some people scramble to get started early, and still find themselves rushing at the last minute. Others *need* pressure to get anything done. Space requirements exert similar pressures: some writers dread condensing material, others dread expanding it. The power relationship between reader and writer intimidates some writers and provokes others. You may not want to write your memo or your paper in Modern Civilization, but you know you must do it. That gives the pressures of deadline, length, and the relationship with your reader real force, and you must respond, either unconsciously by letting them dominate your writing process and product, or consciously by transforming them into advantages. QWP harnesses the writer's energies against these pressures by shaping his or her commitment to the task.

The Writer's Commitment

The writer's commitment is a matter of planning, and is unique with each project. Such planning involves answering the following questions:

- How much does the reader know?
- How much has the writer learned?
- How much does the reader want to learn?
- How much does the writer want to tell?

Each of these questions pushes at the boundaries of the writer's task. I don't know of any method of writing that can turn a half-done, grudgingly written draft into a satisfying product. Readers immediately sense a tone of resistance, a lack of structure, a failure to explain or illustrate. Writing without commitment is the writer's curse. You have to find some way of making every topic unleash your energy and conviction. Otherwise, the process will be remorselessly painful, for both writer and reader.

In our scenario, then, I am writing for the dean of my college, a man for whom I have some respect, who is both creative and timid to the same degree as most of us. We are friendly, but not close; we know almost nothing about each other personally. I'm not sure how he feels about a competency exam, and I don't want to alienate him unnecessarily. Yet he has asked for my opinion as a teacher of writing, and I do have to respond.

My own attitude prevents me from cranking out meaningless writing simply to get the job done. Even if I have reservations, I know from experience that if my initial goal is to figure out what I *think* about this issue, I will learn something. If I can manage the writing process so that I approximate in words what I think and feel, then I will strengthen my writing skills as well. Beyond that, if I can get the reader *engaged* in my words, I will be especially pleased. It may well be that however committed I am to the subject, my reader will disagree and take an action I oppose; no one ever claimed that if you write well everyone will agree with you. But a writer can *learn,* and then tell readers what he or she has learned.

Generating a Provisional Thesis and *Because*-clauses

Having resolved issues of audience, reader's and writer's questions, and my commitment to this project, I am prepared to begin.* My instinctive response to the dean's request is that a competency exam in English is a terrible idea. I'm aware, however, that many people in and out of the profession feel differently, and the dean may be among them, or at least may be depending on such people for his decision. Even though research or the exchange of views might well change my mind, I don't have the time to engage in much reading or discussion. Writing under

---

*With QWP, it is important to have a realistic sense of time and space *before* writing. For the purpose of demonstrating this system the first time through, I want to get to *content* immediately, because QWP is based on discovering and presenting meaning. But at this point in a project I would set out a timetable and a structural plan. Roughly half the time in the QWP timetable is for preparing, planning, and generating material; the remaining time is for producing: moving from raw to final draft. The timetable is also integrated with a structural plan, sketched out against a framework of beginning, middle, and end (see Chapter 3).

pressure, I can do only *some* exploring. Most of the material for this kind of assignment is found in the mind of the writer, not in books, journals, or other people. If this were a research paper (Chapters 11 and 12), writing under pressure would still involve the skill of knowing what to look for in the available material.

My job, then, is to sort out my own views and decide how best to present them to the reader. Generating a provisional thesis gives me the head start I need. I really don't want a competency exam. Committing myself to that position in a thesis enables me to *start* the writing job honestly and energetically. If I discover other arguments, or convincing opposing views, then the material itself will shape my raw draft. I can write this paper without worrying much about the various 'pressures because I will actually discover what I believe and why I believe it, and then present the reader with the truth as I see it.

Writing, after all, is discovery, a way of finding out what I think and what my experience or learning tells me. My own thinking may include conflicting ideas. But writing under pressure will be somewhat less discovery and somewhat more development: you find the site, and you construct the city. You've settled on a decision, and now you make the most of it. Whatever you discover in the process adds to the power and clarity of the writing product, or has to be acknowledged as something you need to study further. Under pressure, some of that will always have to wait until the next writing project.

Here, then, is the start of the thesis which, by the end of the rough draft, will become my full-fledged introduction:

A writing competency exam is a bad idea *because*—

The work of generating *because*-clauses, which at this point should have a strong element of spontaneity in it, now begins. What follows "*because*—" forms the foundation for the rest of the process of writing well under pressure.

The competency exam is a bad idea. *Why?* If I had to create a twenty-minute oral presentation on the subject, I only would have time for the few best reasons. If I were writing a book on the subject, I would need plenty of reasons, covering a variety of related issues. For this three-page paper, I want to generate ten or twelve ideas as freely as possible, without judging their merits or importance, and then focus on three or four to develop into my position paper.

A writing competency exam is a bad idea *because*—

(1) No one would want to read it.

(2) Who would define "competency"?

(3) How will the exam be useful for students who don't pass?

(4) Who would teach the failing students and help them to pass? Where would we get the money to pay these teachers?

(5) What would it mean for the teaching of writing in general if we tailor it to an exam? Many of our students need to be less afraid of writing, not intimidated by an exam.

(6) What will be the impact on younger teachers? How could we convince them that they will not be judged for promotion based on how many of their students pass the exam, but on how well they accomplish the ideal goal of helping everyone to write with power and clarity, on whatever level, as a lifelong skill?

(7) What will the existence of a writing competency exam do to the relationship between the English department and other departments of the college? Will it create a sense that everything we do is "remedial," and that they themselves don't have to work at helping students write well? (Face it, you never hear: "I have such good writers in my history class, the English department must be doing its job.")

(8) What function would the exam serve? To give peace of mind to some people on the faculty or in the administration because so many people don't write well? It's a national concern, after all. How can you ensure that all students will write well as judged by a single essay, or even a series of essays? You might do this with math, or computer competency, or even with a writing inventory to place people in the appropriate level of writing class, but what is our goal here as teachers of writing? (More on this later.)

(9) I define "competency" in writing as the writer's capacity to say with power and clarity what he or she thinks, whether

asked by teachers, bosses, prospective employers, or on his or her own initiative. How will this be tested by a competency exam if in fact a student works hard and moves toward that goal in the fourteen weeks of a writing course? If we insist that students repeat writing courses or get help to pass the exam without getting credit for it, then we can throw in the sponge! It's a miracle anyway when students take the idea of writing well seriously, and the more cynical you become *about* writing, the less you do it. Training people to write in order to pass an exam is one sure way of stopping them from learning *how* to write with clarity and power as a lifelong habit. People get up for a chemistry exam and never want to look at a chemistry book again. But while you may never need to know the chemistry equations once the exam is over, you will certainly have to write, again and again, throughout your life. (More on this, too.)

(10) Actually, I admit the competency idea *is* a tantalizing one. Wouldn't it be nice to draw a line and say, "Yes, Robert writes well enough to get our degree; Judy can't yet, and needs another semester; Tim may need to take writing each semester until he graduates, and even then. . . ." The idea is appealing. The more people working at writing, the better. But no competency exam has ever been accompanied by a whole new series of credit courses in writing, with the hiring of additional staff, and a college-wide commitment to teaching writing. Usually, such exams are meant to "certify" that students have been trained in some way, as we might certify that a student can run a computer program. What should we certify in writing? That a student can write an engaging cover letter for his or her resume? A good essay for a transfer application? A strong, clear letter to the editor of the home-town newspaper? A paper free of mechanical errors that says nothing, and leaves the teacher "no work to do"? I know the sheer dread most people have of writing in the first place, and the fear that a standardized exam might prevent graduation

will kill the whole deal, especially for the very students who need the most attention. (*Much* more on this.)

(11) "Competency" for what? I keep coming back to this: competency for a test, for writing decent papers in a course, or for a lifelong skill? In some ways, these are mutually exclusive goals if all we have is one semester of required writing—an issue that should be at the heart of this argument anyway.

(12) All right, what do *I* suggest instead? A year-long required writing course for everyone? A writing staff charged not only with the theoretical (and now machine-correctable) mechanics of writing, but with the task of helping students express themselves, not only in our courses, but throughout their lives as citizens of a society? Because if we don't do this, we will be responsible for permitting the vacuum that is filled not by real, hard thinking and explanation but, instead, by the incessant yapping of advertisers and photogenic newscasters who smile about weather, drought, starvation, parades, and invasions as if the words they speak have no more meaning than the drone of the refrigerator, and the people of whom they speak no more substance than objects without feeling.

Commentary on Generating the *Because*-clauses

Okay, that's enough. Listing the potential reasons against a competency exam was an intense, exhausting activity, but doing it without stopping to judge these ideas, I felt my energy flowing into several of the *because*-clauses. I even got to the point of realizing (clause 12) that I have the basis for an alternative to a competency exam. I also took a look at the possible good points of a competency exam (clause 10), but I'm not convinced, so those will become part of my own argument against one. I know already, as I prepare to look back over the list (Chapter 4) that I have more than enough good material and some of the actual language I'll use in my raw draft. I've discovered things I would not have thought of before. Even if I don't make use of some of them in the final product, I have learned plenty about myself and my attitude toward these questions by spontaneously listing more than enough *because*-clauses.

For a longer paper, I might have used a directed freewriting exercise first, before committing myself to a thesis and listing reasons why I support (or oppose) it. But although a page of nonstop, uncensored writing on the subject would have helped me find what Peter Elbow, in *Writing with Power*, calls the "center of gravity" of the piece of writing, my list of clauses gives me a convenient way to analyze the various ideas and opposing views, and provides some of the language I will use later. As with all the stages of QWP, this listing is useful in itself and ignites the next activity: analysis of the *because*-clauses. Analysis produces an argument-outline, which leads to the raw draft. This *continuous* process is one advantage of using QWP instead of mapping strategies or brainstorming. It begins with *meaning;* and the provisional thesis, the *because*-clauses, the constant integration of inductive and deductive thinking create a sense of order. A psychologist has a hunch in a family therapy session; a biochemist has an inkling of where to look for something useful. Anyone who writes has a whole storehouse of ideas and feelings that will come into play in the writing process. QWP takes the energy of thinking and turns it into a system for writing.

Of course, no process can assure that I will find the ultimate explanation or the basis of my instinct against a competency exam. Although I fully expect to discover ideas and evidence that I haven't thought of before, it is not always possible to find what Sheridan Baker, in *The Practical Stylist*, calls the "public reasons for . . . private convictions." The process of writing under pressure may or may not force them out into the open before the deadline. Perhaps I'll find them later on, as I think over what I've written and what I've left out, or as people respond to my paper. I may change my mind as I learn new things. But for now, within the deadline, I can approach, clarify, and represent my thoughts, and acknowledge, as part of the power of writing, the things I *don't* know, the things I'm not sure about. There will be more chances, more writing tasks. This particular one may lead to others, or it may be forgotten. All I can do is work with as much intensity and concentration as possible in the available time and get *this* job done.

The more you use QWP and make it your own, the more you will be able to depend on having enough information, analysis, and ideas. There will be ample discovery for the writer and food for the reader's thoughts. It's not a foolproof process. The history of writing is a history of attempts. Think of Faulkner writing the second section of the *The Sound and the Fury* because he felt the first didn't work, and then the third and the fourth, and finally, putting them all together as a book although he still was not satisfied. There's tremendous effort and limited success in most writing. Incidentally, that's not the kind of truth a competency exam would allow—another thought to consider as I move toward integrating meaning and structure in my position paper.

☐

Time and Space

The QWP Timetable Related to Structure

At this point, I have much more than enough raw material for my paper, more than I would have imagined when I began to analyze my relationship to the intended reader, and my own feelings about taking a stand on the competency exam question. QWP delivers you headlong into the writing process, neutralizing the familiar resistances to starting. Besides, in this case study, I don't have *time* to put it off: the paper is due in the dean's office tomorrow morning. My limit is three pages, and I'm about to sift through and analyze the list of freely generated *because*-clauses linked to my thesis, to put them into the order of a balanced argument-outline.

Before the analysis, it's important to understand the relationship between time and structure in the writing process. Managing time is essential to the writer's control and independence. Most writers don't appraise their time realistically and, invariably, the result is a distortion of the writing process. The writing we do under pressure (an exam essay, a memo due in the afternoon, a cover letter for a wonderful job that has to be mailed today) has a last-minute feel to it. We know it as we write, and we feel it as soon as we turn in the bluebook or memo, or send off the letter. We hope it will "get by," or we resign ourselves to a disappointing response. But this familiar experience reflects something more than an unsatisfactory process. The deeper issue is that most people are not fully aware of the way words strike readers, and so they use their writing time inefficiently.

Readers may not have an inkling of a writer's most difficult struggles. Novelists are often fondest of the writing that has given them the most trouble, but readers are grateful for books that seem to have been written "easily," as if without effort. Two different goals are involved here: the writer develops his or her craft; and the reader seeks enlightenment or pleasure, or both. If you explore the diaries, notebooks, and letters of writers, you may be surprised to find that what gave the writer a private feeling of accomplishment may have little to do with a reader's experience of the work. This is generally true, at one time or another, of everyone who writes. But for inexperienced writers, it is usually destructive. "Sorry, I know this is full of comma errors," a student says, even though her essay, read aloud, has overwhelmed the class with its power. Another writer will assure you the meaning's *there*. "I haven't bothered to put in all the connections," he says, "but you know what I mean." Both writers have little perspective on the way readers perceive their work. With some constructive feedback about the effect of their words, they could redistribute their energies throughout the writing process to strengthen communication.

The most common and debilitating distortions in writing result from this failure to manage energy and time. It is extremely common for people who dread a particular writing task to spend most of their available time on one small part of the process, so that the final product is out of balance, lacking any sense of focus, and constructed so loosely that the reader can't keep the message or method in view.

It's easy to see this in terms of structure. Some people will spend almost all their time and energy on the beginning of a piece, getting it "right," but leaving no time for the middle or the end. There are people who never write conclusions even though this ensures that whatever they have to say will spill out and evaporate. (No matter how good the beginning and middle of a piece of writing, if there is no sense of an ending, and of pointing the way to a new perspective, the reader may retain only an example or a vague general impression.) There are people who only write conclusions. They give the reader their general assertions, even a sense of perspective; but the writing is abstract, without the necessary specificity of the argument, the anchored examples, the progression of evidence that supports general statements. The whole essay drifts away.

Most familiar are the writers who have no time, inclination, or confidence for beginnings and endings, and give the reader only a middle, with perfunctory opening and closing statements. The result, for readers, is irritation and disorientation: "What was this really about?" or, "Oh, I see. Why didn't you tell me you were talking about that in the first place? For most of your paper I thought you were writing something else entirely and now I don't have the time to go back over it." Planning time, in QWP, is important for avoiding these writer's distortions. It is done within a basic sense of structure, so that your essay or memo has the balance of an interesting, useful beginning, a solid, specific middle, and an inspiring conclusion.

If I have less than a day for my position paper I must commit myself to a timetable for the whole project even before I generate a thesis and *because*-clauses. Planning time and space is best done even before considering the fundamental questions about the writer and the reader that initiate the writing process. QWP acknowledges the boundaries of a writing project because its goal is to get the job done. Underlying this principle is the inescapable fact that setting these boundaries is the writer's responsibility. It is an act both of independence and of control, and gives the writer motivation and confidence. Deciding *first* the time you can realistically allot to your writing project to get it done thoroughly is essential, and liberating.

Amid the pressures, fears, hopes, excuses, and good intentions, you simply say, "This is what I'm willing to do—set aside this much of my best time and energy to complete this job." You may wish you could do more; you may want to do nothing at all. But the decision about time helps you resolve the issue of your commitment to the project immediately, and that is a tremendous help.

The QWP timetable ensures that none of the writer's energies are wasted, and that each activity in the process initiates the next. For example, although I've written about twenty pages of explanation about QWP so far, I've actually spent less than an hour in writing time on the position paper itself, preparing and then generating a thesis and *because*-clauses. Now I have a quarry of raw material, which assures me I have "enough" to say, and can do a thorough job. The real virtue of QWP is that everything you do counts, in producing the best piece of writing in the given circumstances. The QWP process transforms time from one of the pressures against writing well to one of the writer's most important resources.

Here, then, is the timetable for this project as I worked it out *before* even preparing to write. It is predicated on three hours of writing (an evening's work) for a three-page paper. This timetable includes everything we have done so far to arrive at our list of *because*-clauses, and includes the *producing* part of the process, to be described in Chapters 5, 6, and 7.

Timetable: "Competency Tests?" (3 pages; 3 hours)

(1) Timetable related to structure: Beginning, Middle, End. (*5 minutes*)

(2) Who is my audience? What is my relationship to the reader: power, professional, collegial? What is the appropriate approach, distance, and level of specificity, given time and space pressures? What needs to be put in, and what can be left out? (*5 minutes*)

(3) What is my commitment to the project? How are my instincts, feelings, fears, and energy related to my commitment? (*5 minutes*)

(4) The provisional thesis, up to the "*because-*": the answer I propose, based on experience, learning, instinct, or passion. (*10 minutes*)

(5) Plenty of spontaneous *because*-clauses to support and oppose my provisional answer, enough for a paper three times as long as the one I have to write. (*25 minutes*)

(6) Analyzing the list of *because*-clauses to choose the most important ideas for my argument, the ones that elicit the most from me as a writer and turn the question to my strongest interests. (*20 minutes*)

[*first hour*]

(7) Ordering the most important *ideas* into a quick Beginning, Middle, End argument-outline that organizes the best raw blocks of *because*-clause writing. The provisional thesis, cor-

related with the ordered ideas of the Middle section (including a look at the "other side") becomes the introduction. Note perspectives for the conclusion. (*15 minutes*)

(8) Filling out the argument-outline with the selected material from the *because*-clauses. Rough-cutting this raw draft for coherence and consistency. (*30 minutes*)

(9) Reading the roughly final draft aloud (or having someone give me constructive feedback) to find missed connections, loose ends, flat places where I can polish the draft down to size. (*15 minutes*)

[*second hour*]

(10) Making the roughly final draft *final:* ensuring unity through powerful transitions, and integrated examples and comparisons to engage the reader's thinking. Filling out the End (the conclusion) with a new perspective for the reader. Correlating the Beginning (the full-fledged introduction), with the whole argument, including the final perspective. (*30 minutes*)

(11) Proofreading for a flawless final copy. (*20 minutes*)

[*third hour*]

My *because*-clause list was the first real writing for this position paper. The list took about half an hour to generate (even though many of the clauses, as written out in Chapter 2, for purposes of explanation, are much longer than they need to be). But it was a half-hour well spent. Already I have language I can use in the final draft of the paper. I know the tone and the attitude I will take toward my reader. I have a sense of the places where I have the strongest ideas and feelings to convey, and those where I have less investment. I have my beginning, based on my provisional thesis. Later, in the roughly final draft, I might want to add a few introductory sentences to capture my reader's attention or assure him that I am going to answer the question.

I have no doubt that once I analyze my list of *because*-clauses, I will discover something new that will form the extra sense of perspective beyond a summary a good conclusion needs. In less than an hour, then,

I have the essence of my paper: its psychological underpinnings, a quarry of ideas for my logical argument, and a good feeling for the way the structure will enable me to write efficiently and, at the same time, permit the reader to follow a balanced argument.

If I spend the next half-hour analyzing the main ideas, and then arranging them in the most effective order just as they are, I will be well on the way to the argument-outline that provides the framework for the raw draft.

Writing under pressure doesn't permit the painstaking build-up or filling-in that most writers do when they begin a paper, those first few pages or early paragraphs that exist to get the writer started and disappear in later drafts. The planning stages of QWP serve that function by committing the writer to the project and to the provisional thesis. Generating the *because*-clauses for the thesis locates the argument's center of gravity. Sifting through the *because*-clauses culls the best material into the most coherent structure.

The first hour, then, consists of preparing, planning, and generating a solid thesis, and the central argument of a balanced paper. By the end of the second hour, I will have a coherent rough draft containing more than enough good material. In the final hour, I can ensure unity, anchor examples, and fashion a conclusion that goes one step beyond anything I've said so far to give the reader a broader, deeper, or clearer perspective. Then I can polish, proofread, and type the final product.

This three-hour commitment ensures a paper that gives my answer, in my own voice and style, to the question posed by my intended reader. QWP allows me to *discover* my answer through a relatively spontaneous process within the allotted time, a process over which I exert a good deal of control. I will have done the best I can, not only for the reader but for myself as a writer under pressure.

When I was a student, my first two-page paper took me forty hours. For other papers, I spent ten or twelve hours painstakingly trying to get the beginning ''right,'' without leaving time for the middle or the end. On some assignments, I spent so much time trying to fill in the reader on the rich background for my topic or my thesis that I left no time or energy for the middle—the argument. Most often, this happened on final exams, where I would squander the forty-five minutes allotted to

"Essay Question One" on only one of the issues involved in a full answer (see Chapter 10 for the application of QWP to examwriting). Occasionally, I wrote what I meant in general terms, but could nót discover the specifics or the examples to provide concrete evidence. QWP time-planning prevents you from falling into these ineffective patterns. You discover and commit yourself to your own thesis early in the writing process; you also make a *time* commitment and stick to it. Even if you start your writing project weeks in advance, you still need a process-related timetable to ensure that you will complete the *whole* job through the flawless product, on time.

One of the keys to good writing under pressure is the realization that you must do each part of the process with equal vigor, and as thoroughly as possible within the time limit. You want the good material on the page, not left over in your mind to regret a day later. Moreover, each activity in QWP initiates the next, in a continuous process. We've imagined, here, that I was asked to prepare my position paper on an English competency exam less than a day before the dean needed it. Given such an urgent assignment, I might have sat down at the keyboard and squeezed out three pages word by word. I might have had a hot streak and come out with more material than I needed, giving me the luxury of cutting back. More likely, I would have hit a blank wall, stymied after a paragraph or page, aware that I really didn't know what I meant, or where I was going, or that I simply meant too many things to proceed without planning. I might have found out what I really wanted to say too late to do anything about it; or that everything I'd written up until five minutes to nine (shooing students and colleagues away from my office with a wave of the hand) was suddenly rendered vulnerable by a thought that only emerged as the tower clock rang out the hour.

That's why QWP includes a timetable for each project, within the boundaries of available time: a complete timetable for the forty-five minutes in which you must produce a one-page memo for your boss before he or she leaves for lunch; or the three months for your undergraduate thesis; or the half-hour for the brief speech at your town meeting tomorrow night. You make the time-commitment in order to get the *whole* job done. After using QWP once or twice, with additions or adaptations that suit you in unique ways, you will find that the job of writing under pressure no longer induces a mystifying and debilitating terror.

It took me about a year in college to come up with a reliable process for writing; in graduate school, I adapted the process to longer papers. It was only after I began teaching writing that I realized how few people had any system at all. With QWP, your writing improves because you use more and more of your best energy appropriately, instead of wasting time on the needless anxieties and fears, the dislocations and distortions, that result from a failure to plan time and space.

Developing the QWP Model for My Position Paper

I am a little less than a third of the way through my timetable for this project. I have plenty of ideas in my list of *because*-clauses to analyze for the most compelling blocks of material. Now, I'll order these blocks into a quick, balanced Beginning-Middle-End argument-outline (Chapter 4).

    (A) Preparing and Planning

        (1) Timetable related to structure: Beginning, Middle, End (Chapter 3).

        (2) Audience, level of generality, scope, scale (Chapter 1).

        (3) Writer's own investment in the project (Chapter 2).

                    (*total allotted time:* 10%)

    (B) Generating

        (4) Provisional thesis (Chapter 2).   (5%)

        (5) *because*-clauses (Chapter 2).   (15%)

CHAPTER
4

<hr>

Generating the Argument-Outline

Analyzing What to Keep

Everyone who writes has been told at one time or another to "sharpen the focus" of a paper, a report, or a memo. But that is the kind of advice that often creates more trouble than it resolves. How do you know what to put in and what to take out of a piece of writing, especially if you are writing under pressure?

Most people imagine an analogy between sharpening the image in a viewfinder and focusing their words on a page, but locating the center of gravity in an essay can be far more complicated. Writers do not sharpen as much as they *transform* what they observe into their own perspective. A piece of writing re-creates the world as an hypothesis, whether stated or implied, with proposed evidence to support it. Orwell's essays are good examples of this re-creation. Although it's impossible to tell, from reading "A Hanging," or "Shooting an Elephant," whether they are "true" or "made up," Orwell states, in a straightforward way and in a strategic place in each essay, exactly what question he hopes to discover for us, and what answer he proposes. You might think of Tolstoy's philosophical "Epilogue" in *War and Peace* in the same way, an explanation that makes explicit an underlying principle that has been dramatized in the book. We may not agree with what the writer says, but we can interact with it, weigh it, judge it, and see if we have a use for it.

If there is an analogy between making a photograph and writing, it is more appropriately Cartier-Bresson's "decisive moment," the image

28

taken at the exact moment when it presents the photographer's discovery of meaning. But in writing, as in music, this presentation of meaning occurs not in a sentence or in a single word but in a structure built over time. The principle is to put in anything that can help *reveal* meaning, and take out everything that will obscure it. The writer sorts through his or her own ideas, feelings, and experiences to find the best relationship to the material: that point of view which elicits the most energy and precision, with the least wasted motion. The mystery of "focus," is replaced by control over the process, and the writer's principle of selection corresponds to his or her own wisdom and passion.

With QWP, the writer sorts through the *because*-clauses to find evidence of where he or she stands, and of what matters most. These selected blocks of raw material become the center of the argument. The analysis of the *because*-clauses also yields the "glue," the internal coherence that holds the argument together. You find it as you sift the rough blocks of *because*-clauses, discovering how opinions, experiences, and examples cluster around your thesis. You lay bare the underlying structure, the unifying principle for the whole paper.

I'll begin, then, by testing each of my twelve clauses, asking the same questions about them that any good reader who needed to decide about a writing competency exam might ask. I'll want to see whether each *because*-clause holds up as an idea; whether it elicits more information as I think about it; and whether it should have a place in the raw draft of my position paper.

Clause 1: *No one would want to read it.*

First: Test the *because*-clause as an idea.

"No one would want to read it." Why? (a) Because it's excruciatingly dull to read essays on set questions as if they were real pieces of writing; (b) because it's a dead-end process in which the student gets no useful feedback except a number, or a sheet with inscrutable checkmarks that teach nothing, and which the disembodied reader has filled in perfunctorily; (c) because most of all, no one who teaches writing wants to pretend that one exam question, on any given afternoon, can possibly represent conclusive evidence about any student's writing. The idea of an essay invariably written at the worst possible time for both writer and reader (usually, in the midst of registration, or as classes begin) fulfilling such a function is an illusion maintained by those who don't participate, and a futile exercise for those who do.

Second: Test the implications of the *because*-clause.

"No one would want to read it"—that's undoubtedly true. People can be paid to do it, but then they have to be instructed in what to look for, and their own values and skills have to be integrated with or subordinated to those of the English staff. Or the younger or newer part-time members of the staff can be saddled with the job, which is grossly exploitative.

Third: Relevance of the *because*-clause to my position paper.

For my argument, this *because*-clause is a minor but solid point, to be inserted somewhere in the larger argument as a sentence or two.

Naturally, writing out this analysis took a good deal longer than thinking it through. For a longer paper, with a timetable that allows for it, writing out these analyses will provide you with some of the language that finds its way into your final draft. But with a shorter schedule, you can simply make notes next to each clause, or use a tape recorder. In either case, you are gathering essential ingredients for the rest of the process: ideas, language, and the glue that holds your argument together.

As you become experienced with QWP, you will find you can combine some analysis *with* the first spontaneous listing of your clauses. But to begin, you should allow that initial generating process to go on freely, until you have three times as many *because*-clauses as you need. Then analyze, amplify, and draw connections. Harvesting the *because*-clauses, questioning their assumptions, implications, and relevance, yields blocks and sentences and hints of new ideas for the public version of your private argument.

Clause 2: *Who would define "competency"?*

A philosophical question if taken seriously, but it probably won't be, in view of the narrow-minded pomposity of some academics and the near exhaustion of others. (If we all faced the fact that anyone can and should get proofreading help from a friend or teacher or editor—as professional writers do—or from a machine, the whole question of competency would be seen in a clearer light: not "mechanics," but *meaning* plus proofreading.) What defines "competency" for a person who struggles to write neat, bland essays compared to someone who can think, and who says a good deal in a powerful way, but doesn't know some of the conventions of punctuation? Most college students have elements of both extremes, anyway; and this debate among English teachers is endless, from generation to generation. Having seen that pendulum swing back and forth, I know it won't be one that interests me for this paper, because it leads to dead ends.

Clause 3: *What about the students who don't pass?*
Clause 4: *Who would teach them and where would we get the money to pay these teachers?*

Time and money questions: do we give failing students course after course, in addition to their full load? Or force them to add time to their four years of college until they pass an exam? When would we give such an exam—during their junior year? Do writing skills need *re*-testing in senior year? How would we test 250 seniors, and what would we do with those who failed? We would be valuing the test above teaching and learning. Who would be hired to teach chronic failers? What would this continual focus on failure mean to students and teachers? Sections full of failing students *have* ensured jobs for English teachers at the nearby state college, but no one I've met teaches them with any passion or interest. Besides, at Shakespeare College of the Communicable Arts, we could never support it financially. (This is a little more interesting as an issue, but again, not one for which I want to make room at the center of my essay.)

Clause 5: *What would it mean for the teaching of writing in general?*

Ah, now that's a little more like it. At this point, my teaching experience, in all kinds of settings, with or against all kinds of teaching policies, pushes toward the center of the stage. I've learned that exams invariably become ends in themselves. That would be a disaster for writing—and I see writing as a personal accomplishment, a lifelong skill. Moreover, our students need to be less afraid of writing, not intimidated by an exam, and the same holds true for teachers. With something as difficult as writing well, everyone looks for an easy way out. A competency exam would be a destructive illusion, a way of avoiding the real task: to help people tell each other what is really on their minds. (Definitely hold on to this.)

Clause 6: *Impact on younger teachers.*

This is related to clause 5 but adds the pressure of promotion to the writing teacher's job, which accentuates the wrong things: if you, the teacher, train students to pass a competency exam, you will have done your job. If you fail at that, you will be judged a failure as a teacher. Forget about meeting students at their present level, and helping them to discover and tell what they mean. Get them through the exam. We don't care what happens afterward. But the hitch here, as always, is that passing or failing an exam may have almost nothing to do with the lifelong need to adapt one's capacity to write to a variety of demands and situations.

But . . . here's the discovery of something really important to me, worth struggling to write about: how do you show someone who knows little about the possibilities for learning and teaching writing that an exam is exactly the wrong emphasis to give to a course? My students, even those who start by not being able to write a sentence in an hour, could pass an exam at the end of the term if it were a genuine task (a step in their development toward independent writing with imagination and skill), and if this developing skill were reenforced elsewhere in their school life. But that would also involve educating teachers who give students no feedback on their papers, or who correct only typos and punctuation. The emphasis, instead, must be on giving people the self-confidence to share, to believe in an audience despite their experiences in school that militate against such confidence. A competency exam works against all of those goals. (This may be the heart of my paper, the place where my own feelings, experiences, and ideas ignite the act of *writing* it. Come back to this!)

Suddenly, half-way through reviewing my list of *because*-clauses, I have a vision of how this position paper can satisfy the needs of the reader *and* the writer. I see that clause 7 is related to this discovery as well: of course a competency exam will weaken the image of the English department in relation to the rest of the faculty. We will be seen as a "remedial" department, which is what most academics think anyway ("you writing teachers have the toughest job in the college, why can't you teach those illiterates how to spell?").

So, too, with clause 8. The competency exam is really not for the student or for the English department; it is for the comfort of people who really know nothing about the teaching and learning of writing, but who imagine it as a testable skill that can be measured in increments on a false analogy with computer literacy or statistics. These people are probably skeptical about what students have to "say" anyway.

Clauses 9 and 10 directly address this point, too: what I consider necessary in the teaching of writing is actually *undercut* by the competency exam. They can go right into the middle section of my essay, in their present wording. Clause 11 adds the point that more writing should be required of *everyone,* while clause 12 explains why, and goes on to the more general point about the connection between writing, educational goals, and society, reflecting our larger responsibilities as teachers of writing and as educators. Certainly, part of the twelfth clause will find its way into the conclusion of my paper.

Looking through my original list, then, has yielded lots of information about my essay; not only a sense of the whole design, but the center of the argument, and the feelings, the anger, the ideals, the energy I can bring to a question that at first seemed dreary and futile. I know, too, that I'll be able to use some of the language of the original clauses. This analysis might take about fifteen or twenty minutes, with a few notes sketched on the page as you go along. But you will feel confident that, writing under pressure, you can make your paper, memo, or report clear and powerful.

Discovering What's Left Out

Now I have a good opportunity to make a crucial discovery: what have I left out of my argument that needs to be in it? I have the major argument of my essay, a sense of the overall design, and some usable language, but I still have the nagging feeling that I can't quite get at the heart of the question of a competency exam. I need to look at the "other side" before I make my outline, not only for perspective, but because if I do so now, I'll have time to integrate whatever I discover. This is much more efficient than rearranging, as any writer knows who has discovered a useful or crucial point too late in the writing process.

What would the little world of Shakespeare College of the Communicable Arts be like *with* a competency exam? Aside from the practical problems of administering a test, and forgetting for the moment any philosophical debate about its wisdom, there are, after all, precedents in other areas, such as computer science and swimming. The swimming test is something everyone who graduates from SCCA must pass: up and back the length of the pool. Writing is every bit as important in the modern world as computer skills, although probably not as important as swimming. Dedicated teachers of writing should jump at the chance to reenforce what they do by requiring that students pass a test. Besides, wouldn't it raise the level of literacy throughout the college, and raise the reputation of Shakespeare College at the same time? Why don't more schools, especially the so-called "elite" schools, have a competency exam? We could be in the forefront. Harvard would follow us! A competency exam would put old Shakespeare on the map. We would be known as the school that demanded and got a minimal standard of writing from its students.

I'm not convinced by making a case for the other side, but I am disturbed. Why shouldn't we *try* it? What harm would it do?

But writing is not like swimming or computer skills. For one thing, it is not a life and death matter; people get along without writing easily and well, and society trundles along with a mostly mute citizenry while still managing to put a man on the moon and lower interest rates. Moreover, writing does not lend itself easily to incremental learning; and there isn't any program tailored specifically to the kind of learning writing *does* require. Swimming saves lives; computers get bills out faster. Writing doesn't have an easily achieved, demonstrable payoff. It's not surprising, then, that people are interested in cutting losses: if you can't write, at least learn how to spell. And even that futile goal is unrealizable. The truth is, you can't *make* people write well unless you provide a supportive learning environment over time. No exam has ever created that. Fear, panic, and anxiety do not create a good writer; instead, they discourage risk, and prevent people from tapping their deepest resources and sharing them with readers.

The real problem with a competency exam is that it would work *against* the goal of competent writing, intimidating both students and teachers, rather than invigorating the learning process. A self-paced course in statistics, with good tutors available, is a workable endeavor. But writing is an unfolding process in which practice is linked to the discovery of meaning, and the motivation to reach an audience. Students will not learn to find their own voice for an exam. Teachers will not support students in attempting new, risky kinds of thinking and writing if they are to be judged on how well their students perform on an exam. Teaching people to swim once up and back the length of the pool is a marvelous goal; and students will find computer skills useful in their work lives, and in their personal lives. But writing requires nurturing, patience, honesty, and wisdom. Some people will not write clearly or powerfully without a great deal of help and support. A competency exam completely misses the goal for teaching writing, and will consume everyone's best energies at a time when resources are dangerously scarce. *The exam undercuts the very goal it is intended to achieve.*

Adding an *Although*-clause to the Thesis

Now I feel better. That was the insight, the piece of structure that was missing from my argument, showing that the appeal of the competency exam (bringing everyone up to a standard writing level) is not only illusory but self-defeating. If I incorporate this insight into my thesis:

> *Although* the idea of a writing competency exam is appealing on the surface as a way of strengthening a basic skill, such an exam would be self-defeating because . . .

then there are two benefits for my paper. First, I'll have the opportunity to refute the major argument for such an exam. Second, in doing so, I gain a structural benefit as well, linking my opening thesis statement to my conclusion, where I'll suggest a better way to strengthen students' writing skills while urging writing teachers and others across the curriculum to help. By taking a look at the "other side," I have clarified my own view, and forced my conclusion on the whole question out into the open. (When we integrate the QWP system in Chapter 8, it will make sense to include an *although*-clause in the original provisional thesis.)

Although it has taken some thirty pages to explain each of the QWP decisions so far, I actually have used less than half the time allotted for the whole project in my timetable. Much of what I have written here for the reader takes place in the conscious writer's mind: assessing the relationship between writer and reader; devising a thesis; and analyzing possible material for the raw draft. Actual *writing* time and energy is devoted to what matters most: the spontaneous *because*-clauses that provide the material, the language, the explanations, and the passion for the whole project. Now I can weave ideas I have selected from my *because*-clauses into the main strands of thought by combining, connecting, and rearranging them into the persuasive order of a quick argument-outline.

The Argument-Outline

English teachers have always been fond of outlines, but when people write history papers, or memos, or reports, they rarely use them. One of the lessons I remember most vividly from grammar school is Miss Gaffney's fourth-grade outline lesson, demonstrated at the board and tested over and over again. This outline, in increasingly meticulous versions, was the mysterious treasure map to success that term, demarcated by Roman numerals, large and small letters with prime marks and sometimes even double-primes. Some of our outlines were beautiful; others were chaotic disasters. Nobody mentioned at the time, least of all Miss Gaffney, that outlines could help us order our thoughts. Instead, we took quizzes on what kind of thought went where: did a phrase including a noun and an adjective get a prime or a double-prime? When did a generalization merit a capital letter? There was much anguish about these matters, although a few people seemed to thrive on them. Still, I wish someone had linked those exercises with the act of *writing*, because it took years of frustrating, disappointing work before I understood the usefulness of thorough planning, and years more before I taught myself how to speed up the writing process by using, among other things, a good outline.

The QWP argument-outline is neither theoretical nor abstract. It is a quick Beginning-Middle-End outline that will help you create the best order for the chunks of material you've selected in analyzing the *because*-clauses. Almost all rhetoric books and writer's handbooks spend at least some time listing possible patterns of organization. English courses spend weeks on such exercises: compare and contrast; classification; definition. As a writing teacher, I would be happy to know that everyone understood the characteristics of deductive and inductive thinking and had developed some facility for combining the two, as Orwell does in many of his essays. But patterns and outlining exercises can become ends in themselves, without reference to what the writer needs to discover and then tell the reader. Whatever pattern you might choose, however complex the comparisons or the underlying metaphor, your reader will still expect and seek a beginning, middle, and end, so it's safe and efficient to begin with that sense of structure.

After analyzing the *because*-clauses to select those blocks of material I want to include in my raw draft, I see that the first several clauses point out problems in giving a competency exam. The next few have to do with the negative effects such an exam will have on students, teachers, and the college environment. Then there are blocks of material that express my own feelings and ideas about specific and general alternatives. Almost always, there is this inherent order in the material.

Writers, like sculptors, work to bring the figure out of the stone. First, I can group my selected material under these three main ideas: problems, negative effects, and my own alternatives. Next, I can arrange the three groups in the most persuasive order. I'll put the other side first, setting up the tension to counter it with my own views. If it comes first in the middle section, I'll have the rest of my paper for what *I* want the reader to think about.

Immediately after the other side, I'll want to state my most pragmatic arguments against the exam. These are the least open to interpretation, and may appeal to the reader's common sense. If I remind the reader that such an exam would be too expensive, too inefficient to administer, and labor intensive, my more complicated arguments might not even be necessary.

Next I want to explain the more general idea of how such an exam would demoralize not only students and teachers, but also the environment for learning at the college. Once that is established, I'll combine the problems of giving the exam with the difficulties caused by the exam itself to explain how an exam will not ensure competency. Finally, I'll end the middle section of my paper with my discovery that the exam is self-defeating and destructive to the *goal* of competency. This idea, in turn, provides a natural transition to the conclusion, where I'll recommend alternatives.

With the order of the middle section roughed out, I can rearrange the several clauses or sentences of the thesis to fit it, and this now becomes a full-fledged introduction. The conclusion will provide the reader with my alternatives, and then go one step beyond anything I've said so far to a new perspective on the subject. This, too, will be reflected in a final adaptation of my original thesis.

The quick argument-outline balances material in the best order. Once you're satisfied with it, you can move on to block out and then cut the raw draft.

A Quick [Annotated] Argument-Outline

(I)   BEGINNING [The thesis, condensing the selected, ordered *be-cause*-clauses into the phrases and sentences of a full-fledged introduction, including the other side—the *although*-clause—and the broader perspective of the conclusion.]

> Although the idea of a writing competency exam is appealing on the surface as a way of strengthening a basic skill (*see below,* IIA1), such an exam would be self-defeating and destructive (IIB4). It is expensive, inefficient to administer, and demoralizing for both students and faculty (IIB1). It would destroy the English department's effectiveness in helping all students to learn (IIB2), and will not ensure competency in writing, a skill that cannot be legislated for a wide spectrum of our students in a set period of time, or through programmable steps (IIB3). Such a goal can only be accomplished by a creative, flexible, highly motivated teaching staff making instruction as individualized as possible toward the goal of writing as a lifelong skill (IIIA) and, in a broader perspective, as a citizen's advantage in a democratic society (IIIB).

(II) MIDDLE [Main points of my argument, structured in a balanced order of importance, beginning with a look at the other side and leading toward my most important discovery which, through a transition, opens out to a broader perspective at the END.]

  (A) Arguments *for* the competency exam (the other side)

   (1) [Introduced by the appropriate phrase in the thesis.] "*Although* the idea of a writing competency exam is appealing on the surface as a way of strengthening a basic skill. . . ." (use *because*-clause 10)

  (B) Arguments *against* the competency exam (transition phrases taken directly from the thesis statement to introduce separate, interrelated points)

   (1) *Practical problems in giving the exam:* would be expensive and inefficient to administer (use analysis of *because*-clauses 1, 3, and 4)

   (2) *General negative effects:* demoralizing for both students and faculty (use analyses of clauses 5 and 6)

   (3) *Specific negative effects:* would destroy the English department's effectiveness in helping all students to learn (use clauses 6, 7, 8, and analyses)

(4) *Summarizing issue:* would not ensure "competency" in writing, a skill that cannot be legislated (use analysis of clause 2)

(5) *Discovery and transition:* would be self-defeating, and destructive (use clause 8 and analysis, and new writing from the analysis of the other side)

(III) END [Concludes the argument by recommending alternatives and a broader perspective.]

  (A) Summary and critique

    (1) Exams become ends in themselves (clause 5 analysis)

    (2) But writing is a lifelong skill (clause 12 and analysis)

  (B) Recommendation and new perspective

    (1) *Specific alternative:* a new writing requirement, a new commitment among writing teachers (clause 6 analysis; clause 12 and analysis)

    (2) *Larger perspective:* the new goal of empowering people to have their say beyond the college walls, throughout their lives as members of society (clause 12 and analysis)

So far, I've used about half my allotted time to arrive at a solid outline correlating the language of my thesis with the key points supporting it, and a broader perspective arising from it. Now I can fill out the raw draft on the framework of the argument-outline. The time for generating material is over; the time for *producing* has begun.

The Developing QWP Model

  (A) Preparing and Planning

    (1) Timetable related to structure: Beginning, Middle, End (Chapter 3).

    (2) Audience, level of generality, scope, scale (Chapter 1).

    (3) Writer's own investment in the project (Chapter 2).

                                    (total allotted time: 10%)

(B) Generating
  (4) Provisional thesis (Chapter 2). (5%)
  (5) *because*-clauses (Chapter 2). (15%)
  (6) Analysis of *because*-clauses (Chapter 4).
      Provisional thesis revised to include *although*-clause
      ("other side") (Chapter 4). (10%)
  (7) Ordered argument-outline (Chapter 4).
      Thesis revised into full-fledged introduction to reflect
      order of ideas in argument-outline (Chapter 4). (10%)
      (total allotted time: 50%)

CHAPTER
5

Producing the Raw Draft

Overcoming the Anxiety of Revising

Most of us are trained from our first exercises in school to think of revising as a painful, unrewarding act, changing a word or two here or there without actually improving our writing. We don't really know what to do. Later on, we learn to cut and perhaps to rearrange a few blocks of writing, but often we do this with a draft that is already fragmented, filled with remnants of discarded thoughts and passages that need *more* rather than less explanation for a reader. Writing becomes the easier part of the process, and *revising* a series of unpredictable, disjointed acts that can spoil the whole thing. People often feel they have squeezed the life out of the writing in revising it, or that they can't change one part because then they would have to rewrite the rest. No wonder, then, that some people settle for eliminating passive voice or unnecessary commas from their rough drafts; there isn't time, energy, or confidence to do more. But revising should be as exciting as composing a draft. We may not be as intrepid as Dylan Thomas, who delighted in ''improving'' Shakespeare in his public readings, but we should anticipate some refreshing surprises as we *improve* an essay or memo. The parts of your raw draft are interdependent; they influence each other, and have a relationship to the whole design that gives revising the sculptor's sense of freeing the figure from the stone. In the context of the raw draft, some pieces of the argument will seem too fragmentary, or too isolated; others will call for greater emphasis or illustration. Eliminating the fragmentary pieces scales the draft down to size and provides internal coherence. The improvements for unity on your roughly final draft filter and shift material organically, as would occur naturally over a longer period of maturation. QWP simply speeds up the process by harnessing the pressures of time, space, and audience.

Freeing the Figure from the Stone: An Overview

If I cut and paste, or move blocks on the screen to run together all the selected, unedited material in the order of my quick argument-outline, I have before me the raw draft of my position paper, written under pressure.

This raw draft will be about twice as long as the final product needs to be. Placing the blocks in sequence reveals gaps, shows where connections need to be made, and inspires good examples. I have plenty of room for cutting what isn't absolutely essential, and then polishing the roughly final draft. You needn't be rigid or timid about what to keep or what to let go. Once you know you have more than enough good material, you regain the excitement of discovery and presentation even in the revising process. Now you can be as hard-nosed and creative an editor as you like. Since QWP ensures all along that you select the best of the best material, you're free now to chop away digressions, repetitions, remnants of discarded thoughts, and inconsistencies in content and tone.

You can read the cut raw draft aloud to gain a measure of detachment and objectivity. Try reading into a tape recorder and then listening as if you were hearing your words for the first time. Have someone else read the draft back to you; you can hear flat places, lost connections, undeveloped meanings, and inappropriate emphases. Although a reader's responses at this point can tell you a lot about how the piece of writing works, give some thought to the kind of feedback you need. Too often, people close to a writer either try to "soften the blow," holding back comments that might be really useful in revising, or attempt to protect the writer from imagined criticism by giving stifling or intimidating advice. What you need most is someone who will read over your words with a fresh eye, and simply tell you how they sound, what they mean—without either protecting or disciplining you. For this "second reading," you need feedback, not misguided therapy.

As the roughly final draft becomes the final product (see Chapter 6), new writing adds transitions and examples, and reenforces overall unity in tone and content. But this project, which you planned from start to finish so you would be sure to complete it on time, reaches an end in the polished final draft, the clearest, most honest, most powerful piece of writing you could manage under the circumstances. The last look at the writing, especially if someone else types it for you, should be a flawless proofreading, to ensure that your reader is not distracted from the power of your meaning and the clarity of your style. Many people are not good proofreaders of their own work (even if they are experts at helping others); software programs can do some of it, or you may have a colleague or teacher who will help. But no matter how proficient you are at the process of writing well under pressure, the product must be perfect.

It's important to distinguish between revising and proofreading. Once you've expended your best energies on a piece of writing under pressure, a proofreader's reservations about content or style will probably hurt more than help. At the end of a project, you need someone to perform the simple, friendly act of reading through the final draft for any words left out, word processing idiosyncrasies, or chronic misspellings. By then, your timetable has expired (see Chapter 7). Turning in your paper or memo is the immediate goal. It is also an act of liberation, the one way you can integrate what you've learned about your writing skills on this project and free yourself for the next one.

Cutting the Raw Draft Down to Size

My raw draft (on right-hand pages 47–61) is composed of the fourteen unedited blocks of material I ordered in my argument-outline. Block (A) is the introduction developed from the provisional thesis. Blocks (B) through (O) are introduced through simple transitions taken from either the introduction, or the argument-outline. I include working headings in the text of the raw draft ("Suppose that . . ." or "But . . .") to remind myself of the goal of each section as I cut the draft down to size.

On the corresponding left-hand facing pages are the explanations for keeping, moving, or deleting passages. (Boxed passages with diagonal slashes in the raw draft text are to be deleted; passages to be kept for the roughly final draft are bracketed in the left-hand margin.) I also note questions that need resolving, and places where I will want to *add* new material in the roughly final draft.

Freedom in cutting the raw draft arises from having more than enough usable material. Once you experience the fun and mastery of revising with confidence, you will automatically clarify and tighten your editing skills. You will also feel differently about *beginning* a writing project. You will know beforehand that improving the draft is as creative as writing it, and this will allow you to begin the whole process with less hesitancy, less anxiety, and more energy.

Commentary on the First Cuts

As the author of this position paper I would not have needed, of course, to write out my explanations for the changes in the raw draft. But I wanted to demonstrate the fun of freely cutting a raw draft when you know you have more than enough good material. Without the usual dread and anxiety, you can see your own thinking and writing more clearly, and you are better able to judge what will help or hinder coherence. This leaves more energy for shaping a powerful product for the reader.

What's left, then, is a draft on the screen or in cut-and-paste form, that is now roughly scaled to size, and basically coherent: most of what needs to be there is in; almost all of what should come out is out. Some of the deletions from the raw draft in this first round of cuts were fragments of larger issues, or allusions to broader arguments that would demand much more explanation to make sense to the reader. I deleted a few general passages stating opposing views because they were so far removed from the *immediate* issue that refuting them would be a waste of time and space. Other passages were in the wrong tone for the intended audience. Finally, there were the inevitable passages written for myself as part of the composing process but not for a reader, and these were deleted as well.

(1) KEEP opening paragraph—but CUT to correspond with the selected material in the rest of the raw draft.

(2) CUT—these points are too abstract and require too much groundwork to explain in so short a paper (I'm not going *that* far, am I?).

(3) KEEP—I can use the challenge in this statement to engage the reader.

(4) CUT—don't need these anecdotal examples.

(5) KEEP—this sets up the false analogy I'm going to knock down (but turn the question about certification into a rhetorical one). QUESTION: how to show the goal of a "cover letter" isn't *enough?*

First Cuts on the Raw Draft Text

(A) Although the idea of a writing competency exam is
appealing on the surface as a way of strengthening a basic
skill, such an exam would be self-defeating and destructive.
It is expensive, inefficient to administer, and demoralizing
for both students and faculty. It would destroy the English
department's effectiveness in helping all students to learn,
and cannot ensure competency in writing, a skill that cannot
be legislated for a wide spectrum of students in a set period
of time, or through programmable steps. Such a goal can only
be accomplished by a creative, flexible, highly motivated
teaching staff making instruction as individualized as
possible toward the goal of writing as a lifelong skill
and, in a broader perspective, as a citizen's advantage in a
democratic society.

SUPPOSE THAT...

Arguments for competency exam (the "other side")

(B) Actually, the competency idea is a tantalizing one.
Wouldn't it be nice to be able to draw a line and say: "Yes,
Robert writes well enough to get our degree; Judy can't yet,
and needs another semester; Tim may need to take writing each
semester until he graduates, and even then..."? Usually, such
exams are meant to "certify" that students have been trained
in some way, the way we might certify that a student can run
a computer program. What should we certify in writing? That
a student can write an engaging cover letter, or a letter to
the editor of the hometown newspaper? Or a paper free of
mechanical errors that leaves the teacher "no work to do"?

47

(6) CUT—I'm only going to have room to allude
to the computer analogy later.

(7) KEEP—might provide a transition to my own
argument.

(8) CUT—what I say here might be so, but it's
too far afield for this paper. Besides, its
sarcastic tone is inappropriate in a position
paper for the dean.

(9) KEEP—each states an important practical
objection in a simple, straightforward way.

(10) CUT—these objections need more explanation
than I can give in so short a paper, and
without fuller explanation, they *seem* less
troublesome than they really are. I would
rather not cut this, but pressures of time and
space force my hand.

(C) There are, after all, precedents in other areas, such as computer science and the swimming test. The swimming test is something everyone who graduates from SCCA must pass: up and back the length of the pool. Surely we should do the same for writing. Writing after all, is every bit as important in the modern world as computer skills....Dedicated teachers of writing should jump at the chance to reenforce what they do by requiring that students pass a test.

Besides, wouldn't it raise the level of literacy throughout the college, and raise the reputation of Shakespeare College?...We could be in the forefront. Harvard would follow us! A competency exam would put old Shakespeare on the map. We would be known as the school that demanded and got a minimal standard of writing from its students.

BUT...

Arguments against the competency exam

Practical problems in giving the exam: would be expensive and inefficient to administer.

(D) But who would teach them...and where would we get the money to pay these teachers?

(E) No one would want to read it...People can be paid to do it, but then they have to be instructed in what to look for, and their own values and skills have to be integrated with or subordinated to those of the English staff. Or the younger or newer part-time members of the staff can be saddled with the job, which is grossly exploitative.

(F) And what about the students who don't pass? Time and money questions: do we give failing students course after

(11) KEEP—these are hard questions to ignore, so they're good for my argument.

(12) CUT—seems *too* specific here. Moreover, stated simply as questions, without adequate explanation, they lend themselves to simple-minded answers like "Oh sure, we could do that."

(13) KEEP—ah, a good, focusing phrase for the reader to consider.

(14) CUT—don't need; will *seem* an irrelevant example to someone not in the field.

(15) KEEP—within the politics of the school this all pretty much counts, but CUT unsupportable generalization (and old grudge).

(16) KEEP—points to the shortsighted pragmatism of some administrators and teachers on so important an issue.

course, in addition to their full load? Or force them to add time to their four years of college until they pass an exam? When would we give such an exam--during their junior year? Do writing skills need re-testing in senior year? How would we test 250 seniors, and what would we do with those who failed? We would be valuing the test above teaching and learning.

Generally demoralizing for both students and faculty.

(G) Who would be hired to teach chronic failers? What would this continual focus on failure mean to students and teachers? Sections full of failing students _have_ ensured jobs for English teachers at the nearby state college, but no one I've met teaches them with any passion or interest. Besides, at Shakespeare College of the Communicable Arts, we could never support it financially.

Specifically, it would destroy the English department's effectiveness in helping all students to learn.

(H) Of course a competency exam will weaken the image of the English faculty. We will be seen as a "remedial" department, which is what most academics think anyway. And an exam excuses other faculty from doing their part of the job of teaching writing across the curriculum.

(I) Impact on younger teachers...accentuating the wrong things: if you, the teacher, train students to pass the competency exam, you will have done your job. If you fail at that you will be judged a failure as a teacher. Forget about meeting the students at their level, and helping them to discover and tell what they mean. Get them through the exam.

(17) MOVE to (20)?—could be the conclusion to the several points about the difficulties in the concept of a "test" for writing competency, because it transfers attention from the student in school to the citizen in society.

(18) KEEP—more valuable questions for my side, leading to the *real* situation in teaching writing, as I see it.

(19) CUT—don't need (stated in a less threatening—and less self-righteous—way elsewhere).

We don't care what happens afterward. The hitch here, as
always, is that passing or failing an exam may have almost
nothing to do with the lifelong need to adapt one's capacity
to write to a variety of demands and situations... My
students, even those who start by not being able to write a
sentence in an hour, could pass an exam at the end of the
term if it were a genuine task, a step in their development
toward independent writing, with imagination and skill....The
emphasis, instead, must be on giving people the self-
confidence to share, to believe in an audience despite their
experiences in school that militate against all of those
goals.

THUS...

Summarizing issue: Will not ensure competency in
writing, a skill that cannot be legislated for a wide
spectrum of our students in a set period of time, or through
programmable steps.

(J) Who would define "competency"? A philosophical
question if taken seriously, but it probably won't be, in
view of the narrow-minded pomposity of some academics and the
near exhaustion of others. (If we all faced the fact that
anyone can and should get proofreading help from a friend or
teacher or editor as professional writers do, or from a
machine, the whole question of "competency" would be seen in
a clearer light: not "mechanics," but meaning plus
proofreading.) What defines "competency" for a person who
struggles to write neat, bland essays compared to someone who
can think, and who says a good deal in a powerful way.

53

(20) MOVE passage from (17) here?

(21) KEEP—the basic premise of my argument as
I discovered it in the process of analyzing the
because-clauses and the other side.

(22) CUT—some of this is offensive, some of this
is unnecessary, and all of it will seem off the
track without further elaboration.

(23) CUT—certainly the wrong tone for a dean's
request for a position paper.

(24) CUT—basically repetitive.

54

but doesn't know some conventions of punctuation? Most
college students have aspects of both these extremes, anyway.

MOREOVER...

Discovery and transition: Such an exam would be self-
defeating and destructive.

(K) What function would the exam serve--what is it
really for? To give some people on the faculty or in the
administration peace of mind because so many people don't
write well? It's a national concern, after all. How can you
ensure that all students will write well, judged by a single
essay, or even a series of essays? You might do this with
math, or computer competency, or even with a writing
inventory to place people in the right level of writing class
(if you had the different offerings in the curriculum), but
what is our goal here as teachers of writing?

(L) The competency exam is really not for the student or
for the English department; it is for the comfort of people
who really know nothing about the teaching and learning of
writing, but who imagine it as a testable skill that can
be measured in increments on a false analogy with computer
literacy or statistics. These people are probably skeptical
about what students have to "say" anyway.

(M) Why shouldn't we try a competency exam? What harm
would it do? Writing isn't like swimming or computer skills.
For one thing, it is not a life and death matter; and it does
not lend itself easily to incremental learning. Teaching with
only an exam in mind would destroy any possibility of
creative teaching as soon as teachers realized, once again,

55

(25) KEEP—as is: holds together as the underlying
thrust of my argument, and embodies some
of my deepest feelings on the subject.

that you can't _make_ people write competently unless you provide a good, open, supportive learning environment. Fear, panic, and anxiety do not create a good writer. Instead, they discourage risk and prevent people from tapping their deepest resources and then sharing them with readers. The real problem with a competency exam is that it would not accomplish the goal of competent writing and, in fact, would work against it, intimidating both students and teachers, rather than encouraging the learning process. A self-paced course in statistics with good tutors available is a workable endeavor. But writing is an unfolding process in which practice must be linked with meaning, motivation, and the belief in an audience. Students will not learn to find their own voice for an exam; teachers will not support students in attempting risky and new kinds of thinking and writing if they are to be judged on how well their students perform on an exam. Teaching people to swim once up and back the length of the pool is a marvelous goal; students will find computer skills useful in their work lives and even their personal lives. But writing requires nurturing, patience, honesty and wisdom....A competency exam completely misses the goal for teaching writing, and will consume everyone's energies at a time when resources are dangerously scarce. _The exam undercuts the goal it is intended to achieve._

THUS...

Summary and critique: Exams generally become ends in themselves.

57

(26) CUT—has the passion I need to draw the
reader's attention to my conclusion, but I've
said all this elsewhere.

(27) KEEP—my alternative, clearly stated, to
balance my criticism of the competency exam.

(28) CUT—offensive, the wrong tone for the
intended audience (this passage really was
written for *me,* not for a "public" audience).

(N) It would be a disaster for writing...which I see as a personal accomplishment, a lifelong skill. Moreover, our students need to be less afraid of writing, not intimidated by an exam, and the same holds true for teachers. With something as difficult as writing well, everyone looks for an easy way out. A competency exam would be a destructive illusion, a way of avoiding the real task: to help people tell each other what is really on their minds.

INSTEAD...

Recommendation and new perspective: specific alternative.

A new writing requirement, a new commitment among writing teachers.

(O) A year-long, required writing course for everyone. A writing staff charged not only with the theoretical (and now machine-correctable) mechanics of writing, but with the task of helping students have their say, not only in our courses, but throughout their lives as citizens of a society. If we don't do this, we will be responsible for permitting the vacuum that is filled not by real, hard thinking and explanation but, instead, by the incessant yapping of advertisers and photogenic newscasters who smile about weather, drought, starvation, parades, and invasions as if the words they speak have no more meaning than the drone of a refrigerator, and the people of whom they speak no more substance than objects without feeling.

59

(29) ADD—example of the summer writing project
at Bard College that impressed me so much as
a potential solution to the problem of writing
competency . . . a perfect lead-in to my
conclusion.

(30) ADD—summary and crystallization: "no
competency from an exam if the exam
enforces incompetent teaching."

FURTHERMORE...

Larger _perspective_: the new goal of empowering people to
have their say beyond the college walls throughout their
lives as members of society.

Once freed of inappropriate, inconsistent, or irrelevant material, a cut raw draft clearly reveals what is *missing*. This is usually the moment when you think of your best examples. At the end of my raw draft, it was obvious that I needed a larger perspective for the reader, some example that would illustrate in a concrete way many of the issues I had raised so far: money, time, academic goals, ideals. I needed something to help convince the reader of the *reality* of improving writing instead of the illusion fostered by the advocates of a competency exam. Thinking about what other schools have done, I realized that the example of the Bard College writing project was exactly what I needed. I can develop that example in my roughly final draft. Moreover, thinking about that summer writing program at Bard, where the best teachers of writing were chosen to work with all incoming freshmen, I realized that although an exam won't inspire competency among students, it *will* permit incompetent teaching. I'll want to work that simple statement into the *conclusion* of my roughly final draft, too.

The concentration and freedom of these first cuts, then, gave me the opportunity to trim away the fat, emphasize the coherence of my argument, and to see what was still needed. All along, I have been selecting the best material, knowing I'll have more than enough, and that I am working within a system that will enable me to get the *whole* job done well, on time. As I removed from the raw draft what did not fit, settled on what *did,* and discovered what was missing, the argument crystallized for me organically. In my roughly final draft I want to ensure that the reader will similarly experience the organic unfolding of meaning.

When and How to Use Feedback

It is almost always worthwhile to get the perspective on your paper that only another reader can provide, but if you write under pressure, waiting for or wading through a reader's comments may seem impractical and risky.

Talking through a topic as you begin a writing project can be the most useful feedback of all. You needn't find an expert in the field, only someone who can ask good questions: Why are you interested in this topic? What are you really hoping to say? What do you think you will conclude? Such questions inspire you to clarify your own intentions, and it's likely that if you record the talk-through, some of your answers will find their way into the draft of your paper, article, or memo.

You can also make direct use of feedback before you polish the roughly final draft into the final product. When writing under pressure, you concentrate on finding out what you really want to say: integrating ideas, feelings, experiences, and learning into a clear, powerful argument for the reader; and finally, improving your capacity to write well. Feedback is most useful if it is related to these three issues. The timetable for writing under pressure will not accommodate (and your increasing experience as a writer will not tolerate) the comments of people who insist that you revise your writing in their image: either the way they think about a question, or the way they approach an audience.

In most cases, you will be able to include feedback in your timetable for completing the project. The benefits of feedback make the allocation of time well worth it. A few minutes of discussion about your cut draft can save you untold misery in trying to crystallize an argument or demonstrate its coherence to a reader. With my position paper, due at nine in the morning in the dean's office, I'm going to have to trust my experience with the issue and the QWP sequence of conscious decisions that have led me to the carefully selected material of my roughly final draft. But I know from thirty years of writing experience that I will need someone to give the final draft a thorough proofreading before I print it out for the dean.

Producing the Roughly Final Draft

Unifying the Cut Draft

Rough cutting the raw draft is an absorbing activity commanding the largest portion of the QWP timetable. Now, a second reading will reveal the gaps left by cutting repetitious, tangential, undeveloped, and inconsistent material, and combining related points. You will discover the need for transitions, a good example to crystallize an idea, and new material to reenforce the internal coherence of your argument. This second reading, especially with the constructive feedback of an objective reader, gives you the opportunity to develop a broader perspective for the conclusion, and then to transform your thesis into the complete introduction. You establish the unity of your paper, memo, or report. Finally, you attend to the vitality of the language, integrating style and argument on the word, sentence, and paragraph levels.

The roughly final draft (on the following right-hand pages) has only a little more material than I'll need for my position paper for the dean. There is still some room to cut, combine, and add new material, as in passages (A) through (J) on the left-hand facing pages that follow. But the goal is to use all the remaining time, except what you have reserved for proofreading, to unify the clearest, most powerful product.

Improving the Roughly Final Draft

Now, almost all the familiar, often frantic work of revising is completed. With QWP, it has been part of a developing process, selecting what to keep and what to leave out, to engage the reader in the unfolding structure and meaning of your argument. Cutting the raw draft down to size to make it coherent and consistent, and then unifying the roughly final draft, is intensely absorbing, bringing out the passion and personality of your thoughts and words. In this way, QWP makes a necessity of a luxury most writers under pressure cannot afford: having done the hard work of thinking through, setting out, and analyzing your argument, you become your own sympathetic, totally committed editor, bringing fresh energy to the task of improving words, sentences, paragraphs, transitions, tone—anything that has to do with the clarity and power of your product.

I've had thousands of students and writing clients who have never felt this kind of involvement in the process and product of writing. The ultimate reason may have to do with our view of education and communication, our vision of democracy; but it is clearly not a matter of talent. Any writer or teacher can tell you about talented people who never wrote. QWP assumes that *anyone's* writing is worth doing well, for both the writer and his or her readers. This second reading, then, in which you improve the roughly final draft, is not the usual slapdash substitution of words and phrases, or a last-minute attempt to impose order on chaos, but the natural efflorescence of an organic process, bringing ideas to life before the reader's eyes. People may experience this organizing, shaping, vivifying intensity in their work, in their relationships, in their recreation; but they don't expect it from writing. Creating a piece of writing *should* be exciting and fun, challenging your deepest resources, and making you vulnerable in a healthy way. These are, after all, the qualities we admire in our favorite writers.

Improvements

The introductory paragraph of my roughly final draft carried the assertions I intended to develop for the reader throughout my paper. On a second reading, the next two paragraphs seemed to meander, to imply rather than state, depending too much on a sympathetic reader. They

(A)   *Combine, fill in gaps left by cutting, and tie in more directly with overall tone and theme:* "The competency idea *is* a tantalizing one, but it raises serious questions. Usually, as in computer literacy, such exams 'certify' that students have been trained in some way. But what could we or *should* we 'certify' in writing? Who would make that decision? What are we prepared to do for the students who fail? What kind of teacher could we hire to read such uninspired essays, or to preside over a competency-test prep course? Generally, what would the mechanization of writing instruction mean to our students and teachers of writing? These and other questions suggest how a competency exam will force us to put our resources into a test rather than into teaching and learning."

The Roughly Final Draft

Although the idea of a writing competency exam is
appealing on the surface as a way of strengthening a basic
skill, such an exam would be self-defeating and destructive.
It is expensive, inefficient to administer, and demoralizing
for both students and faculty. It would destroy the English
department's effectiveness in helping all students to learn,
and cannot ensure competency in writing. Such a goal can
only be accomplished by a creative, flexible, highly
motivated teaching staff making instruction as individualized
as possible toward the goal of writing as a lifelong skill.

The competency idea _is_ a tantalizing one. Usually, such
exams are to "certify" that students have been trained in
some way, the way we might certify that a student can run a
computer program. What should we certify in writing? That a
student can write an engaging cover letter, or a letter to
the editor of the hometown newspaper? Or a paper that leaves
the teacher "no work to do"?

Surely we should do the same for writing. Writing is
every bit as important in the modern world as computer
skills....Dedicated teachers of writing should jump at the
chance to reenforce what they do by requiring that students
pass a test. But who would teach them...and how would they
be paid? No one would want to read it....And what about the

67

(B) *Combine, substitute more consistent, formal language for* ———
    the casual metaphor: "Of course, a competency exam will
    reduce the role of the English department to that of a
    remedial service, with no realistic expectation that faculty
    in other departments would take up the responsibility of
    teaching writing."

(C) *Combine passages, and change to a "public" tone, more* ———
    consistent with the overall texture: "The impact on new
    and younger faculty will also be devastating, accentuating
    just the wrong things: training students to pass an exam, a
    narrow skill that may have almost nothing to do with the
    lifelong need to adapt one's capacity to write to a variety
    of demands and situations."

(D) *Set off with transition that ties this point to the overall* ———
    theme and establishes progress toward the conclusion: "In
    short, the exam creates an untenable paradox for writing
    teachers and, by extension, for the college as a whole: what
    is 'competency' for a person who struggles to write a neat,
    bland essay, compared with someone who says a good deal
    in a powerful way, but doesn't know some of the testable
    conventions of grammar? What problem would a
    competency exam *solve*?"

(E) *Cut: inconsistent, private, not public tone.* ———

students who don't pass? Time and money questions: do we give failing students course after course, in addition to their full load? We would be valuing the test above the teaching and learning. What would this continual focus on failure mean to students and teachers?

Of course a competency exam will weaken the image of the English faculty (we will be seen as a "remedial" department) and excuse other faculty from doing their part of the job of teaching writing across the curriculum. If new faculty's students fail the competency test, they will be judged failures as teachers. Forget about meeting students at their present level and helping them to discover and tell what they mean. Get them through the exam. We don't care what happens afterward.

Who would define competency? What defines "competency" for a person who struggles to write neat, bland essays compared to someone who can think, and who says a good deal in a powerful way, but doesn't know some conventions of punctuation? Most college students have aspects of both these extremes, anyway.

But the hitch here, as always, is that passing or failing an exam may have almost nothing to do with the lifelong need to adapt one's capacity to write to a variety of demands and situations... My students, even those who start by not being able to write a sentence in an hour, could pass an exam at the end of the term if it were a genuine task, a step in their development toward independent writing, with imagination and skill... The emphasis must be on

(F) *Make transition more dynamic, and make language more*
active: "In many ways, such an exam is self-defeating and
destructive. You can't make people write well unless you
provide a supportive learning environment."

". . . but, instead, will intimidate both students and
teachers. A self-paced course . . ."

(G) *Cut: the analogy seems indirect at a point where I want*
to bring in my alternative sooner, directly after the key
last line of the paragraph.

giving people the self-confidence to share, to believe in an audience despite their experiences in school that militate against all of those goals.

Such an exam would be self-defeating and destructive. Teaching with only an exam in mind would destroy any possibility of creative teaching as soon as teachers realized, once again, that you can't _make_ people write competently unless you provide a good, open, supportive learning environment. Fear, panic, and anxiety do not create a good writer. Instead, they discourage risk and prevent people from tapping their deepest resources and then sharing them with readers. The real problem with a competency exam is that it would not accomplish the goal of competent writing, and in fact, would work against it, intimidating both students and teachers, rather than encouraging the learning process. A self-paced course in statistics, with good tutors available, is a workable endeavor. But writing is an unfolding process in which practice must be linked with meaning, motivation, and belief in an audience. Students will not learn to find their own voice for an exam; teachers will not support students in attempting new kinds of thinking and writing if they are to be judged on how well their students perform on an exam. Teaching people to swim once up and back the length of the pool is a marvelous goal; students will find computer skills useful in their work lives and even their personal lives....But the competency exam completely misses the goal for teaching writing, and will consume everyone's energies at a time when resources are dangerously

(H) *Transition, and development of language:* "Instead of a competency exam, we need a new writing requirement, and a new commitment among our faculty to teach writing in every department. We should have a year-long, two-semester writing course, and a writing staff charged not simply with teaching the theoretical (and now machine-correctable) mechanics of writing but also with the task of helping students have their say, and helping teachers in all departments improve the writing of their students."

(I) *Save this largest generalization for the conclusion.*

(J) *Add example and new material to anchor it in overall theme; move passage (I) to end and broaden its implications to unify the whole paper:* "For perspective, consider the summer writing program for all incoming students at Bard College. Teachers, tutors, and students read, analyze, and give feedback on writing. The results have been gratifying: students produce better writing, and teachers are more engaged with their students' ideas and words. The initial program's budget was $192,000. As far as I know, Shakespeare College can't afford that. But neither can we afford the illusion that we will achieve anything really useful for our students by spending a little money on an inefficient, demoralizing exam that doesn't test for the skills and wisdom we all hope to inculcate in our students.

"Competency cannot be achieved through the incompetent teaching an exam will elicit. Instead, we should work for writing competency in a larger perspective of empowering our students to think and write with clarity and power, not only while they are in college, but throughout their lives as members of a democratic society."

scarce. The exam undercuts the goal it is intended to achieve.

Specific alternative: a new writing requirement, a new commitment among writing teachers. A year-long, required writing course for everyone. A writing staff charged not only with the theoretical (and now machine-correctable) mechanics of writing, but with the task of helping students have their say, not only in our courses, but throughout their lives as members of society.

New conclusion, beginning with the Bard College example, and including the summary and crystallization....

were too *personal*. I combined them in (A) so that my objections, my series of "serious questions" about giving a writing competency exam have a cumulative effect, leading the reader through time, money, and pedagogical concerns to my conclusion about the inevitable danger of putting the test before teaching and learning. I wanted, also, to tighten the language and make it more straightforward (less dependent on an understanding or on an identity in values between writer and reader), and to make the passage more consistent with the overall tone and texture.

The next paragraph, too, wasn't as clear as I wanted it to be: the "image" of the English department *is* important in the politics of the college. Although there may be people who would like to see the English faculty immersed in "remedial" instruction, it is important to identify the dependent relationship among the different departments of the college. To diminish the role of the writing faculty alters the responsibilities of *other* faculty members, especially those who see themselves as conscientious teachers. Improvement (B) made this clearer.

The fourth paragraph also needed to be stated in a more direct way, in a "public" tone consistent with the rest. The irony, and the hypothetical arguments were holdovers from the moment when I first felt really engaged in this topic and my own experiences and frustrations and ideals about teaching writing suddenly became relevant to the writing of the paper. However useful in *generating* material that led me to an important private discovery, casual irony is not appropriate for the reader. More important, in so short a paper, I need each point to serve as many purposes as possible, so I re-created this passage about new teachers by combining two others to reflect a sense of my overall theme: that we should teach writing as a lifelong skill (C).

Then, with a simple transition, I brought all the separate points to bear on the ultimate question about an exam: how do we define "competency"? Instead of another series of questions, I illustrated this issue in the form of one of the many paradoxes it raises, hoping to prompt the reader's thinking (D).

There was certainly no need to include the next passage, filled with self-righteousness, drawing attention away from the argument and toward the writer. Although the passage was fun to write, and stated what

I believe, it was not consistent with the overall tone of the paper. The second reading saved me from myself there (E).

The next paragraph was really the turning point of the paper, moving from a criticism of the details of giving an exam to the larger issue of teaching and learning writing in a liberal arts college, a larger *context* in which I wanted the reader to consider the exam. I needed a stronger transition (F). I already had the example of a self-paced course in statistics, so I could delete the others at the end of the paragraph. But I did want to keep the last line as a summary, an indication that there is a better way to achieve the goal (G).

A new transition turned the reader's attention away from the competency exam (H) to a better idea, the alternative I briefly wanted to develop. The last line of the original paragraph was one that kept coming back throughout the whole writing process, and I retained it for my conclusion.

The flaws and illusions involved in the competency idea could be typified by comparing it to the comprehensive Bard College program. (The play of words on "affording" an illusion may help draw the reader's attention to these basic issues as well.) The final sentence of the paragraph expressed my deepest conviction about teaching writing, and offered a perspective in which a competency exam at best is trivial and irrelevant, and at worst is a black hole for our scarce resources and energies (J).

The new conclusion, including the sentence retained from (H), matched the order and feeling of the introduction, and went one step further than a summary, to leave the reader with a larger context, a comprehensive perspective for thinking about the question posed (I).

Proofreading

No matter how many times I read over the final draft of a paper I miss some typos, a chronic misspelling, a word left out in the cutting or polishing process. Sometimes I fail to see a difference in number between subject and verb, or a cumbersome adverb between the parts of an infinitive, a habit I've found hard to break. On the other hand, in publishing jobs, in committee work, and as a teacher, I've been an excellent proofreader for other people. The only satisfactory solution I've found for this irritating paradox is to barter: I will proofread their work if they will proofread mine.

The fact that when I type I sometimes reverse letters (a learning disability) doesn't help. When I was in college, I could never show my rough drafts to anyone because they looked as if they were in a language only vaguely related to English. The papers I turned in were readable only because I put in twice as much time as anyone else making them so. I knew that some teachers equated proofreading with intelligence. But I had no idea how little inclined *any* reader is—whether conscientious, exhausted, or jaded—to put faith or energy into what seems at best the work of a sloppy writer, or at worst, an illiterate one.

Any reader feels jolted by a typo, whether in a *New Yorker* article or in a personal letter. Professionally, this natural recoil can be devastating: grants are not granted, jobs are not offered on the grounds that "if they don't care enough to proofread their work then we don't want them here." Beneath this impatience is the myth of the perfect draft springing as if by magic from the hand of the perfect writer. Yet a look at the drafts of some of our favorite authors corrects this notion. Thomas Wolfe delivered his chaotic manuscripts in boxes, to be shaped into novels with the help of his editor. Most professional writers and most professional people who write have editorial help: an editor, an executive secretary, the "secretarial pool." Writing well does not necessarily mean proofreading well. If you can't proofread flawlessly, find someone who can do it for you, and view their help as a conscious necessity that some other people can take for granted.

The goal is to make sure the spelling, punctuation, grammar, the whole final draft is free of distractions: typos or word processor quirks (a sentence missing inexplicably, a word repeated or placed out of order). You simply want to ensure that the reader *reads* what you have written, as you intended it. Unfortunately, the myths about final drafts (if they are not perfect the writer is unintelligent, sluggardly, illiterate, defiant, insulting) are shared by many people who are trying to write:

"If I can't get it perfect myself," they say, "then it doesn't deserve to be read."

Teachers often are not much help. If they fail to transform a student into a good proofreader through their red marks and exclamation points in the margin, they find it hard to suggest that the student get help elsewhere. In fact, the myths of the final draft preclude getting help ("Your father read this over for you? That won't do you any good in the real world!" or, "You should take a course in grammar and clear up this mess before you expect anyone to read your writing!"). Some people back off: they write less or not at all. I've had countless students who dreaded writing because they equated it with the act of proofreading; and I've known too many teachers who find that equation appropriate, and continue to apply it.

Writing is difficult enough. Few people do it at all, except under pressure. Fewer people do it with any pleasure. Fewer still do it well. If the myths of the final draft prevent you from writing, or from writing with pleasure, or from writing well, you have to permit yourself, slowly and with determination, to give them up. Proofreading is the *end* of a complicated process of finding out what you think, and how to *say* it with clarity and power to readers. It's a process people need to call on throughout their lives. Proofreading is absolutely essential; but it is a simple act that neither measures intelligence nor talent. It is an act devoid of morality or nobility. It involves no more wisdom than putting the stamp on an envelope in which you mail your application or the article you've spent six months writing. Without the stamp, you won't get *any* reading; without proofreading, you may get a poor, or irritable, or unenthusiastic reading. If you need help, have someone proofread your work for you.

Burnout is clearly evident in those teachers who can only respond to typos and "mechanical errors." I don't blame them, of course. After reading thousands of *un*-proofread papers, who would want to crank up energy again to try and find the ideas and personal style in someone's garbled writing? But the advent of software that corrects spelling and even grammar puts the whole question in a new light. For the writer, correcting the final draft by machine is a blessing. Without the burden of circling errors and weeding out good writers (or applicants, or subordinates) from bad based on the myth of proofreading, some readers will have to retrain themselves to look for meaning in the words on the page.

And that is all to the good. Writers have been doing the hard part long enough. Writing done well, on time, and flawlessly proofread, shifts the burden to the reader. Let the reader be *vulnerable;* that makes communication possible.

Producing the Final Draft

The Final Product

Here is the final draft of my position paper.

Teachers, Not Testers

Although the idea of a writing competency exam is appealing as a way of strengthening a basic skill, such an exam would be self-defeating and destructive. It is expensive, inefficient to administer, and demoralizing for both students and faculty. It would destroy the English department's effectiveness in helping all students to learn, and cannot ensure competency in writing. Such a goal can only be accomplished by a creative, flexible, highly motivated teaching staff making instruction as individualized as possible toward the goal of writing as a lifelong skill.

The competency idea *is* a tantalizing one, but it raises serious questions. Usually, as in computer literacy, such exams "certify" that students have been trained in some way. But what could we, or *should* we "certify" in writing? Who would make that decision? What are we prepared to do for the students who fail? What kind of teacher could we hire to read such uninspired essays, or to preside over a competency-test prep course? Generally, what would the mechanization of writing instruction mean to our students and teachers of writing? These and other questions suggest how a competency exam will force us to put our resources into a test rather than into teaching and learning.

Of course, a competency exam will reduce the role of the English faculty to that of a "remedial" service, with no realistic expectation that faculty in other departments would take up the responsibility for teaching writing. The impact on new and younger faculty will also be devastating, accentuating just the wrong things: training students to pass an exam, a narrow skill that may have almost nothing to do with

in a larger perspective of empowering our students to think and write with clarity and force, not only while they are in college, but throughout their lives as members of a democratic society.

The Writing Blues

With QWP, it took me about as long to arrive at the final draft of "Teachers, Not Testers" as it has taken you to read this far in *Writing Under Pressure*. After two rounds of revision, what remains must stand for a lot of thinking and some writing that is no longer visible in the final draft. But the paper has a substantial feel to it, representing more than it actually states as information or opinion. QWP's organic process enabled me to select for the best material, and to strip away excesses and inconsistencies in tone, style, and meaning. I see the main argument clearly, and it is expressed with some of the passion that I feel about it. It is also a paper I've worked hard over, and that gives me a sense of satisfaction as well.

Any piece of writing can be improved, and most of all the one the writer has just finished. That's part of the writing blues. Years ago, a writer told me there was nothing worse than coming to the end of a book. She had never experienced writing blocks (certainly not the kind that John Steinbeck described with such anguish in *Journal of a Novel: The East of Eden Letters*). But whenever she came to the end of a project, she suffered the writing blues. For some people, the intensity and involvement of writing makes *finishing* unbearable. Having devoted so much energy to the challenges, compromises, and struggles, the writer can't let go, even when other projects call for attention. It's not surprising that some people finish the job, turn in their paper or memo, or send off their manuscript, and then experience an awful restlessness that the writing could have been better.

Writing is never "finished." Every generation "rewrites" the Bible to make it more "readable." We've seen with fascination the original version of Eliot's *Waste Land* before Pound's editing. And there are classic examples of the writing blues: Kafka, Dickinson, and Hopkins repudiated or hoarded their writing, work we might never have known except for the intervention of the people involved in complicated relationships with them. Readers add, truncate, and interpret from one generation to the next, and then that accumulation of wisdom and folly is mixed into subsequent readings of those works as well. Can a writer *know* his or her intentions fully? Can intentions be imparted to readers, or is writing a kind of Rorschach test in which readers see only what they are prepared to see? What were the unfulfilled possibilities in the draft of Melville's unfinished *Billy Budd?* As readers, we are grateful for what we have. As writers we can take some measure of reassurance from the fact that writing is one of the few good ways we have of finding out what we think, and of informing (or learning from) others. Writing doesn't give us a final answer; but it helps both writers and readers think, and so it is worth the imperfect effort, again and again.

Finishing a piece of writing we have worked hard over and intend to present to an audience of one reader or many is scary. We may not want to do it. We may put it off; we may hope to make it better, first. But for certain kinds of writing—writing under pressure—it *is* necessary to finish and let go, in order to begin again with increased knowledge and skill, and higher expectations for both the process and the product.

You can never do enough. Having finished my position paper on time, I wish now that I had transformed the opening paragraph to read more simply, more directly—something like this: ''Although we can improve our students' writing skills if we use our resources wisely, an English competency exam would be destructive and self-defeating. . . .'' This would immediately give the paper a positive, practical tone, rather than the slightly defensive tone it takes. I would also want to think about changing some of the questions in the second paragraph into statements, or at least explanations. I would make more room to explain my conviction that we must teach writing as a lifelong skill rather than a school-bound ''competency.'' I would want to draw out the distinction between a neat, dull piece of writing, and a powerful ungrammatical one. (*Is* that clear? How many people can understand that point without having taught writing?) How can I hope to convince inveterate testers that there is another, perhaps better way to educate people? And who can I reasonably expect to agree with me that writing well is worth the extra effort and commitment, not only for each individual student, or for the papers in all our courses at Shakespeare College, but for all of us, as we try together to make our lives less disconnected, clearer, more empathic, more powerful?

Too late. The paper is on the dean's desk. As I walk toward my office, I see some ways it might have been better. Perhaps it is only *because* it is done that I can see these possibilities. But it is as good as it could be under the circumstances. And what I've learned from writing it will help make my next piece of writing better.

These writing blues are inevitable, but they diminish as your writing experience grows. One alternative was not to do the paper at all. There are people who have left projects unfinished, or abandoned finished ones not because of flaws in their reasoning, or weaknesses in their skill, but because they have made too much of what is really the natural recoil, the vulnerability following intense, personal effort. I prefer the example of Trollope: if he finished one novel in the middle of his daily writing session he began another.

Real and Imaginary Readers

Of course, on the way to my office in the humanities building I wonder, too, about the readers. Not only my intended reader, the dean, but the people with whom he will discuss the issue of a competency exam, and then those others I would *want* to reach: people I know or know of, people I will never meet, people who don't agree with me, people who or may not be capable of persuasion. These are such black and white times, with a craving for simple answers to enormous questions, that I wonder if I could convince the parents of our own students, or even the students themselves.

My position paper acknowledges some of the possible uses of competency testing rather than avoiding them; won't that lessen its appeal? Still, I know what I'm talking about; I've thought it through in writing the paper, and had my say. It's up to the reader to sort out his or her own view.

When I was in college, it was fashionable to say, about one student or another, ''He won't ever write, he feels Dostoevsky looking over his shoulder''; or, ''Who cares about her mountain people? She should be writing about Cambridge love affairs.'' Those comments, from teachers and students, always struck me as bizarre, mixing ignorance with envy and perhaps cruelty as well. People trying to write are particularly vulnerable to readers' pronouncements. But one need only listen to the comments at a public reading or the questions in a writing class to know how foolish this can be. Readers' comments are revealing more often for what they say about the *reader*. I try hard in my writing classes to make sure this is clear to everyone. Feedback *is* invaluable to a writer, but readers often adopt those criteria which were applied to them in school, and they often act as if they were now the teachers who put *them* off writing in the first place. Some writers, too, aren't comfortable unless they get the ''tough,'' pre-manufactured comments they were used to in school, the comments which are the easiest for readers to make and, in the long run, the least constructive for the writer. If the reader hopes to sound profound, then the comment will probably push for some ideal the writer clearly has failed to reach. The reader may have good examples of that ideal, and sound advice for the writer about the ''correct'' way. If, on the other hand, the reader decides that the writing is a pale reflection of what the writer ''wanted'' to say, then the reader may back down, to spare the writer's feelings (''That's nice, dear, but I think you must have a lot more to say'').

There is a familiar repertoire of such double messages, based on the relationship between writer and reader. But, as I've indicated in discussing feedback, such comments are less useful to a writer than the plain, energetic explanation of what works and does not work. Such readings are hard to come by. Good editors are as rare as good writers, and far less appreciated. Yet, over time, it's extremely healthy for a writer to replace anxiety over possible failings in the finished piece of writing, or misinterpretations by readers, with a sense that he or she has a large measure of control over the writing process, including a reasonably good sense of how readers will respond.

The writer delivers the product and moves on, learning something new each time. Developing your own skill is a private matter. You take pleasure and some pride in it, and these feelings are mainly private as well. The completion of a painting or a musical composition or a piece of writing is not usually cause for widespread celebration. I first understood that when a professor came into the office of my freshman advisor to announce that he had just finished a book he had been working on for five years. My advisor looked up with a faint smile: "Well, now you can begin the next one," and looked down again at my study card. It's like that more often than not. Writers need to build in their own rewards, and not get caught up in longing or perfecting themselves for acceptance, gratitude, or honor. The writing that you do is, finally, your own, however clear you are about your reader's needs.

Moreover, talking about or responding to writing is extremely difficult to do well. First responses can be quite different from more mature ones. The insights that matter to writers are often time-locked. They come privately to the writer during a later rereading, or long after the project is done, when he or she is working on another. Such insights (the real point of an essay, the more appropriate structure) seem to be a product of persevering, wrestling with the problems intrinsic to writing over and over again. Real readers may or may not help; imaginary readers may or may not hinder. But I would feel more confident of my position paper if I had gotten some useful feedback. However pressed for time, I will definitely include feedback in my next project. The reassuring thing is that the fear of *being read* loses its power to block or stifle, or to give you a bad case of the writing blues. Writing becomes, instead, a realistic possibility for expression, for connection, and for communication.

8

The Integrated QWP System

The QWP Model

Here is the QWP model as we have developed it through the first part of this book.

(A) *Preparation*. The writer's point of view on the question posed; his or her relationship to the reader; the writer's own commitment.

(B) *Timetable*. From planning the process to completing the final product.

(C) *Plan of overall structure*. The intrinsic characteristics of beginning, middle, other side, and conclusion.

(D) *Provisional thesis*. The writer's commitment as a challenge to the reader.

(E) Because-*clauses*. Completing and supporting the thesis, and generating the other side.

(F) *Analysis of* because-*clauses*. Selecting the best material for the argument, and for developing the conclusion.

(G) *Analysis and integration of the other side*. Extending the provisional thesis to include an "*although*-clause."

(H) *Argument-outline*. Arranging the best material in the best order.

(I) *Full introduction*. The amplified thesis incorporating the other side, and the main ideas in the sequence of the argument-outline.

(J) *Filling in the raw draft, and first cuts*. For internal coherence and consistency.

(K) *Feedback.* What works and what does not work for a reader.
(L) *Second reading: improving the roughly final draft.* Overall unity: powerful transitions, integrated examples, a polished introduction, a conclusion that gives the reader a new perspective. Vital language.
(M) *Proofreading* the final draft.
(N) *A writing policy.* Integrating new knowledge into the writer's version of QWP.

Integrating the QWP System

QWP allows you to prepare, plan, and generate writing in a continuously self-correcting process that allots adequate time and energy to improve the final product. The system focuses on both meaning and the most compelling structure to clarify and enrich meaning. Since QWP emphasizes efficiency, your writing continuously serves two purposes simultaneously: moving toward meaning; and providing the basis for the next stage in the evolution of the project. The logic and concentration of QWP, driven by the writer's alliance with time, produces good writing under pressure.

Preparation

The writer's analysis of the question posed and of his or her commitment to answering it initiates the process. Imagining different approaches, you seek the one that will teach you the most, and elicit as much of your own experience, knowledge, and feelings as possible. You clarify the power relationship between you and your reader. How much and what kind of information does the reader want? How free are you to provide that information? What point of view enables you to set the appropriate scope, scale, and style for the project?

Part of this initial preparation is to get started by immersing yourself immediately in the material. You want to transform the topic or the question into a subject you can write about with a sense of discovery.

You may need to explain your divergence from the question, or your interpretation of it, and this explanation also becomes part of your argument. The boundaries of a writing project often seem narrowly defined by format or convention. You may need to redefine those boundaries for learning, and for teaching the reader.

Timetable

In beginning a writing process that drives toward a balanced, clear, powerful product, the writer makes an early decision about time. In some cases, you will have little choice: a memo is due the same day; a case study must be in the teaching assistant's mail slot by nine in the evening. In other instances where the writer can schedule work for a deadline a week or several months away, a realistic timetable ensures thoroughness and balance. Skill in making this productive alliance with time increases as you use QWP, ensuring that *each* part of the writing process will receive your best energy.

QWP insists on a realistic timetable for completing the *whole* project before you write the first word. Once you create a schedule for writing under pressure, you abide by it. You may have more to say. There may be unresolved issues. There may be no ultimate, convincing proof for your thesis. But your job is to acknowledge what you know and what you don't know *now,* and to complete the project on time. Perhaps you will have an opportunity later on to continue, or to return to the project, or to use what you've written under pressure as the stepping stone to a longer, more comprehensive work. But with QWP, about half of your time is for planning and generating material (the *because*-clauses that elaborate your thesis, and their analysis), and the other half is for transforming the selected, ordered raw material into a coherent, consistent, unified product. The timetable will ensure a comprehensive, balanced piece of writing under pressure.

Plan of Overall Structure

After considering the reader's needs and the writer's commitment, and setting a timetable for the whole project, make a quick sketch of the overall structure. Structure enriches, at times even conveys content. The beginning of a piece of writing should set the context and deliver a challenge to the reader; the middle should provide a sense of comprehensiveness, including a view of the other side; and the conclusion should offer the reader a new perspective, a new context in which to view the question. This underlying structural plan facilitates both the process of writing and the act of being read. Some writers expect their readers to fill in gaps and find the lost connections in their writing. That is not a promising expectation. A good reading is rare enough; an intense, symphathetic second reading is almost nonexistent. The clearer your concept of structure at the beginning of a writing project, the greater the likelihood your reader will follow you through that first and only reading.

Provisional Thesis

A thesis takes a stand on a question, commiting the writer to provide an answer, with reasons, explanations, and examples. It is the internal energy driving good writing: the writer's struggle to make sense, and to engage the reader in the quest for meaning. The answer must be clear before it can be persuasive. In scientific writing, the answer must be clear enough for other researchers to duplicate the work in their own laboratories. In an action memo, the reader must first understand the writer's judgment of various options before adopting or rejecting the recommendation for action. A provisional thesis, then, immediately moves the writer and reader toward *meaning*. The thesis may become more elaborate as the project unfolds, or it may lead the writer to an unexpected point of view, but its continuous influence is to point the writer to careful selection, explanation, and illustration in organizing his or her thinking.

Because-clauses

The spontaneous list of *because*-clauses, the "reasons" for a thesis, works on a number of different levels. By freely generating ideas you move beyond the anxiety of a blank page toward the center of gravity of your commitment to the project. Moreover, *because*-clauses provide some of the actual language for the raw draft. It's important to generate far more *because*-clauses than you need, before judging the material. If you are writing a three-page position paper, give yourself at least a dozen ideas to analyze; for a term paper of twenty-five pages, you might want to sketch out twenty-five ideas. The goal is to explore widely, discovering specific and general issues, and perhaps some examples. You also want to uncover your own feelings toward the material, those issues about which you have a good deal to say, and good reason to say it. You want to gather as much raw material as possible for discovering the project's center of gravity, and for the breadth, power, and clarity of your final draft.

Analysis of the *Because*-clauses

Once you have generated more than enough reasons to support your thesis, you can begin to select the best material to order into an argument-outline. Test each clause to see if it is true or useful in your argument, making notes to develop the ones that mean and matter the most. Analyzing the *because*-clauses reveals the writer's passion for a topic, the aspect of the thesis that will teach the writer the most, or that the writer most wants to explore. You see this clearly in the clauses that run on to paragraphs or pages. *Because*-clauses are the heart of the QWP system. They lead to meaning and they provide the language of fresh ideas. And fresh ideas are what you need. Imagine, for example, what it is like for a teacher to receive yet another set of papers on a topic given year after year. Most of the papers cover familiar ground, in familiar language. Even the selection of quotations from sources will be predictable. (Of course, the teacher *could* ask new questions, or think of new approaches to basic material; but, under pressure, many do not.) When you analyze your raw material, respect the fresh ideas which impart your own way of looking at familiar issues. Just as generations of students have been trained to evade responsibility for their words by the prohibition against using the first person singular pronoun, so, too, most people have been trained to neglect the very material that best represents their own thinking. The drabness in our private and public discourse to some extent reflects the devastating effects of this training. Take some risks in proposing your ideas; stretch *too* far, at least in generating raw material. Vision depends on it. Your own learning demands a larger context, linking your present argument, your struggle for meaning in a paper or memo or article, to the ideas and concerns that will endure beyond it.

Analysis and Integration of the "Other Side"

When writing under pressure, many people feel they should ignore evidence that runs counter to their argument: "Why should I spend valuable time making the case for the opposition?" But writing is different from face-to-face debating before an audience. If you can think of opposing arguments, so will your readers, even those inclined to agree with you. It is far better to acknowledge the other side and to integrate it into your own argument. Moreover, by generating some reasons *for* the other side, you discover fresh ideas for your own.

If you leave too many unanswered questions about what you have failed to consider, or if you make a show of knocking down a straw man, you undercut your reader's trust. A convincing, honest look at the other side strengthens your argument, engaging readers, even those who may not agree with you. By including the other side in your introduction and exploring it at the beginning of your argument, you demonstrate on a structural level a confidence and power that naturally attracts readers.

If you think of the extended provisional thesis as containing an *although*-clause first, and then several *because*-clauses ("*Although* _____, it is better to _____, *because*: _____"), then you will have the structure for a reasonably comprehensive argument written under pressure.

Argument-Outline

Once you've analyzed the *because*-clauses, and selected your raw material, make a quick outline of your argument to establish coherence among your ideas. It often makes sense to put your most convincing point last in the middle section, where it will naturally lead the reader to your conclusion. (You may, of course, want to *write* the most important point first to develop it fully, and devote less time and energy to subordinate points.) Moreover, by grouping related ideas under headings in the quick outline, you create the separate phrases that combine to become the introduction and the transitions of your raw draft.

Full Introduction

Selecting and ordering the best of the raw material into an argument-outline, including the other side, enables you to adapt your provisional thesis to the sense and sequence of your argument. Read the thesis back from the headings of your argument-outline, and then develop it, if appropriate (as in an article), into a full-fledged introduction.

Filling in the Raw Draft; First Cuts

Filling in the argument-outline with the blocks of selected material creates a raw draft much longer than you need. This gives you the freedom to cut freely, deleting passages that are inconsistent, tangential, or too abstract. This freedom to choose only the best material, one of the great benefits of QWP, is the exact opposite of what most people experience in revising. When an anxious writer cranks out a draft in a headlong rush, or piles up evidence and examples, revising almost always reveals the lack of planning. Usually, it's too late at that point to do anything about major weaknesses. With QWP, you keep only the best material from your ample raw draft, balancing the specific with the general, and ensuring coherence and consistency among the parts.

Feedback

The questions a reader asks invariably elicit answers from the writer that ought to have been included in the writing. Such questions often challenge parts of the argument the writer has taken for granted, or conflicts the writer has failed to resolve. Readers can also tell you where your language is unclear or lacking in vitality. This kind of feedback is a gift; one's writing is always stronger and clearer because of it. It's important to ask for it, and to make use of it.

Second Reading

The first cuts that transform the raw draft into a leaner, coherent, consistent roughly final draft leave some loose ends and missed connections. Although each piece of evidence in an argument should relate to what comes before and what comes after it, it also must be linked clearly with the overall thesis. At every point in the final product, the reader should be able to sense this unity: the clear relationship between evidence, example, and thesis. Moreover, unifying the elements of your argument, you gather momentum for a powerful conclusion that not only summarizes but creates a new perspective for the reader.

Proofreading

It is the writer's responsibility to present his or her reader with a flawlessly proofread document. In some cases, the absence of proofreading can put a whole project in jeopardy. There are readers who confuse proofreading with intelligence, who are more concerned with "errors" than with meaning. Such readers should be required, at the very least, to read what you have written as you meant it.

A Writing Policy

The *process* of writing well on time has a beginning, middle, and end in itself. You need to assimilate what you've learned and move on. You may or may not get a response from your intended reader. There may or may not be more work to do: more reading, writing, and talking about the subject; another essay, memo, or research article to write. But under pressure, a writer works with intensity and concentration, and it is important to let go once the project is finished. It takes skill and experience to permit yourself to do this, but it is crucial in fashioning a strong writing policy that allows you to approach each new project with increasing confidence, energy, and pleasure.

Adapting QWP to Your Own Needs

Writers have to bridge the reader's natural defenses, the natural reluctance to take on still another job of making sense of someone else's words, and testing their authenticity.

Adopting a new system for something as complicated and personal as writing may seem more trouble than it is worth. You have to give up some old habits that, however unsatisfying, work well enough, and are familiar. But applying new strategies that organize your work more efficiently and make it more consistent opens up exciting possibilities for change. The clearer you are about your commitment to a project and what you can learn from it; the clearer your relationship to the reader and your anticipation of the reader's legitimate questions; the clearer your sense of how to select the most representative material, the more confident you become. If your writing process is based on solutions to problems rather than on compromises with writing blocks or a sense of powerlessness induced by the old mythologies of pedagogy and culture, then your writing will become more flexible, resilient, and powerful. You will discover fresh, elegant ways to represent the complexity of your thoughts.

Unfortunately, most teaching dwells on small details, without attending to the continuous activity of writing. QWP emphasizes discovery, flexibility, and mastery. Adapting the QWP model to your own needs, you will discover a stronger voice, a sense of structure reenforcing meaning, and a new energy in choosing words.

QWP also changes the way you *read* (see Chapter 9). You examine whether the writer has set the context for his or her argument in the beginning of an article; whether he or she has embedded a challenge to the reader in the texture of an essay. You look for coherence in reasoning, and comprehensiveness in scope. You come to expect new perspectives, a broader context at the end. Everything you read with this heightened awareness contributes to making your own writing clearer and more powerful.

Knowing that you can write well under pressure changes your policy toward any kind of writing, from exam essays (Chapter 10), to research papers (Chapters 11 and 12), to memos and reports (Chapters 13 and 14). Runners speak of the "right" feeling in their muscles; painters and pianists tell of the confident feeling "in their hands," because their art is the disciplined physical expression of a mental process. Writers also experience this intense clarity and elegant simplicity. For them, too, the feeling arises not so much from the mystery of talent as from attention to process, to perfecting their own form and style for accomplishing work.

QWP depends on common sense and writing experience, not on magic or talent. It is a system for resolving writing questions, and a policy about writing. With QWP, the writing product is suffused with the writer's own style and substance, and that may well be the real magic.

PART TWO

The Writer Under Pressure

9

Applying QWP: Reader and Writer

Enriching Writing

The first part of this book presents a system and a general policy for writing well, on time, under pressure. This second part applies QWP to common writing situations in a public context.

QWP is constructed out of common sense and writing experience. The essence of the system is discovering and presenting meaning. The emphasis, then, is on a planned, *continuous* process, from thinking about the subject and the audience, to structuring more than enough raw evidence for a provisional thesis, through fashioning the most powerful ideas and examples into the clearest language. With this kind of preparation and planning, anyone can write well under pressure. Moreover, each time you use QWP, you make personal adaptations in the system until it becomes an instinctive approach to any writing assignment.

The next several chapters consider writing under familiar pressures that emphasize different aspects of the writing process. The pressure in examwriting, for example, is to *demonstrate* what you have learned clearly and efficiently to the reader. QWP examwriting focuses meaning through the lens of time. The pressure in research writing, however, is to *discover* and then *present* material effectively. Using QWP in research writing helps you find, select, and structure material clearly for your reader. At work, the pressure is to integrate the qualities of good writing with the demands of different kinds of readers: subordinates, supervisors, clients, colleagues. QWP allows you to accomplish two jobs simultaneously: to approach specific issues from a comprehensive perspective; and to increase the flow of authentic communication in the environment.

Every piece of writing, whatever its subject, is also about writing; every completed memo, letter, or essay reflects the writer's view of what it is possible to accomplish in writing. This concentration on the writing process enriches the product: reading an essay by James Baldwin or George Orwell, we learn about the subject from the author's point of view and are set to thinking about our own ideas. But we are also prompted to think about the writing as well: how and why the essay is so clear, so powerful. The result is a renewed awareness of good writing, and the inspiration to improve our own.

The way we write inescapably tells our reader a good deal about what and how we mean. If we adopt conventions or approaches at school or at work that are inappropriate for what we know, or what we want to say, then our writing necessarily suffers: meaning diminishes, style and tone reflect internal conflict, and the reader is confused. It is inappropriate to write an answer that only obliquely refers to the question on an exam. But it is equally futile to repeat what the lecturer has said in class if your own reading and thinking have led you to other perspectives. It is inappropriate to begin a research paper or article without a survey of the work that has been done before. But it is equally self-defeating to gather material without continuously integrating it with your own thinking. It *is* inappropriate to prepare a position paper when you've been asked for a one-page action memo. But it would be irresponsible to allow any sort of organizational code to prevent you from telling the reader what you think is most important about the issue. So many pressures work against our sharing our knowledge that we continue to write with a low level of satisfaction in school or at work, meeting arbitrarily narrow conventions and expectations. QWP helps us to shape what we know into our own words.

Transforming the Pressures Against Authentic Writing

The pressure of school, work, or profession often forces us to hold our common sense and experience in abeyance. We *know* better, but we don't have the time or the confidence to struggle for authentic writing. Part of the skill in applying QWP is to overcome those pressures within each situation that make good writing on time so difficult. Some of these pressures against writing are internalized. We think that we don't have anything to say, or that someone else can say it better, or that it really doesn't matter because no one is listening. Other pressures, imposed by an environment, result not from wisdom about communication, but from unexamined habits, false analogies, or illusions about efficiency. Every writer needs to acknowledge these pressures in order to transform them. By analyzing them from a *writer's* standpoint we provide a strong foundation for improving writing.

Writing courses may make improvements possible through a structured environment and constructive criticism. But once a course is over, the *intensity* diminishes unless something has caught the individual writer on his or her deepest level of motivation. (That is why I encourage students to form subgroups during a term and ongoing groups at the end of a semester, as one way to ensure that the struggle to write well continues in a supportive environment.) If a student wrote one hundred pages in a semester, without attending a class, he or she would learn something useful about writing. But the solitary writer misses a public perspective on his or her writing, and loses the benefit and inspiration of exchanging views with others engaged in the same struggle. The discipline of writing three pages every day for three months would be instructive; but not many people can systematically improve their writing on their own.

Books about writing usually gather rules and conventions. A few may even inspire a new attitude toward writing. But there are thousands of such books, and most disappear before anyone makes much use of them. Part of the trouble is our stance toward the whole process of learning to write. Writing is a frustrating business. It is also one of those skills we want everyone else to have, instantly. Teachers, supervisors, and colleagues all want better writing from their students, subordinates, and peers. But once we conceive of good writing as good thinking made public, and as a claim on our reader's thinking, we pull back, scaling down our expectations: "No one should be allowed to graduate until he or she can write a good cover letter"; or "Tell them what they want to hear"; or "Just get them to spell correctly." Yet writing is not learned in increments, and thinking clearly is not simply a matter of good intentions or hard work. That is why handbooks, courses, and workshops have not been widely successful in improving writing. Some people can learn that way, but even so, the question remains of whether they will continue to struggle to write well once the workshop, course, or book is finished.

In writing classes, I ask people to read the chapters on the internal structure of an essay in Sheridan Baker's *The Practical Stylist*. It is hard-edged in an old-fashioned sort of way, and seems to me a good summary of what the best handbooks have had to say. But I also ask people to read about freewriting (a timed, nonstop exercise to circumvent self-censorship), sharing, and the two different kinds of responses to writing ("reader-based," and "criterion-based") which Peter Elbow describes in *Writing with Power*. His focus on the struggle to write well, beyond the conventional (and more "teachable") issues of grammar and patterns of organization, invites the writer to take control of his or her own process. Writing is much more difficult to teach and to learn than most people realize because it represents so much that is personal about the individual writer: not only the capacity to share, but the many different reasons we might have for *not* wanting to make our thoughts and feelings clear. *Writing Under Pressure* is not a handbook for reference, but a *system* for application to familiar writing assignments. As the writer makes the system his or her own, it becomes an instinctive approach to process and product. Experience in using QWP makes it a personal skill, and that is why it can help where courses, workshops, or handbooks may not: the emphasis is on becoming an independent writer.

There are, of course, other applications for QWP than those I will consider in the next several chapters. Speechwriting, for example, is vulnerable to the same pressures that complicate writing in organizations: there is the same pressure of audience, the same need for compression in form and content, and the same intention to elicit an immediate response. QWP can prepare and focus the speechwriter's efforts appropriately. Group writing projects, in gathering and integrating material efficiently, need to make the most of the differences among the people involved. The QWP timetable and agenda can organize such projects effectively. The basic system, as demonstrated in the first part of this book and applied in this second part, can be adapted to *any* kind of writing project. But before moving on to applications, there is a less visible, yet equally important benefit: by using QWP in your own writing you transform the way you read.

QWP Writing and the Enlightened Reader

As you use QWP in your own writing, you will discover almost immediately that you read with an increased awareness of how a memo, essay, or report was put together and how the writer struggled to express meaning clearly and powerfully. For some people, this new perspective has the power of a revelation: taking control over writing enables you to understand the writing of others from the inside out.

At times, this new understanding will make you impatient. You sense the writer at work, but you cannot ignore the clear signs of *lack* of work, the lost opportunities to think through and present an issue. This more realistic view of writing restores authenticity to the process: it enables the reader to see writing as a struggle to communicate rather than as a service or magical talent. While you may be more aware of what does not work in writing, you will also appreciate what does: those qualities of good writing that result from concentration and experience.

Readers often view writing through a paradoxical vision. On the one hand, they attribute magical powers to the writers they admire. On the other hand, they cannot imagine the writer consciously struggling to achieve those magical qualities. "She couldn't possibly have *meant* all that," a reader says, as if the writer were capable only of magic, and not the thought and feeling that the writing inspires in them. This paradoxical combination of awe and lack of insight is a familiar response. We may admire a tennis player, but find it hard to believe that placing a ball down the line just out of reach of an opponent is the instinctive result of daily practice. Such an act seems to be all the more a magical gift because it appears effortless. The idea that such skill results from and depends on conditioning and hours of drills seems farfetched. But to a large extent, talent *is* consistent preparation. Writing in general, and our own writing in particular, cannot be understood fully until we write several hours a day, for months at a time. Some readers may view the words on the page as if they were placed there through grace or good luck. But if you write with control, you read with a new understanding of how much more the writer had to say, and of how carefully the writer arranged his or her words to represent as much meaning as possible in as few words as possible.

A QWP Reader's Agenda

This sharpened awareness enables you to detect writing that short-changes or misleads the reader, obscures meaning, or presents frag-mented, incomplete, or distorted information. You will weigh the writer's thesis and evidence more deliberately in textbooks, memos, and reports. As you employ QWP in your own writing, you inevitably apply it as a standard, and a key to your reading.

The Writer's Preparation

As you read, assess the decisions the writer made about his or her relationship to the reader. What tone, style, or level of intimacy did he or she consider appropriate? How did the writer understand the power relationship: are you, the reader, dependent on the writer for instruc-tions or crucial information? Does the writer need you to feel or think the way he or she does?

The Writer's Planning and Generating

As a reader with experience at writing consciously, be aware of how well the writer has defined the issue. Is the underlying or explicit thesis appropriately supported in the material? Is the thesis comprehensive? Does it include a view of the other side, and generate a coherent essay, in a balanced structure?

The Writer's Product

Test what you read to see whether the writer has illuminated the most important issues with the richest details and examples. Do the examples really *represent* the point the writer is trying to make? Do the analogies hold true, or do you sense the writer groping for authority through them? Are the explanations adequate? Are general statements and underlying assumptions anchored in specific facts, experiences, or analyses? Is the *order* of the argument appropriate to the task of making it clear and interesting to the reader? Is there, throughout, a sense of the writer's awareness of vulnerability: is he or she straightforward about the distinction between evidence and opinion? Can you disagree with some of the ideas or explanations and still trust the writer?

Experienced in the struggle to say as much as you can in a clear, powerful way under pressure, you will expect the writer's conclusion to integrate the whole argument, as reflected in the most vivid examples or evidence. Does the conclusion broaden your perspective, and give some sense of what comes next, or of what direction to follow? Did the writer acknowledge the problems inherent in his or her argument, while putting forth a new way to look at the issue? After you've read a piece of writing, can you state the main argument in a sentence that contains a sense of the best evidence and the opposing side?

Finally, can you appreciate, even admire, the freshness of the words, the way the sentences hold your attention, the way grammar and punctuation facilitate meaning? Do you feel the writer understands the work the reader must do in completing the transfer of meaning?

As readers, we ask such questions because, in a common sense way, the answers govern whether the writing enlightens and entertains us. But we ask them because we have asked them of our own writing, in the hope of moving the reader to feel, to think, to understand, to change. As the whole process of communication becomes more conscious, we imbue it with more meaning and more possibilities. Reading and writing become inseparable: we read with an awareness of how good writing is possible, and we write with an awareness of what the reader will need to make good communication possible.

Examwriting

The Eighth Bluebook

Imagine that you are an undergraduate taking a three-hour final exam in Renaissance Literature. You have done fairly well on two short papers in the course, but your hour exam was disappointing. If they ask the right questions on this final, you should get at least a B in the course.

You have reviewed your lecture notes. You did the readings carefully, and distilled your notes into a comparison of the major themes and methods of the important writers. You've browsed through one of the books on the recommended reading list, a study supporting most of your lecturer's interpretations. Taking five courses, playing intramural basketball, and trying to discover who you are and how you fit into the world beyond home and family, you are about as well prepared as you can be for a final exam. But literature is not your major, and you don't speak the language of literary criticism the way English majors do. You chose the course because you have always found reading a comfortable way to compare your own ideas and experiences with those of others, but you would hate to have a class that you took for fun pull your average down. A good exam will make a difference; a poor one will do some damage.

You read the first question on the exam and find it incomprehensible. Perhaps you missed a phrase, a verb, some sort of key. But the second time through is no better. It's as if the question were for a different course entirely, on a graduate level. You can't figure out what they are really asking. It's so frustrating! All term long the lecturer has emphasized the readings instead of the technical details. But this question is much too abstract, phrased in such a way that only a literature major could understand. You race through the other questions, but there's no comfort to be found there, either. This is going to be a disaster, one of those exams you write in panic, trying to pile up enough information to show that at least you have done the reading.

Now consider this exam from the point of view of an instructor in the English department who is serving as section head in an undergraduate survey course in Renaissance Literature. You suffer the anxieties and frustrations of an interminable apprenticeship as a junior member of a department which, once you finally get your degree, will escort you to the door of the job market. No matter how conscientious and supportive you want to be with your students, and how devoted you are to the poetry of Andrew Marvell, you don't look forward to reading your allotment of sixty rushed, chaotic essays about the material you know and love best. One or two students will have gotten a sense of the poems and essays in their intellectual, social, and political contexts, but most people take the course because it is a distribution requirement that no one ever fails.

This year, the professor has asked you to prepare rigorous questions to weed out the good students from the bad, because he hopes to raise the reputation of the course. You set aside the weekend to do the exams, dividing up the work to give each one as fresh a reading as possible. You'll look for familiarity with the texts and lectures, and an appreciation of the professor's approach to the material. You hope the essay questions challenge the best students and inspire the weaker ones. The professor is your thesis advisor, and you don't want him to be disappointed with his undergraduate course.

The first two bluebooks are sketchy, leaving out most of the examples you had in mind when you wrote the questions. You are shocked at how little undergraduates retain from their readings and lectures, how unreflective they are about applying the course *method* to the texts. You're beginning to feel the burden of repeating comments from one bluebook to the next. But after four more uneven ones, you read an exam that is so complete it makes the others seem abysmal. Why was this student able to put it all together when the others could not? Now you have a clearer sense of what is possible, and you feel you were right all along: the exam questions *are* good. Maybe at the end you can go back over the first few tests and do a better job of explaining what's missing.

Then you begin the eighth bluebook. It's unreadable. You try the first sentence over again. Maybe you missed a verb, a phrase, something left out that might make sense of it all. But you can't track the meaning in these scrambled words. Is this student pulling a fast one because he didn't do the work in the course? Your doubts make you guilty. What if he is really trying to say something? And then you get angry. The bluebook is twenty pages of disconnected, illegible scribbling, and a second bluebook is nested inside it with more. You look up at the clock. If you reconstruct this exam word by word, idea by idea, you won't get any of the others done. Why should one student deserve so much of your attention at the expense of the others? You skim through it one more time hoping to find some of those touchstones of a good answer, something you can hold onto and grade. If the rest of the exams are anything like this, the professor is going to feel the course was a failure.

Exam Pressures

There's no doubt that exams can make fools of us all. They are often unpleasant to make, unpleasant to take, and unpleasant to read. They may not show what the student has learned, or what the teacher wanted students to learn. They may not teach the writer or the reader anything, and instead may reenforce those experiences that work *against* teaching and learning.

Examwriting is too important to leave to the classroom or to a particular course. You can transform examwriting from an exercise in meeting arbitrary pressures into a process for discovering and presenting meaning; and no matter how narrow an assignment, the act of writing can teach the writer something useful about his or her own writing process. With QWP, the task of examwriting is twofold: first, to do a good job of writing under pressure; and second, to improve your process of matching ideas, perceptions, and feelings with words on the page.

People not used to planning writing almost always upset the balance of an essay in some way: too much middle, not enough beginning, or no conclusion. They may have too many specifics and not enough generalizations. They may fail to develop their most convincing piece of evidence and give too many subsidiary points, or they may leave out the best examples to illuminate their argument. On exam essays, without the benefit of a second draft, a lack of planning can be disastrous.

Exams foster anxiety and increase the writer's sense of helplessness. Some people put their faith in *quantity,* and race through an exam at top speed. But you can write a lot and *say* very little; and you can think a lot, write only a little, and still answer questions fully. Sprinters *prepare* for the few moments of their race through months of continuous work with weights and training runs, and by shaping their diets and work habits around competition; examwriting requires the same kind of preparation. Other people envision examwriting as entirely unpredictable. Their sense of helplessness is reenforced by the common notion that you cannot study for some tests: "Either you know the stuff or you don't." Finally, there are the loopholes in the exam process. A few students who have done little work in a course will do well on an exam because, understanding the inherent weaknesses in the system of making, writing, and reading exams, they substitute style for substance.

But once you understand the *context* for an essay exam, the relationship between the method and the material in a course, the unarticulated agenda and the stated expectations of the exam maker, then essay exams are not very mysterious at all. Good examwriting depends on preparation before and during the exam, and a determination to make the exam useful not only in itself but as a way of improving your skill as a writer.

The essay exam is an immediate challenge within a larger context of common-sense preparation: attending lectures and taking good notes to review for the test; keeping good notes on the assigned readings; reviewing course handouts and the papers and comments you get back during the term; reading beyond the required list; and discussing questions throughout the term with your instructor. In some schools, you can review previous exams on file in the library. A few students do all these things as preparation for an exam; some do nothing at all. But a good deal of what takes place on the day of the exam depends on consistent preparation. In this context, QWP examwriting helps you plan, generate, and produce good essays, while constantly improving your capacity to discover and present ideas.

The QWP Examwriting Model

Planning takes place on two different levels: *structuring* each essay around a thesis supported by an argument with a beginning, middle, and end; and managing the *time* to produce a good essay—from analyzing the question to proofreading the final product. Essay questions usually ask the student to demonstrate facts, compare theories, or extrapolate from facts or theories. The facts may be as simple as dates, or as complicated as the steps in a complex argument; comparisons may simply require familiarity with material, or may involve interpretation; extrapolation may involve applying a theory, or may challenge you to critique the assumptions and methods of the course. In each case, you can plan a framework and a procedure for making the most of what you know in the available time by taking a project view of the whole exam.

A Project View of the Whole Exam

Exams generally are created as projects, either by an instructor or by a dedicated staff working under a professor. The unarticulated goal is to epitomize the course method (the instructor's approach to the material) and to reiterate the significance of typifying events, ideas, and shifts in thought. When you read through the whole exam as if it were a coherent work, you find out how the exam makers value the course material. You also discover what they anticipate from good essays: the level of generality, the intensity of focus, the breadth of references and examples.

You will see, too, of course, where your preparation is weak or strong: the events or issues or theories about which you could write for hours, and those about which it will be difficult to write anything at all. To plan effectively, it is better to know this at once, so that your mind can begin retrieving as much information as possible. Everyone has had the experience of remembering a fact or an example *after* an exam is over. Acknowledging a difficult question at the start prompts recall.

A good exam will inspire you to think hard about the material. It will cover a lot of ground, and ask you to draw on the books, articles, and cases you have studied. Each question requires a full answer; you cannot hope to make up for what you don't know on one question by doing a wonderful job on another. Essays on the two questions may be read by different readers, who may not have written those questions in the first place. Viewing the whole exam as a coherent piece of writing, however fragmented it may appear, allows you to use time and space efficiently, and to do a thorough job.

Time and Space

If the exam allots time or credit to each question (''fifty minutes, forty-five points''), you take this into account as a signal of the importance of the material. But your *personal* timetable should be realistic and practical in harnessing your energies.

For every exam, of whatever difficulty or importance, set aside time for planning at the beginning, and proofreading at the end. Most people do far too little of either, under the mistaken notion that the main goal is to pile up as many words as possible. There may be exams in which that is a temptation: you might have learned from experience in a course that your readers only skim your exam book, and give little or no comment except a letter grade. These signals of poor readings are clear enough, and you and your friends may decide that ''they'' are less interested in what you know than in how much you can get down on the page. But a writer's assumption that the reader is irresponsible is self-defeating, the counterpart to the illusion that a reader will supply the words and ideas the writer meant to say, the steps left out of an argument or remembered after an exam is over. Exams are a burden to most people, too serious as a symbolic ritual to be taken lightly, and not serious enough to be transformed into an opportunity for learning. The danger for the examwriter is that such illusions about audience undercut control over the writing process. Instead, imagine your audience as a willing but pressured reader fully capable of understanding what you have to say, as long as you say it with reasonable care.

Focus briefly on what each question asks. It is not necessary to pin down everything at this point. You will return to each question once you've created a quick set of argument-outlines for the whole exam. But the overall perspective you build by working quickly through the whole exam enables you to rearrange or change the emphasis of material in each outline once you begin writing. Moreover, you will be able to integrate new material prompted by thinking through each question, instead of trying to squeeze it in later on or leaving it out entirely.

Write out a provisional thesis, condensed *because*-clauses that represent evidence in support of it, and one or two good examples. If you start your provisional thesis with ''Although . . .'' you will be sure that you sketch out at least one opposing argument as well. Review these condensed *because*-clauses, number them in order of importance, and you will have a miniature argument-outline, an efficient design for a structured essay. If you do this for *each* question first, before you begin to write, you create a safety net for the whole exam.

Concentrating briefly on all of the questions first gives you the confidence that you are going to make the most of what you know. Planning twenty minutes or so at the end to proofread what you've written ensures that you will get a decent reading. (A simple way to keep the writing process open for new ideas even as you actively proofread is to write your exam essays on the right-hand pages of the exam booklet, and use the left-hand facing pages for words, phrases, examples, or qualifications to be inserted into the text with arrows.) Your preparation, planning, and structuring enables the reader to follow your argument; the proofreading allows your reader to make out the words on the page. Papers can go through successive drafts; exams cannot. Your essay exam proofreading brings what is essentially a rough draft closer to the level of a finished product.

At first, planning may seem to take more time than you can afford. But consider this typical case.

> A final exam lasts three hours. There are three thirty-five minute questions, each worth twenty points, and a seventy-five minute question worth forty points.

With the QWP approach, you know at once that you need about twenty minutes at the end for proofreading, and a little more than that at the beginning to review the whole exam, planning each question to the point of a quick argument-outline. This will leave about two and a quarter hours for the actual writing. No matter how you do on the shorter questions, you will want to do a good job on the long one because it carries so much credit. If you plan twenty-five minutes for each of the shorter essays, you will have about an hour for the long one, a timetable corresponding roughly to the amount of credit allotted to each question.

You may want to do the long question first, to give it your best energy; or you may want to save it for the second hour because you know from experience that as you write the shorter essays you will remember more facts, examples, and theories from the lectures, readings, and handouts. Whatever your decision, if you take control of the examwriting process consciously, you will ensure your best product, and learn the most about your own writing under pressure.

Pace and Intensity

One advantage of starting with an easy question is that the process of writing with confidence goes a long way toward diminishing the anxieties inherent in the examwriting situation. Once you have an idea of what you want to say, hesitation in committing words to the page only increases fear and panic, making clear thinking difficult. Writing what you know first gets you started. Invariably, writing what you do know helps you recall other things that can be applied to the questions about which you have less to say.

But there is another benefit as well. As you select the most representative material for an easy question, you store up fresh energy and time for the more difficult ones. This is a simple matter of pacing. If you expend all your energy writing out everything you know about one question, you will feel drained, and find it all the more difficult to turn to the questions for which material is harder to recall. People don't often think of *pace* as a skill in writing. But consider the analogy with a runner in a mile race who goes out too fast in the first quarter, opens a lead, and then begins to fade as the other runners who have maintained a steady pace draw on their reserves of energy for the last turn. An examwriter's race against time requires pacing; each part, the difficult as well as the easy, must be done with appropriate energy to ensure a thorough job.

Whether you've chosen the easiest, or the most important question first, you will have taken into account pace, time, and structure in shaping your answer. When the time is up for one question, move on to amplify the outline for the next.

Answering the Question

Essay questions are not often clearly stated. Few are challenging. Translating each question into a working outline before you write gives you a perspective on the goals of the whole exam. Then, you will have a better sense of how to come to terms with what each question really asks, and what kind of answer it anticipates. This may be evident upon the first reading with some questions, such as those framed to elicit facts or steps in a sequence. But with others that are abstract, obliquely worded, or deliberately open, you can begin by clarifying, narrowing, or transforming the question in order to write what you know in the most effective way. This isn't as difficult or as risky as it sounds.

Every question on an exam is the result of compromises in the exam creator's mind. There is too much material in a course to cover in a few questions or in a few hours. Conscientious teachers may try to cram too much into a question, or use the exam to make a point they feel was neglected during the term. They may work very hard on their exam questions and overdo it, as in this example from a core science course:

"How does energy on the earth flow to and through living organisms?" This might make an interesting title for a lecture to a sophisticated audience, but it represents an inappropriate level of abstraction for freshmen in a survey course. In this particular case, the instructors were appalled by the lack of details in their students' answers. Even the most conscientious students were appalled by the lack of specificity in the question.

Here is a more concrete example, a take-home essay in a graduate course: "Trace the history of American foreign policy from George Washington to Ronald Reagan. Limit your answer to one page." What can one say for a question that asks for so much in so little space? The exam maker could not reasonably anticipate a chronological recital of doctrines and manifestoes. The question calls for a conclusion only, perhaps a restatement of the underlying principle of the course. However puzzling (or absurd) this question seems, you can respond by translating what you know into the appropriate level of generality. Writing good exam questions is a skill few people possess. Questions that evoke learning or elicit knowledge are rarer still. The examwriter should take this into account in transforming poorly posed questions into issues worth writing about.

The standard advice about exam essays is to begin by simply restating the question: "Energy on the earth flows to and through living organisms in many ways." Perhaps this may help you get started. But there is usually some latitude, and even an obligation to *take a stand* toward the question, to show not only that you have covered material, but thought about it, too, making comparisons, and drawing connections: "Both plant and animal cells . . ."; or, "Although American foreign policy has fluctuated between. . . ." Turning the question to your own interests is an essential first step in the writing process. If you diverge from the main thrust of a question, or redirect it, your explanation for doing so becomes your introduction.

Keep in mind how dull it can be to read a set of bluebooks. A few essays approach an ideal in the reader's mind; most others fall far short in one way or another. Any reader welcomes a piece of writing that is fresh and full of energy. If your reasons are interesting in themselves, and if the turn you take allows you to demonstrate the kind of knowledge called for in the question, you will produce one of the few essays that can be read with pleasure.

Focusing Meaning through the Lens of Time

Moving from one question to the next, you abide by a timetable and quick argument-outlines. You've reserved time for proofreading at the end. Once you have shaped a question in a way that will bring the best out of you under pressure, you can write clearly and efficiently. Since examwriting is so concentrated, examples from the course material are especially important.

The issue here is representation: choosing material that can reflect more than itself by crystallizing an explanation or standing for related material. This selection is a skill in itself. If you read an abstract essay, you feel relief at every concrete example, not only for the specificity but as a resting place to gather your thoughts. Concrete examples keep the reader involved in the work. To the reader of an exam, the examwriter's views are a set of abstractions which can only be made clear, in the limited time available, through good examples. If a philosopher develops a theory in one or several works, then an exam question about that theory will usually ask the student to apply it to a particular case. The examwriter's job is to make an example represent as much as possible about the whole theory: its origins, strengths, and weaknesses. Again, as was the case in transforming the question, the writer does best to choose examples that reflect his or her own interest in the material. Careful selection of examples that represent your own understanding of the material extends your essay beyond its severe space and time boundaries.

Internal coherence and overall unity in writing ensure clarity. Since you won't have a second chance, your transitions must alert the reader to the connections among the specific points, and the relationship of each to your overall thesis. It is possible to find a striking, fresh example and neglect to show how it illustrates or explains an idea, yet such connections are rarely self-evident. If your example represents your point of view, anchor it in the context of your whole essay.

Finally, you have to *write out* the conclusion of your essay for the reader, rather than hoping to save time by implying it, or depending on the reader to interpret it for you. A reader may or may not know what you are driving at; your job is to explain, to let the reader in on your thinking and then present your conclusions. A good conclusion returns the reader to the question, but seen now through the lens of the examwriter's thesis and examples.

Proofreading

Nothing is harder to read than a piece of writing chaotically produced under pressure. Most students and professionals at every level fail to realize this simple fact. Some students assume the professor in a large survey course (instead of the teaching assistant) will read their papers. In the same way, people who take exams imagine a meticulous reader capable of making sense out of what is usually little more than a rough draft, annotated between the lines, with crossings out and perhaps a note or two added ("misspelled?"; "sorry—no time to fill in here"). But professors and teaching assistants rarely have the time to decipher illegible writing, fill in gaps, forge missing connections, or rectify errors in spelling, grammar, punctuation, and diction. Instead, the reader's difficulty in understanding your words on the page will be reflected in his or her evaluation of your writing. QWP examwriting insists on a timetable, structured essays, a practical distribution of the writer's energy, and a sustained level of intensity. But this practical system for organizing your thinking efficiently is rendered useless if your essays are impossible to read.

Students rarely understand why an active proofreading is so important. But if they are asked to read their own writing or that of a fellow student aloud, they discover words missing, ideas truncated by revision and not completed, repetitions, and unresolved paradoxes. A typewritten paper usually can be *read;* handwritten exams can be largely illegible. More often than a student imagines, at least part of what he or she wrote in a bluebook contributed nothing to the grade. As with all writing, but particularly on an exam, the writer needs to take responsibility for every word to ensure a decent reading.

An Examwriting Policy

Exams almost always disappoint the exam makers: "We prepared them for a whole week. We gave them handouts. We even gave them the questions the night before, and still they did a lousy job!" Students usually don't realize that the *way* they take an exam is a message in itself: "What!" the insulted reader thinks, "Five pages without a single example. . . . How can I read this!" But carefully planned, comprehensive essays with appropriate examples show a familiarity with course material and an appreciation of method, integrating facts and ideas in clear, fresh ways. They allow writer and reader to transform the ritual of testing into a learning and teaching experience.

I began this chapter on examwriting by describing the futility for both writer and reader of an "eighth bluebook" that is written frantically and read grudgingly. But an exam that is undertaken as a writing project, with an awareness of both the reader's need for clarity, examples, and comprehensiveness, and the writer's need to learn something useful and increase skill and confidence in writing, is bound to be satisfying.

An Examwriting Agenda

(A) Preparing
   (1) Reader's questions
       (a) The relationship between method and material
       (b) The exam maker's unarticulated agenda
       (c) The exam maker's expectations
   (2) Writer's questions
       (a) Examwriting habits, grade pressure
       (b) Preparation throughout the term
       (c) The writer's short-term commitment and long-range goals

(B) Planning
- (1) A project view of the whole exam
- (2) Timetable for the whole exam: planning the process from preparation to proofreading
- (3) Structure: provisional thesis, including an *although*-clause for the opposing side; condensed *because*-clauses as evidence; argument-outline in beginning-middle-end form for each question
- (4) Write the essays on the right-hand pages, and use the left-hand pages for insertions while proofreading
- (5) Pace and intensity: distributing energy appropriately throughout the exam

(C) Generating
- (1) Answering the question; explaining the divergences
- (2) Representation: typifying examples

(D) Producing
- (1) Focusing meaning through the lens of time: coherence, consistency, transitions, and unity
- (2) Conclusion: a new perspective
- (3) Proofreading

(E) An Examwriting policy

Preparing and Planning Research Writing

Resolving a Research Writing Block

Imagine that you are writing a political science paper on back-channel diplomacy. The topic is ''Nixon's Opening toward China.'' Your professor admires that act of diplomacy, and she has presented material in the course to indicate that it depended a good deal on back-channel negotiations. There's not much time so you decide you won't need to plan extensively; once you get going, there will be plenty to say.

How far back should you start? What about Nixon's long history as an anticommunist, and then his sudden turn toward China? You realize immediately you'll have to push these questions aside until you've compiled a basic narrative of the steps that led to the change in policy. You've read Nixon, and now you read Kissinger's huge book. Much of the opening *was* back-channel diplomacy, but you can see that your narrative, based on different, sometimes contradictory versions, is becoming increasingly complex. In five rough pages you have only the barest outline. By the time you have to start writing, you realize you have too much to say, and time is running out.

What should you do? Is the narrative itself going to have to be the paper? But the professor already knows the sequence of events. What is *your* paper about? You started with admiration for the policy, but something is missing. You're blocked; you need to talk through the topic with an objective listener.

Even (or perhaps, especially) a person unacquainted with the subject can help you rediscover your subject by asking you to explain it in simple terms. Such an objective listener might ask, for example, why the policy worked, and whether back-channel diplomacy was the only way to accomplish the opening toward China. Were there any long-range disadvantages? In answering these questions, you recall that you were fascinated by how well back-channel diplomacy fit the personal styles of both Kissinger and Nixon. Shouldn't this be included in your paper? And why that policy at that moment in time? As you think aloud, the listener may ask you if you really need to know why, in order to explain how. In the ensuing silence, the question seems unnervingly rhetorical. But how much of the opening to China was a domestic political move? Could Nixon have done anything that wasn't political? Was it a strategic way to shift attention from other issues? Was it an attempt to form an alliance against the Soviet Union? Was it a combination of all those things?

"How are you going to work all of that into a twelve-page paper by Friday?" your objective listener asks.

Suddenly, you're impatient to get back to work. What about other examples of back-channel diplomacy? How much background do you have to give? You didn't think all of this through because you took the success of the opening toward China for granted. If you spend too much time on the background and the narrative, you won't have any space left over for your own ideas.

"When doesn't it work?" your listener asks. "Has back-channel diplomacy ever backfired?"

"Of course. Selling arms to Iran, for example. When it goes wrong, it's really a disaster! Maybe I'd better include a section on that. I don't know. She's never mentioned it in class. Maybe back-channel diplomacy has to be conducted with the right people at the right time. Maybe there are times when the risks aren't worth taking."

"Well," your listener says, "I don't know anything about back-channel diplomacy, but if it worked for Nixon and not for Reagan, that's interesting in itself."

"Yes," you say, clarifying at last a thesis about the relationship between personal style and back-channel diplomacy, "but Reagan is not my topic."

This is probably as much help as you need. Your objective listener is missing the point, *your* point. Although you're grateful for his time and effort, now you have to get back to work. The conversation has returned you to where you began, only now you have a better sense of direction, based on active ideas of your own that you lost sight of as you became immersed in conflicting details and interpretations. In a long-term study, a scholar might try to resolve the conflicts about what actually happened, and when. But *that's* not your paper either. You want to write about the relationship between policy and style as it is exemplified in back-channel diplomacy. Talking through the block with an objective listener has given you back the paper you wanted to write.

Three Schools of Disenchantment

Research is an opportunity to build personal and professional discoveries on the foundation of previous work, and to present those discoveries to readers who, in turn, may build upon them. How is it that such an exciting prospect can come to seem so exhausting and futile to so many people? How is it that research writing so often reduces confident writers to confusion and helplessness?

It's useful to understand the larger context, the political and economic pressures that have transformed research writing from a skill into an exclusive privilege. These pressures are brought to bear early in the educational process, when the burden for teaching research writing falls on English teachers. Caught between the chaotic, unsupported opinions of students, and the insistence on certifiable competencies by the culture, high school English teachers have few truths to grade objectively. After all, their colleagues in chemistry or math can give multiple-choice quizzes, draw up grade curves, and rank students. English teachers have to make up the criteria and the curve in full knowledge that the teacher down the hall might judge the same student's paper very differently. As class enrollments rise, it is not surprising that teachers settle on the conventions of grammar, punctuation, and documentation as their "subject." Rather than encouraging students to discover connections and present interpretations, they shift attention from thought to appearance. The effect on writing in general has been devastating; on research writing, it has been catastrophic. Teachers drill and quiz their students on footnote form, or train them to build fantasy bibliographies. A student who spends weeks on his or her research paper and sees little or no comment on the content, but instead finds points deducted for incorrect footnote form, gets a clear message about the teacher's priorities and grading policy. These priorities are by now so familiar that teaching genuine research writing on almost any level seems impossible.

When the student gets to college, he or she finds that expectations for research writing have broadened to include the proper use of sources. But although plagiarism is clearly condemned in the college catalog and a cause for immediate dismissal, students are rarely instructed in how to *integrate* their own experiences, opinions, and ideas with what they read. Because they do not consider their students mature thinkers, college writing teachers emphasize not thinking but a series of steps in a research writing process that gives the appearance of thought. This sequence of steps has little to do with learning how to think clearly or write powerfully. Once again, the English teacher's job is to prepare students for writing "real" research papers later on, in graduate school.

How much more helpful it would be to train people from the start to look for meaning, gathering, filtering, and assembling material to present ideas to readers. In high school, for example, we would have been fascinated to know that accurate documentation instructs a reader, or that an annotated bibliography introduces a reader to works he or she did not know. We would have been able to see research writing in the context of communicating our own discoveries and interpretations, and we might have come to expect an exchange of ideas with our teachers. In college, it would have been gratifying to be liberated from the restrictions against taking a stand, and encouraged in our research writing to integrate our own views as we gathered material. We would have seen research writing in the context of developing and presenting a more precise yet more comprehensive perspective than we (and perhaps even our readers) had before.

Finally, in graduate school, we are expected to say something. But instead of coming to the end of our training, we have only begun an apprenticeship. We apply our knowledge of the conventions of research writing to carefully circumscribed topics: the minor poet, the marginal issue, the footnote to history—an endless series of flat subjects which we are expected to bring to life with the energy of ambitious disciples. The opportunity to write about a subject of our own choosing is a long delayed, jealously guarded reward, despite the fact that professional success depends more and more on publication. In the same way that training for law school and medical school involves adopting appropriate attitudes, training in research writing involves obedience in choosing humble topics.

Thus, the long development of research writers begins with empty forms, and all too often ends with empty content. We lose the pleasures of genuine research writing; as readers we are deprived of discovery, new knowledge, and the inspiration to learn more. Trained to list a certain *number* of sources, and *how* to list them, we lose sight of what sources are for, and how to weigh one against another. We compile information in the absence of a compelling principle or a unifying interpretation. Under pressure to demonstrate dependability, we substitute tenacity for intensity as if deeply felt thought were wholly disconnected from the pursuit of knowledge.

It's easy to see, in this three-stage progression, why many people come to dread research writing. It is not a matter of organization or editing. If you want to do authentic research writing, you have to free yourself from the culturally induced inhibitions against the vigorous exchange of ideas. High school students should know that footnotes and bibliography are editorial matters that emerge from the active process of gathering and filtering material. Proper form can be found in any number of handbooks, consulted in the same way mathematicians consult logarithm tables. In college, taking a stand should not be considered a luxury or a reward. An essay that sticks too closely to sources results from the writer's failure to develop a thesis of his or her own to test in the material. Moreover, because the very selection of material is bound to be subjective, an independent, honest writer must work to distinguish his or her own point of view. In graduate school, writing should be an authentic challenge engaging the student's enthusiasm for learning. This is the risk and the value of research writing: being fair and comprehensive while having something important to say. Why else would we want to make a claim on a reader's time?

The Personal Discovery of Meaning for a Public Audience

The pulse of any research writing project is the writer's passion for the subject. Writing assignments on demand in school or at work without the personal commitment of the writer often become chaotic or mechanical. A project that reflects the writer's conscious decision to discover and present meaning takes on a life of its own, with internal coherence and sustained energy.

Readers are especially sensitive to this commitment, or the lack of it. If, out of confusion or conflict, the research writer avoids defining the connection between material and commitment, or else comes to it too late in the writing process, then he or she can only give the project a superficial unity. The reader is left with the impossible task of assembling the fragmented argument around a latent or unarticulated thesis. This is especially true when there is an obvious tension between the writer's selection of material and his or her conclusions. Whatever the subject, however limited or complex the presentation, your reader needs a sense of your active connection to the life of your argument.

Some people are interested in everything they read. Others habitually skim, or look through the introduction and conclusion before deciding whether to continue. At times, the writer knows his or her fate in advance: a research report at work will be filed away; a term paper will not be returned with any comments. Even so, the most efficient, effective way to do clear, powerful research writing on time is to discover the inner, *active* life of your project—why, and how you care about the subject and the reader— and to integrate those concerns into your work.

This is where the skill, craft, and art of writing coincide. Recording data for an accident report requires discipline; evoking the power of a dream requires intensity. But research writing combines the clarity of analysis with the power of conceptual thought; and the relationship between details and concept must be demonstrated thoroughly in a process that continuously engages the reader. Experienced research writers have a feel for their own kinds of question, the level of generality at which they are comfortable drawing inferences, establishing connections, and making assertions. This kind of confidence arises out of experience with the organic development of ideas: how one piece of

information settles and reacts with others. Research scientists and disciplined artists proceed with patience and concentration, as evidence
gathers into explanation. The issue in research writing is not an unattainable conclusiveness, but an enduring passion for inquiry linked with
the satisfaction of communicating discoveries.

Most people do not have much experience linking the two. Instead,
they fall victim to one of the familiar anxieties: "I have so much
material that I don't know what to say!"; or "I have so much to say I
don't know where to begin!" To avoid this, research writing requires an
initial working principle for filtering facts, opinions, theories, and raw
data. With QWP, the provisional thesis provides that principle, enabling the writer to sift through readings, interviews, or experimental
data for support or refutation of an idea or concept. Then, this material
is constantly tested *as because*-clauses. Does the passage support or
refute the provisional thesis? Does the evidence elaborate the thesis or
point to a more clearly defined one?

Research writing requires this kind of prior commitment. In theory,
the scientific method instructs us to gather as much information as
possible without, for the moment, ruling out things that cannot be
explained. But most works of art (Henry James' "germ" of a novel,
for example), and most scientific experiments begin in hunches, theories, or images of a product that establish direction. Good research
writing capitalizes on what the writer knows as he or she searches for
meaning. The process requires conscious choice about scope, scale, and
style, time and space, and the efficient gathering, sifting, and integrating of material.

Preparing: The Writer's Choices

Scope

The goal of research writing is seldom to state the "final" answer to a
question. Instead, the writer provides *an* answer, with supporting evidence. Each research writing product is part of a continuum that em-
braces past and future work on the subject. The best way to begin is to

consider the point at which you are entering that continuum. Is the larger context of your project that of a student coming to the material by request from an expert in the field who will grade the product? Or is your approach that of a worker expected to produce a report on an assigned topic by supervisors who need to evaluate or initiate a new project? Are you a researcher whose audience is limited to the members of your profession, or a translator of technical knowledge for a general audience? Clarifying the context for your work ensures an appropriate fit between your goals and the reader's expectations. For example, applying decision-making models to a particular problem might make a satisfying undergraduate paper, but without a critique of the models themselves, it would fall short of the expectations for a graduate paper. A professional article might go a step further and test the critique against other cases. The scope of the first paper is to demonstrate learning; the scope of the second is to demonstrate analytical skill; the scope of the third is to create new knowledge through analytical skill. Each format has requirements of its own, and it is often not possible to do a good job in one format by meeting the expectations of another. It might seem reasonable to transform your undergraduate paper into an article for a general audience, for example, but the intrinsic requirement that students demonstrate "coverage" of material will not make the transformation to a general audience an easy one.

Begin a research writing project, then, with a realistic appreciation of your reader's expectations, and of how much you can accomplish given your experience and your present knowledge of the subject. A high school student writing a research paper about William Lloyd Garrison might be able to convey a general description of the man, of other abolitionists, and perhaps of the power of the press by reading histories or biographies. A college student developing research skills might place Garrison in his historical context. A graduate student, after a thorough review of previous research, might consider a small point that has not yet been discussed thoroughly. A historian, on a personal timetable, might explore the undercurrents or themes in American history that, in his or her view, are reflected in Garrison's life and work. An article for a general reader might see Garrison in the light of subsequent history, drawing comparisons with the present time. We don't expect a high school student to see the range of possibilities inherent in a topic, or a college student to move easily from specific to general, or a graduate student to cover an important topic fully, partly because we have been so busy with forms, conventions, and prohibitions that we have neglected the reason and method of authentic research. But with so much material available from laboratories, journals, books, and computer-assisted searches, it's easy to feel swamped, to have so much material that you don't know where to begin, or where you fit in. Some realistic view of the *scope* of your project is essential preparation for writing.

Scale

Once you have a sense of the kind of project you are engaged in, you want to know how much ground you can reasonably expect to cover. How will you illuminate the *specific* issue you choose within the context of all the available material? How can you make the most of your time and energy immediately, as you begin to read through your sources or analyze your data? Mathematicians estimate an answer before doing a calculation. Research writers need to do some of the same sort of estimating in order to settle on the *scale* of their project.

An experienced writer senses patterns and relationships in any material that would be valuable in a potential project. This research wisdom results from knowing your own capabilities, and experiencing what research projects can and cannot accomplish. Working through a series of projects enables anyone to develop these instincts. But all research writing benefits from thinking through the scale of your work before you begin, so that you can select material from the start either for its intrinsic interest, or for its direct relation to your project.

Imagine, for example, that you are near the end of a course in Soviet Foreign Policy and it is time to pick a topic for the term paper. You've been interested in decision-making in the Soviet Union, and although there has not been a great deal of information about it during the course, wherever it has come up you have gone beyond the assigned chapter or article to find more. You feel there is probably a strong political bias submerged in the conflicting scholarship on the subject. There is a definite split between those scholars who apply models for decision-making in the United States to the Soviet Union, and those who hold that there may be significant differences. There is another conflict, as well, between those scholars who see decision-making as immutable, linked inextricably with the past, and those who see the potential for change. You know you have something to say about all of this, but you're not sure what. The professor in the course is very skeptical of change in the Soviet Union, and you are aware of her hard-line approach. But you also know that she is interested in what her students have to say, and you feel pretty secure in approaching the project with the goal of finding out something for yourself and then articulating it clearly for the reader, without worrying about whether she agrees.

The point is to do a comprehensive, careful job of gathering and sifting material in support of an active idea, one that is interesting in the way it explains actions, events, or apparent conflicts and contradictions: an idea that looks to the past and present and has something to say about the future. But as exciting and worthwhile as this would be, you realize you are a student pressed for time, with only so much background in the subject and experience in research. You wish you could take a month off and do only this project. After all, the paper might turn out to be the most important thing you do as a student. But then you come back down to earth. The scope of the topic is intimidating. You are far from being an expert, and the experts disagree violently. Your paper is due in three weeks. You don't see how you can possibly say it all in twenty pages, so you'll need to ask the professor if she's serious about the page length and the deadline. But more than that, you really want to do a good job for yourself and for the course, which has been inspiring.

You realize you have to scale down your expectations. This paper will not be the ultimate answer, the resolution of all the conflicting evidence on the subject. You won't be able to do as complete a job as you wish, and even in what you do you will have to make room for demonstrating in some way that you have understood the course method. You're not going to be able to read everything that has been written on the subject to get a thorough background. You may not get much beyond the standard works and a few current articles or books. You'll need to be efficient as you gather material, discarding work that simply •

reiterates what others have written instead of adding anything new. (You've learned from previous research that there are both seminal writers and imitators or elaborators.) You're only going to have time for the main ideas, the best evidence. You may want to read your professor's book as well, but you're not sure it will be relevant to your specific topic.

Scaling down, then, you need to choose a *specific* instance, a representative example of your topic, decision-making in Russia, for which there is some reliable evidence that adequately reflects the larger questions and conflicts. If your specific example is good enough, then what you say in a few pages can stand for a lot more: you will represent a good deal of thinking and reading by focusing everything you know on the details of the specific case. In the past, when you have been clear and thoroughly prepared, your writing has been charged with active information, encompassing specifics but imbued with a general perspective. You are going to do something small, but make it reverberate with larger overtones. That's the best you can do in your present relationship to the material. (Later on, for a thesis, or a research report as an intern in Washington, you might do a deeper, wider study.)

An issue that has captured your attention is the peculiarly amiable agreement of the Soviet Union to set up two agencies—one in Russia, one in the United States—to monitor nuclear emergencies. You read about it in the newspapers, and you know there is some sort of disagreement among the experts about how and why the Soviets have been so willing to agree. You wonder how that decision was made; and you wonder, too, if you have enough background to offer some reasonable comments on the subject. Couldn't you show what you've learned about Soviet foreign policy *and* come up with some fresh ideas by focusing on this issue? You know there's no final answer on the question: it's a matter of current debate; within that context, you may have the room and the right to put forth your own idea.

It seems a good way to start. If you do write this paper, what do you think you might discover—that the monitoring agreement is a trick, a propaganda device? Or that it represents something new in the way the Russians look at the issue or make decisions? You've read that the pressure for the decision came from a group in the Soviet Union known for generating propaganda, yet Gorbachev seems to have made the decision to go ahead suddenly, as if on his own. Could this be possible? And if so, what might it mean for other issues? Is the United States prepared to take into account changes in the Soviet Union, or are we too committed to our present views? Musing, you become enthusiastic again: suppose it is a change? Suppose we need to understand it and respond to it in a fresh way? This paper might actually help. . . .

Exploring the scope and scale of a paper delivers you to your research with energy and interest. Clear about what you are looking for, yet open to discovery, you are fired with a desire to enlighten the reader. If you begin your research writing this way, you can judge immediately whether your readings, data, and evidence hold potential for your project. Everything you do from the start will count. You won't be caught in the middle changing your mind about the kind of project you're doing, or what you can reasonably expect to accomplish, and then having to go back over all the material you've already considered.

There are countless ways in which research writing is sheer drudgery. You can thrash around in the material without confidence or patience, attracting the sharks of research anxiety. But to enjoy research writing for what it can teach you, and for what you can teach your reader, carefully think through the scale of the project to get a realistic sense of what you can include, and what you will have to leave out to finish the job on time.

Style

Even careful preparation cannot resolve the problems in organization and consistency created by fragments of previous attempts at a subject, or data gathered at different times for different reasons. In addition to scope and scale, you should give early consideration to *style* as a function of meaning. What organizing principle locks the pieces of your argument together? What gives the whole paper consistency and balance, from introduction to conclusion? Scope and scale may undergo refinement throughout a project, but style is so thoroughly embedded in structure that the writer must make choices at the start to avoid frustration later on.

Organization

Imagine you are an architecture student who becomes fascinated with the fragments of the past that are evident in the buildings in your neighborhood in London. Each day, on your walk to school, you study those remnants of earlier design that were once functional but, in the mid-nineteenth century, became ornamental. Through the winter of your first semester in London, you explore other neighborhoods to see if this pattern of structural memory applies. By early spring, you are convinced that there is a fascinating article to be written, not only for students and practitioners, but for a wider audience: people who would be interested in seeing more of the world around them. Here is a chance to explain what structures reveal about social change and values, through a familiar example. And the example is perfect: reflecting the need to move country life to the city, even if in image only, as the wretched, closed-in quality of city life, with its loss of privacy and sunlight, came to be accepted as a compromise with the need to find work.

You begin your article one night sitting at your new word processor. Before sunup, you have fifteen pages about the economic and social influences on design. You take the draft off to school, but a few weeks later your professor still hasn't returned it, and exams have begun. You shelve the project. In Design Seminar the next semester, the assignment is to choose a structure at least a hundred years old, and describe it as a person of that time would have seen it. Perfect! You know exactly what to say; in fact, somewhere on one of your floppy discs you have the beginning of this paper already. The assignment is supposed to be three pages long, but you have enough to fill an article.

When you retrieve it, you're surprised at how abstract the writing is, how turgid the prose. You'll need to rewrite, adding a practical section about the surface of a building, simply describing the details of the ornamentation. It's a silly requirement, but that will take only a page. A few photographs would be better. The real issues are the shift in population, the transfer of wealth, the change in the manufacture of goods, all of which can clearly be seen in the attempt to make row houses look like miniature castles.

But through draft after draft—certainly not in three pages, and probably not in fifty—you cannot bring the two chunks of material together; you cannot reconcile the abstract and the concrete. They are pieces of two different arguments, one of which interests you, and one of which is a requirement; and they simply do not fit together. They don't even belong together. You approve of one question, and would like to dwell on it; you reject the other question, and don't want to waste your time on it. Months go by and neither the fascinating article nor the required paper gets written, until finally the director of second-year studies gives an ultimatum, a final deadline. She's not interested as much in the political and social influences on design as she is in the reason why you can't produce your work on time.

Under pressure to get the paper in, you would have been better off if you had read and thought about the subject first, without writing that enthusiastic fragment. It is often easier to start fresh than to try to mesh different pieces of writing from different times. But there they are, fifteen pages representing real interest and excitement. Facing the deadline, your only recourse is to use the early material as an invisible foundation for a less exciting paper. It's possible; it does not need to be deflating, or disappointing. But it requires a decision about style as it relates to coherence and unity, and as it facilitates process and product, under pressure.

Consistency

Sometimes, as you read through a paper or an article, you are struck by a quotation, an example, or an assertion that seems entirely out of context, or of a texture different from everything that has come before. You notice it in compilations, of course: those articles and books put together out of lectures or essays from different times. But it is also evident in self-contained projects. Such inconsistencies, representing a disjuncture in the process of assembling a piece of writing, diminish the force of any argument.

Let us say, for example, that you are having a hard time finishing a journal article on welfare. Your thesis is that welfare must be separated from politics because the real needs of human beings cannot be met by programs that fail to take those needs seriously. It's not a popular idea to propose in a time when people have turned away from social or community goals. But you have done your homework. You've studied various programs that have worked, and you've found a recurrent theme: the ones that do the most good are run by the wisdom of the people themselves, not by bureaucrats whose careers depend on cutting costs instead of helping the hungry, the abandoned, the battered people in misery and despair.

You have also gathered evidence about another cause of failure, those programs run by well-intentioned, highly educated people who do believe in help, but who see themselves as the indispensable arbiters of what is best. Even with willing administrators programs have blocked or diverted the very help that people need. This is your subject, your issue. Part of your responsibility is to try to educate people. In your article, you want to focus on two community-run programs that the bureaucrats expected to battle over scarce resources, but that agreed, instead, to share the money equally.

Part of the difficulty with the article is that your perspective, rooted in thirty years of experience, reaches past the immediate issues toward the larger questions of hierarchy and inequality. These certainly are not issues you can cover in a thirty-page article that would convince any but a few like-minded readers. Had you known from the start that you would end up with such vast issues on your mind, you might have done your research in a different way, or you might not have agreed to write the article in the first place. But you weren't thinking about scope or scale; you had an idea, a good example, and an urgent desire to communicate it immediately. Now you know, as you come to write the conclusion of your draft, that your experiences and examples have led inexorably to a general question: how to empower the powerless. You should really start all over, but the article is due in a week.

For relief, you browse through a classic text on institutions by one of your favorite writers. Now, as if for the first time, you understand some of the things he says, and see how they relate to your article. In an hour of browsing you find at least ten good quotes, exactly what you need to illustrate what you want most to say. You choose one and draw a conclusion from it. The language of the quotation is so graceful, the thought so powerful that you want more. You quote him again; and then you realize the best way to end your whole article is with a really good quote from his last chapter. Your article will be the stronger for it, more powerful than you imagined when you started, leading the reader to the larger issues.

But it's really three-quarters of a paper, with a broken ending, and interesting quotes tacked on. The reader who has followed your argument may see the relevance of the quotes but wonder where you've gone, what you had to say in conclusion. The reader who had trouble following will be left in the dark by the introduction of quotations from a book which is apparently about a much more general subject.

Changes in the texture or established style of an argument are often difficult for the writer to see, especially in a draft written under pressure. Feedback helps. What's at issue is a balance between the specific and the general, between experience and learning, and between your own thoughts and feelings and those of others.

Balance

Consider the example of a nursing student's report of an OB-GYN rotation that suddenly stops to give a textbook description of the birthing process, and then ends on a dramatic personal statement about the beauty of motherhood. For a reader, such shifts in balance need to be accounted for in the writing itself. Otherwise, no matter how powerful the writing, they raise questions about how well or thoroughly the writer has thought through the project.

Organization and consistency determine balance as an element of personal style. Each writer strikes this balance in his or her own unique way. Research writing, whether meticulous in its objectivity or passionate in its advocacy, carries a message through its style about the intensity and thoroughness of the writer's approach. There are clear, persuasive monographs in biochemistry; there are interesting, active technical manuals; there are papers, theses, and articles on every conceivable subject that have made a difference to their readers. A balanced style in research writing is crucial because it clearly conveys a message about the quality of thought. It is important to sort through these issues before you begin, and to resolve them as you proceed.

Planning: Time and Space

Time

As in any writing project, QWP research writing engages time as an organizing principle. Some people spend months, or years, or their working lives painstakingly tracking meaning through a particular body of material. Ideally, we all would want this freedom to read, experiment, ponder, read and experiment again. Writing without the immediate pressure of time is a blissful experience: the writing process takes whatever turns it wants, day after day, until the material teaches you what you need to know. It is a unique and self-affirming experience. If you ever have the good fortune to work that way, you will gain confidence and pleasure in the process and product of research writing that is difficult to achieve in any other way.

This is one reason why Peter Elbow's concept of freewriting is so wholesome and liberating: it gives the writer, no matter how inexperienced or lacking in confidence, a taste of the researcher's freedom to discover the content of the material from his or her unique point of view. After a week of freewriting exercises (writing non-stop, ten minutes a day, without self-criticism or an audience) you find that you have indexed thoughts, feelings, and connections you may not have been clear about or even aware of before. Recapturing this material is thrilling, and the freewriting process, as a perspective on your writing habits, inspires confidence. Although freewriting is a timed exercise, the liberating sense of writing simply to see what turns up, without categorizing, censoring, or sorting for a particular goal or audience, gives you a hint of what it is like to work toward a product without time limits. Long-term, open-ended research writing is similar to freewriting in emphasizing exploration. You cast your net wide to see what you will catch, without regard to usefulness. Some long-term projects, in fact, proceed as a series of freewritings: disposable drafts, one after the other, moving ever closer to the writer's real subject.

Of course, working without the pressure of time is not an unqualified blessing. There is the danger of amorphousness, or of such obsessive thoroughness that the desire to connect with a reader is blunted past the moment of presentation. This meticulousness is reflected in Camus' "obscure hero," Joseph Grand, in *The Plague*, who spends years on the first sentence of his novel. Open-ended projects present a tremendous challenge to the writer's motivation, especially if some of the questions inherent in the material become unmanageable as the project opens out. There is also the possibility that such immersion in the material becomes an end in itself, supplanting the writer's desire to communicate it.

To adapt the QWP timetable to a research writing project, make a reasonable estimate of the time you can allot to the whole project, from thinking through scope, scale, and style, to gathering and ordering, building the raw draft, and refining the final product. Because each research project is unique, it is not possible to establish an abstract notion of the balance of the research process—reading, interviewing, or experimenting—with the writing timetable, but it *is* essential to estimate the proportions of time and energy to complete the job. It may be that you need to balance research time for a term paper with a writing timetable so that you can finish the project in a month. In that case, you might set aside two weeks for reading, and then two weeks for writing: process and product. Balance in planning your time is necessary from the start.

As you will see in the next chapter, however, the QWP research writing model transforms much of the discovery process into usable material for presentation. You can make an early decision, about when to stop collecting material so that, by the time you embark on the *writing,* you will have accomplished most of the preparation, planning, and generating.

Space

Even though research projects are longer and more complex than other kinds of writing, they still require the most fundamental attention to readers' expectations for order. You need to have a sense of the structure of your project before you begin. Keep in mind a basic conception of beginning, middle, and end, with the functions appropriate to each. The beginning invites the reader in and clearly states the thesis. The middle presents your evidence in the best order, beginning with arguments for the other side, and ending with your most persuasive point. This leads the reader to the conclusion, where you push beyond summary or restatement to a broader perspective. Of course, as in all writing, there may be an important reason or a format requirement to adapt or diverge from this fundamental order. But before you embark on a sustained project, it is essential to think through the structure not only as it relates to meaning, but as it provides a safety net for the writer at work.

CHAPTER
12

Generating and Producing Research Writing

Generating: Thesis and Themes

The Direction of Your Interest

Clarifying your particular approach to the reader and to the material gives you a sense of the scope of your project. Estimating the scale gives you a sense of how to organize general ideas, specific information, and representative examples into a consistent, balanced structure within your timetable. On this foundation, a provisional thesis, including *because*-clauses, provides a framework for gathering and sifting the material.

The provisional thesis for a research writing project is rougher and more malleable than that of an exam essay or an urgent position paper: it is less an *answer* and more a means of driving questions into the open for your consideration. In research writing, concepts mature over time. Generating a provisional thesis and condensed *because*-clauses is an efficient way to begin your search. Whether you have been assigned a topic or choose your own, whether you already know a good deal about it or nothing at all, it makes sense to put your own interest in the subject into words at the start. You have at least three major aims: getting the project done well, on time; finding answers to questions and finding new questions to consider; and enlarging your capacity to discover and present information. A provisional thesis reflects each of these goals, providing the basis for both the research writing process and product, *and* your own developing writing policy.

You may begin, as many scientists and artists do, by transforming a hunch, or an instinct, or a conviction into a provisional thesis to be tested or transformed in the material. You might want to discover where your interest leads by freewriting about the question. You might want to talk through the topic with a careful listener, to discover what you do and don't know, and what you think you need to know. The object is to take direction from your own interests and your own style of learning so that you can make use of every interview, every experiment, every book and article you read. With QWP, the process of doing research is not divorced from the process of writing. Gathering and filtering material is done within the framework of assembling it for presentation to the reader. On the foundation of preparation and planning, generating a provisional thesis enables you to proceed on two levels simultaneously: you are constantly learning as you prepare to teach the reader what you have learned.

Given the rarity of authentic writing at every level of education, it is not surprising that few people pursue research projects with any sense of pleasure or confidence. They rush into research with a kind of anxious energy and then bog down in confusion or lack of direction. Or they may discover in the middle of a rough draft that they have gathered information on an aspect of the topic that is tangential to the meaning they have begun to read back from their analysis. But a provisional thesis gives you a system for comparing material with your *because*-clauses on two levels: what you are looking for, and what you want to say about it. In the course of your research, the original clauses may transform or deepen or be replaced by new ones specifically related to your developing interests. You may find that your provisional thesis is incomplete or too narrow. Clarifying the direction of your interests at the start makes these discoveries an organic part of the research process itself. You make the most of what you know in order to find out what you don't know, collecting and comparing data to construct meaning.

The provisional thesis is disposable; it exists to give direction and a basis for comparison. But the process of generating the provisional thesis and *because*-clauses strengthens the whole project. Transforming a hunch, a feeling, or a conviction into a provisional thesis is at the heart of almost everything we write. The provisional thesis simply brings these hunches, feelings, and convictions to a conscious level and fits words to them to be tested in the material. The thesis is not a magnet that inexorably draws material to it, but a compass to orient the writer in relation to it. In some cases, the thesis need serve only that function; in others, it will sustain the whole project and lead to new ones. You may find that your original feelings on the subject were based on too little evidence, or were too fragile to withstand analysis. You then construct a new thesis, adapted to the material you are gathering. Or you may find that your original hunch, although stated too simply, was a good one, leading to interesting questions which direct your search more efficiently.

If your assigned research topic is dull or narrow, do a directed freewriting on the subject, generating thoughts and ideas for about ten or fifteen minutes without regard to the quality of the writing. Don't edit, don't go back, and don't stop writing. Freewriting helps you avoid self-censorship: when you read over what you have written, you will find ideas, feelings, and questions that point you in the direction of your own interest. If your topic is overwhelmingly interesting, a directed freewriting will protect you from feeling swamped, and will lead you to a manageable thesis to be tested in the material.

Open assignments or articles can be especially difficult. Free to choose a topic, you may draw a blank, or feel overwhelmed. In that case, talking through the requirements of the project will help you establish reasonable boundaries. Talk-throughs serve some of the same purposes as directed freewritings, but go further. If you discuss an assignment with a willing listener for an hour, you will find some issues that interest you: you'll *hear* the interest in your voice, or your listener will point it out to you. You will hear, too, your own frustrations and confusions. As you answer your listener's questions you will begin to shape your own ideas, and your own method of investigation. If you discover paradoxes and contradictions early in the project, these will lead you to new paths of inquiry, rather than blocking your way in an avalanche of anxiety later on. If your listener takes notes for you, or if you use a tape recorder, you can keep track of the questions, the connections, the examples you uncover during your conversation. Although the goal of a talk-through is to come up with a provisional question and some possible answers, you may find, in addition, that you have generated some material for your beginning, middle, or end, and perhaps a sense of a larger perspective for you and your reader.

Freewritings and talk-throughs are absorbing and effortless, and their yield is disproportionately large for a writer's investment. Both furnish the inexperienced writer with some of the advantages of professional writers who keep journals, or establish lifelong files of notes, or habitually air their work in discussions with friends and colleagues. Most people are cut off from this kind of nourishment. They don't keep track of their thoughts and interests on a wide variety of subjects systematically, and they don't get much help from others. Often, the most vivid insight people have into the research writing process is when it goes wrong: the dishonest work of an investigative journalist, a research scientist, or a staff writer. Overburdened teachers give less and less feedback on student papers; insecure students don't often seek out busy professors in office hours; and supervisors at work don't know what to say to help improve the draft of a report. But without such nourishment, research projects can be isolating and exhausting, depriving the writer of the pleasure of learning and teaching. Watson's *The Double Helix* shows how productive the continual dialogue between colleagues can be. Talking and writing about the questions and tentative answers will sustain the energy of any research writing project.

Research writing that is free of the pressure of time, and not directed toward a grade, a promotion, or a grant is a unique pleasure. It is a kind of freedom that has become increasingly rare, not only in school but at work and in the scholarly professions. Usually, only those people who have made the choice (with all its attendant sacrifices) to devote their lives to research experience it. There are people who do research but who have not been able to write about it with satisfaction. There are people who can write about other people's research, but who have not been able to sustain their own. Yet anyone can link a sense of discovery with the power of presentation if they proceed with common sense in a systematic way. Once you have constructed the QWP provisional thesis, you can begin gathering material efficiently.

Discovering and Developing Material

Let us say that you are under pressure to write about research on an assigned topic. By transforming a hunch, or adapting the question to your own interests through a focused freewriting, or talking through the subject first, you have found a direction into the material that you have put into words as a provisional thesis.

People usually begin to gather sources through a course reading list or a bibliography in a text. They may get an overall sense of a subject by reading an encyclopedia article. You will certainly need to search the card catalog, the entries in the reference bibliographies in the field, and the guides to current periodicals and books. A reference librarian can direct you to these standard sources, and there are handbooks that list them in useful categories. But your provisional thesis enables you to begin your research in a more personal way by conducting you, through the card catalog, to the appropriate library shelves for some purposeful browsing.

Experienced research writers are inveterate browsers. They constantly gather information, whether or not it is immediately useful, and develop their own systems for keeping track of it. A science writer, for example, will have file cabinets full of notes and articles on several subjects of interest which may or may not lead to articles or books. This kind of specialized, personal research file (easily indexed now with a personal computer) gives the professional writer a place to start, the impetus to explore a subject thoroughly, and a profile of his or her own interests. But such a file is built up over time as a lifelong professional resource.

For almost everyone else, getting started is hard, and taking down specific titles at the card catalog and from indexes to literature is a tedious, often frustrating way to begin. If, instead, you use the catalog to discover where to look, in general, you have at your command a library within a library, devoted entirely to your subject and topics related to it. This focused browsing gives you some of the freedom of the unpressured, experienced researcher. You will discover articles and books that help you make connections and direct you to other helpful books. Because you can survey a whole range of related readings at the shelves, you deepen your sense of your subject. By examining the most recent books, you will get an immediate sense of the present context for the issue, and this sense can be augmented by browsing through the latest journals in the field.

As with talking through a subject, there is no great investment of time or energy in this purposeful browsing among the stacks and periodical shelves. A few hours will put you in contact with the subject in a personal way that allows you to make practical choices. You will find books you want to read thoroughly. You will take direction from the bibliographies of some of the books on the shelves and immediately seek out references that seem especially interesting. You can pass over books that are derivative, sketchy, or too general for your purposes. Later on, as your knowledge of the subject matures, you will come back to the library with specific needs in mind, books or articles you want to consult, questions you need to answer. But to start, you will find interesting passages, approaches, and writers; and finding them on your own through unhurried browsing increases your involvement and confidence. It's an unexpected pleasure at the beginning of a research writing project, more efficient and more wholesome than tracking down a few specific titles from a fistful of call number slips that may or may not be what you need.

Reading the material you have gathered, you will note immediately what intrinsically interests you and what is related to your provisional thesis. These may be two different functions at first, but both are important in efficient research writing. Since you have an *idea,* a hypothesis to test in the material, you can pursue evidence and examples that support or refute some part of your thesis. But in the long run, you are discovering and refining the *because*-clauses of your eventual final thesis in the material. You want to be open to new ideas, or a more useful approach, or a more appropriate thesis in light of your developing interests. Record whatever strikes you as really interesting, whether or not you can see its immediate relevance.

This is not as complicated as you may think. You want to establish your own level of discourse for the project: the boundaries for your hypothesis, evidence, generalizations, and conclusions. Although you have a sense of the scope, scale, and style of the project, you can't be precise about these boundaries at the start. As you read, you continually adjust them to include as much as you can find of the richest material. This continuous concentration on a thesis, and your openness to new ideas, provides more than enough good material. A simple way to keep track of it is through a research journal, on the model of a scientist's lab notebook or an artist's working diary.

The Research Journal: Emerging Themes

Students required to write research papers are often asked to deliver a series of products to their teacher: the selection of the topic, an outline, a bibliography, note cards, and the rough and final drafts. In theory, each stage is both a demonstration of work done, and an opportunity for guidance, for mid-course corrections. But in practice, each stage be-comes a hurdle in itself rather than part of a continuous process of discovery and presentation. The criterion for success at each stage might be the length of an outline, the number of sources, the thickness of the pack of note cards. But there is usually little interaction between student and teacher about how ideas mature into an argument. The whole process degenerates from a formal exercise to a frustrating ritual.

That's why it is important, when teaching yourself to do efficient, authentic research writing, to look for ideas—both those of the writers you read, and your own. Make your reading notes as simple as possible: not full quotations, or underlined paragraphs, but key words or phrases, a couple of sentences, the essence of what interests you either for itself, or because it relates to your thesis. Record the location so you can go back to it for a quotation or a reference. If you are reading an article, head a note card with the bibliographic material you'll need later, and then keep track of what interests you page by page, in as few words as possible.

As they accumulate, your notes become a running record of your interest in the subject. Review your notes periodically. This will enable you to discover your own unique insights, the concepts and connections you will want to shape and order for your reader. As you gather more and more material, these periodic reviews will reveal emerging themes that support, deepen, or redirect your provisional thesis. If you add a commentary in your notes about these emerging themes (again, in a few words), you build the inner structure of your argument, the blocks of language for your raw draft. In essence, this series of periodic reviews is the research writer's version of generating *because*-clauses, only now you construct them from the material itself, *and* from your ideas about that material.

Most people have some sort of a system for taking notes, but this periodic review to build emerging themes allows you to filter material in a way that broadens and deepens your understanding of your subject and sharpens your own thesis within it. Instead of simply gathering as much material as you can within the limits of time and space, you track ideas as you go along, and develop your own view of the importance of those ideas. This makes the project intensely personal and unique, and this intensity constantly renews your momentum and confidence.

Analyzing the Research Journal: Theme-Families

You have notes and comments keyed to a list of emerging themes in your research journal. As you read, some of these themes gather a large number of references, others only a few. Analyzing these theme-families is analogous to analyzing the *because*-clauses freely generated to fit a provisional thesis. You select what counts, what keeps your interest, what draws the most energy and feeling out of you for the project. You leave behind whatever is tangential, indirect, or dull. You will discover gaps in your argument, and issues that need further examination. This is the time for focused research to balance, amplify, or illuminate an idea, or to develop the argument for the "other side." Now you return to the library with specific needs: a book or an article that may be useful; an idea that must be tested; interpretations you need to consider or reconsider for your own argument.

When you have completed this final research, you have comprehensive notes and comments selected for their relevance to the main ideas of your project.

The Argument-Outline: An Index to Meaning

Some of the theme-families can be combined to make a major point in your argument. Others can be grouped together to form a whole section of interrelated ideas, evidence, and interpretation. The range of material will run from specific to general. This range will contribute to shaping your argument-outline around the progression from the simplest to the most complicated ideas. Once you have established the order of your raw draft, you read back from it a fuller thesis, which becomes the full introduction of your paper.

With research writing, the argument-outline is not only the pattern for the raw draft but, with the attendant references in your notes for each theme, an index to meaning. You can order the notes and comments for each separate point in your argument, and then write freely with confidence. You do not need to move through your outline in order. If you choose to write the most complicated part of your argument first, to give it adequate time and energy, you can arrange the notes for that section and start writing. The outline is at hand; you needn't worry that you will lose your way or bog down. The goal of the raw draft, after all, is to get a sense of the whole paper. Then you can begin to cut.

Producing: Amplifying Meaning

A "Treatment" First

Before you compose the raw draft, write a compact "treatment," a narrative of the whole project in not much more than a page or two. Begin with your thesis, then describe the progress of your argument in the middle section, and propose your conclusion, the new perspective your paper or article will offer the reader.

This treatment serves several good purposes. It provides a project narrative for reference; it forces an early examination of the balance and emphasis of your presentation; and it may well become the actual transition from the introduction to the middle section of your paper. (It could easily be the source for the abstract, or a proposal for publishing your article.) But there is a simpler, practical benefit that has to do with the anxiety of beginning a research paper. Usually, in whatever we write, the first few paragraphs (or in longer works the first several pages) actually serve to start the writer. We almost always abandon these pages later as we find the project's center of gravity. A treatment serves this catalytic function.

Drafting the Product

Raw Draft

Your notes are a quarry for excavating facts and examples. Research journal commentaries provide the internal glue of your argument in explanations and interpretations. It is a great pleasure to begin a research writing project with a stack of notes, commentaries, examples, and quotations sorted under the major themes in your outline. You can move confidently through the raw draft with the single purpose of developing meaning.

Your project may require a particular format, but you will confront that issue in transforming the raw draft into the roughly final draft. Since the goal of the raw draft is to set out the material you have selected in the order of your argument, it will be longer than your final draft. Write freely, without unduly worrying about repetition, or consistency. For the raw draft, you simply want to get the selected material down on paper or up on the screen. Take delight in translating your notes and comments into solid evidence for the ideas that reveal your thesis to the reader.

First Cuts

The raw draft will be too long, somewhat inconsistent, perhaps repetitious here and there, and lacking some transitions. But this raw draft will contain your whole argument, in its most powerful order. Those two qualities ensure that you will get your project done well, on time.

You can cut freely, without the familiar timidity or remorse people feel at "losing all that material." You have more than enough. You want to take out *everything* that doesn't count, that doesn't add to your method and to your meaning. If you have always been unsure about editing, this round of free cuts delivers you from the usual pressure and anxiety and allows you to see your writing from a *reader's* standpoint. You can distinguish between what was really a shorthand or truncated explanation for yourself, or a defensive argument with an imaginary reader, and what advances your argument in a publicly comprehensible way. You may find evidence scattered throughout the draft for another subject entirely, which could provide material for your *next* paper. But you remove it all from this draft without hesitation.

In writing this book, for example, I have taken out far more material than I have left in. Some of the material I cut was really process-oriented, resolving questions of method. But much of it was about the teaching of writing, a subject which has concerned me for almost thirty years (either as a victim or as a practitioner). I have kept only those teaching issues that seemed to me to illuminate the reasons why most people find writing such a struggle, especially under pressure. Although QWP lends itself easily to a supportive classroom setting, it is a system for anyone to adopt independently. This was the principle underlying my first cuts: to keep whatever material was necessary to enable any reader, whatever his or her background, or experience as a writer, to acquire QWP and adapt it to any writing situation. In what I deleted I have accumulated almost enough material for a book about the teaching and learning of writing in America. This is characteristic of QWP: shaping the research writing process around finding, developing, and presenting meaning yields enormous benefits beyond the present project.

The Objective Listener

You have been working with great concentration, discovering and developing meaning, and then playing it back in your own mind to see how clearly it sounds. At this point, it is extremely helpful to have someone else tell you how it *reads*.

You may have talked through the topic even before you started to plan, or you may have had the consistent help of a teacher or a colleague throughout the process. But the cut raw draft is still malleable, and a reader may be able to tell you something useful about the inadequacy of an explanation, the need for an example, or an inconsistency in tone that helps you shape the cut draft into the roughly final draft in a clearer, more powerful way. The objective reader's function is to freshen your perspective on the writing. Almost always, the reader can help by telling you where transitions are missing, an idea is not linked to the thesis, or a particular passage would have more impact in another place. A reader can also tell you what is missing in the presentation of your method, the framework of your project. These questions about content or structure will give you insights into the way the writing *reads* that would be difficult to achieve on your own. In writing the drafts of this book, I sought feedback at several stages. Each time the reader's observations and questions sharpened my perspective.

Of course, if you bog down in the writing, feedback from an objective listener is essential. I began Chapter 11 with an imaginary conversation that resolves a research writing block. But any piece of writing can stall for a time while ideas mature, or the writer's approach crystallizes. We all have been trained to write in artificially compartmentalized stages in which it is easy to lose concentration, instead of in a continuous process. Talking through a problem saves you time and energy and sharpens your perspective on the continuous struggle to link specific issues to overall themes.

When my tennis teacher calls out, ''Turn your shoulder!'' after I've hit a weak forehand, it restores my concentration. The same sort of thing holds true for writing. Even if you know better, from experience, you may convince yourself your notes are good enough and you don't need an outline, or that you really know the subject and can just sit down and write. Someone outside the process can remind you of what you have neglected, or where you took a short cut and got lost. Talking your way through a problem with an objective listener is a reality test: if he or she can follow you, then you have a good indication of an appropriate fit among your ideas, your written words, and your reader's understanding.

Improving the Roughly Final Draft

Once you have removed what doesn't fit in the draft, and gotten feedback from an objective listener, reconnect what does fit. In effect, you have freed the figure from the stone, the essential argument from its more amorphous context in which there may have been confusions, undeveloped or unexplained fragments, repetitions, and distractions. Your cuts have ensured coherence and consistency among the parts of the argument. Each part fits what came before it, and leads clearly to what comes after. Now you want to bind the parts of your argument together into a unified, smoothly flowing draft.

At every point in your paper or article, your reader should be able to relate the specific fact, explanation, or example to your overall thesis. This means writing solid transitions which explicitly refer to the framework underlying the whole project. Anchor every example in the overall argument, and make sure that the importance and relevance of each illustration, diagram, and table is clear. You may have to enlarge, or alter the tone of the introduction, or strengthen your conclusion. Readers of papers habitually skim beginnings and endings before they read through the argument, and you want to be sure everything is there, in compact form.

You need to test the relevance of every piece of structure in this roughly final draft, from the introductory section, to the development of themes, the running commentary on method, the examples, the specific conclusion, and the broader perspective you hope to provide your reader. Furthermore, you want ample paragraphs, clear and varied sentences, vital language, and helpful punctuation. Finally, you need to be certain that your references and quotations are accurate, extending to the reader the opportunity to use your research as a springboard to further study. Then, of course, have someone proofread the final draft.

Product and Policy

The QWP method should be especially useful for people who have viewed research writing as chaotic or unmanageable. The underlying principle is that research writing is a *continuous* process in which the writer simultaneously discovers meaning and prepares it for presentation. Finding facts, ideas, and examples, the writer shapes the most effective product for sharing them with a reader.

This synergism at the heart of research writing explains why some people find such pleasure and excitement in every project, no matter what the format, or how much patience or meticulousness the work requires. Even the painstaking introductory section of a long paper, chronologically reviewing earlier research and providing a clear rationale for taking up the subject within that context, can be imbued with the style and tone of your argument, as any article in *Scientific American* demonstrates.

Most people never experience the enlightenment and stimulation of authentic research writing; those who do usually discover it on their own. The QWP research writing agenda leads the writer to discover meaning, and to present that meaning *as* discovery for his or her readers through an improvable, adaptable system. It is one of the most powerful accomplishments of an independent writer.

A Research Writing Agenda

   (A)   Preparing: the writer's choices
       (1)   Scope
           (a)   The writer's commitment to discovery and presentation
           (b)   The writer's relationship to the material
           (c)   The active inner life of the project
           (d)   Choosing the context
           (e)   What you can and cannot accomplish
       (2)   Scale
           (a)   Choosing the issue
           (b)   Representation: the specific for the general
           (c)   Getting *this* project done
       (3)   Style
           (a)   Organization; consistency; balance

(B) Planning: time and space
  (1) Timetable
    (a) A research timetable correlated with a QWP writing timetable: preparing; planning, generating, and producing the writing
  (2) Space
    (a) Structure: beginning, middle, end—including a broader perspective.
(C) Generating: thesis and themes
  (1) Provisional thesis: the direction of your interest
  (2) Purposeful library browsing
  (3) Discovering and developing material
  (4) Periodic review of notes
  (5) The research journal: emerging themes
  (6) Analyzing the research journal: theme-families
  (7) Argument-outline: an index to meaning
(D) Producing: amplifying meaning
  (1) A "treatment" first
  (2) Drafting the product
    (a) Raw draft
    (b) First cuts for coherence and consistency
    (c) The objective listener: when in doubt, talk it out
    (d) Improving the roughly final draft
    (e) Ample, varied, vital language
    (f) Synergism: product and policy

CHAPTER

13

Writing for Organizations: The Communication Problem

Colonel North's Memo

What could be more exciting for a writer than discovering ideas and words that help people make decisions or take actions? Yet no other writing assignment is more difficult, more vulnerable to external and internal pressures, or more likely to cause mischief. Take, for example, the text of Colonel Oliver L. North's "Action Memorandum for the President," printed in the *New York Times* on January 10, 1987.

This top-secret memo outlines an Israeli plan to "bring about a more moderate government in Iran" by "unilaterally . . . selling military matériel to Western oriented Iranian factions." The Israelis are convinced these resources will result in "long-term changes in personnel and attitude within the Iranian government." The Israelis' only requirement is "that they will be allowed to purchase U.S. replenishments for the stocks they sell to Iran." Although U.S. law requires that recipients of U.S. arms notify the government of transfers to third countries, Attorney General Smith determined that a presidential finding could authorize the CIA to sell arms "outside" the reporting provisions of the law. (The President's "Covert Action Finding," of the same date, is attached to the memo.)

The heart of the memo is that the "Israelis are also sensitive to a strong U.S. desire to free our Beirut hostages and have insisted that the Iranians demonstrate . . . good intent by an early release of the five Americans. Both sides have agreed that the hostages will be immediately released upon commencement of this action. . . . [T]his approach . . . may well be our only way to achieve the release of the Americans held in Beirut.'' The memo notes that ''since this dialogue began with the Iranians in September, Reverend Weir has been released and there have been no Shia terrorist attacks against American or Israeli persons, property or interests.'' After describing the details of the arms transfer, the memo calls for an ''urgent response,'' and notes that while Secretaries Shultz and Weinberger ''do not recommend . . . the plan,'' Attorney General Meese and Director Casey believe the ''objectives of the plan warrant the policy risks involved.'' The memo ends with a recommendation to ''withhold notification of the Finding to the Congressional oversight committees until such time that you deem it to be appropriate.'' A handwritten note by John M. Poindexter, for whom the memo was prepared to present to the President, reads: ''President was briefed verbally from this paper. VP, Don Regan, Don Fortier were present.''

My purpose here is not to view this memo with the advantage of political hindsight, or to argue policy. Instead, I want to focus on the *writing* for what it can tell us about the problems involved in writing for organizations.

The memo at first seems to present a no-fault plan, conceived by another country whose own interests happily coincide with those of the United States, requiring nothing more than the purchase and shipment of arms, but with the potential for immediate, and perhaps long-term benefits for the United States. On a second reading, however, the memo, on so sensitive an issue, bringing together such disparate groups with so many different needs and values, seems disturbingly seamless. Even for an action memo, it is *too* perfectly constructed around its recommendation. There is no consideration of potential dangers or drawbacks, except for the unexplained reference in the next to last sentence to undefined ''policy risks.'' The legal problem about reporting transfers of arms to third countries is brushed aside by Attorney General Smith's ''determination'' that the President can authorize the CIA to sell arms. Conveniently, the Israelis ''well understand our position on not making concessions to terrorists.'' Finally, ''If all of the hostages are not released after the first shipment of 1,000 weapons, further transfers would cease.'' With no apparent political, legal, or strategic drawbacks it seems an offer *too* hard to refuse. Furthermore, although the memo proposes an action in direct conflict with the President's stated policy, the *style* is earnest, energetic, and decisive, as if free of internal struggle.

But there is something troubling here, and since the brief memo is so important it is worth reading again. Now, questions arise. Why, for example, is there no concern anywhere in the memo that the Israelis might be wrong? "The Israeli plan is premised on the assumption that moderate elements in Iran can come to power if these factions demonstrate their credibility in defending Iran against Iraq and in deterring Soviet intervention." *Are* there "moderate elements?" If so, who are they? Is the Israeli premise sound—that moderate elements will come to power because they acquire arms? What is the connection between this purported Israeli strategy and freeing the American hostages: what *connection* do "Iranian moderates" have with the fate of the American hostages? How is it that the Israelis "understand" this arms sale, with the hope of freeing hostages, is "not making concessions to terrorists?" Is their "understanding" in any way related to how other parties (terrorists, for example, or our allies enlisted in the struggle against terrorism) might interpret it?

On further analysis, the memo begins to unravel. What *were* the objections of Secretaries Shultz and Weinberger? What were the undefined "policy risks"? Why does the memo have no sense of the other side to the argument? Surely, on such an emotional and politically sensitive issue as the hostages, there were other perspectives or reservations. Why does the memo contain no evaluation of the *moral* and *political* implications? For example, any reader might ask: Should we be involved in an attempt to bring "moderates" to power in another country? What would the political impact be if our role became known? An action memo for the President should necessarily represent the most thorough analysis as the basis for a decision. Is that what we have here?

The memo raises many of the familiar questions about any poorly written argument. It substitutes energy and urgency for evidence; it distances the writer and the reader from responsibility; it attempts to fill in the gap left by lack of analysis with an unquestioning faith in its own efficacy: "Both sides have agreed that the hostages will be immediately released upon commencement of this action." Reading this memo simply as a piece of writing, it is hard to believe that anyone could take it seriously.

Could all of these questions have been discussed earlier? Were they resolved when Mr. Poindexter "verbally briefed" the President? If that were the case, what is the status of so deficient a memo? Why write it as if it were the basis for a decision if it is entirely lacking in analysis? (What if Mr. Poindexter added his own interpretation when he briefed the President "verbally"?) Perhaps the memo simply confirms a policy already agreed to by the President, and merely records the grounds for his presidential finding. But if that is the case, how could a presidential finding be based on such precarious evidence?

The function of the memo could not have been to inform; it may have been to persuade or confirm. But despite its seamless surface, the low-cost, fault-free appearance of its recommendation, there is chaos and fantasy within. Beneath the concise, decisive exterior, there is defensiveness, inflexibility, and desperation in the writing. In its context, among consenting adults in the White House, the memo may have made perfect sense. Perhaps it really was meant to justify one policy by linking it as if through good luck to another: ransoming the hostages as a fortuitous by-product of an Israeli plan to reach moderate elements in Iran who, with the acquisition of 4000 TOW missiles as a demonstration of their "credibility in defending Iran against Iraq and in deterring Soviet intervention," will topple the Ayatollah. A need for justification, even after the fact, might explain the unsettling indications that it is a fantasy memo: too perfect; too eager; and totally lacking in analysis and perspective.

There are other possible scenarios here, from North's point of view as the writer. Perhaps the memo presents what North thinks the President wants to hear. But by writing what you think a superior wants to hear, you may be ignoring the very data he or she needs the most; and in making such a decision before writing a memo, you may be putting him or her in jeopardy. This may have been the case with North. If the President depended almost entirely on North's memo, without countervailing arguments or more comprehensive discussions, then, in a broader context, North's interpretation of what the President wanted to hear would be extremely dangerous.

Another possibility is that North is presenting his own view in the most persuasive way he can, deliberately omitting an assessment of the possible problems and alternatives. This, too, would leave the President in a vulnerable position, and his vulnerability would extend far beyond the walls of the Oval Office. In the end, your reader may not want to know the truth; but as the writer, you must provide the basis for an informed decision.

Any writer in an organization faces such questions. He or she does not want to write a memo that will displease a superior, or is afraid to write a memo that includes the "cons." But as we have learned with North's memo, the results can be disastrous, putting in jeopardy not only individuals or agencies, but even the nation. To write well in an organization is not simply a matter of writing clear, concise, persuasive prose. The writer's ultimate responsibility, especially when making the best case for an argument, is to give the reader a comprehensive, analytical view of the problems and alternatives, a balanced perspective, and an honest look at the other side. As a writer you can not assume the responsibility of the decision maker. Your own point of view must not be allowed to distort the data. Your role as the person who culls the material and presents a point of view is crucial, because the decision-maker depends on your honesty and thoroughness.

It may never be clear whether North's memo was intended to create, affirm, or obscure policy, but his presumed dedication and loyalty backfired: the memo left the President blind-sided. North's responsibility in constructing his memo, whether he was assuming or supporting a goal of the President or presenting his own strategy, was to keep the President fully informed.

North's memo reads as if it were written in a political and moral vacuum, an extreme case of the problems inherent in writing that is intended to help other people decide or act. Writing in an organization can be exciting or frustrating; it can strengthen or undermine policy. But to be able to do it well, it is necessary to understand some of the causes of our communication problem.

The Uniform Writing Code

Many of the persistent problems for writers and readers in organizations can be understood within the framework of what I call the uniform writing code. This code, explicit or implied in the structure of almost all organizations, confronts the writer with choices that often bring loyalty and responsibility into conflict. Such conflicts create ethical questions that, unresolved, make good writing under pressure impossible. Successful writing in organizations depends on transforming the code.

The Organizational Point of View

The writing code includes a number of injunctions about tone and attitude. Organizational writing is expected to be *positive;* to present, internally and externally, the best case for the organization's conduct. At times, this may mean not writing what the writer knows. Organizations blessed with consensus have an energy that turns problems into prospects. Labs, departments, and new or revivified organizations have this spirit when everyone feels they contribute to a common goal. But organizations in trouble almost always begin by attributing their problem to a failure in "getting their story out." There is an inevitable tension between reality and appearance, and one pressure on anyone who writes in an organization, in good times or bad, is the need to consciously avoid any subject, fact, or analysis that might shake the reader's confidence in either the organization or the writer's role within it. A writer who wants to survive adopts the organizational point of view.

For many people, this positive attitude is a learned behavior that eases daily life, especially in relation to colleagues and supervisors. In some organizations, there may be only a rare occasion when the organizational story does not fit the writer's experience or knowledge. In others, such a conflict can be a constant of daily life. A staff assistant or a manager may see the difference between stated policy and actual practice. A professor may be surprised by the story the college catalog presents to the outside world. But for the writer, there is the pressure not only to affirm the organization's conduct but to avoid communication that might cast doubt upon his or her *capacity* to affirm it. A writer's inquiring or deliberative tone might be interpreted as "indecisive" or "negative." Your intended reader may want to know what to *do,* not what to doubt or ponder.

Communication is greatly simplified if both reader and writer share the organizational point of view. But the price for such consensus is often high. While the formats for writing in organizations—different memo forms, models for correspondence, guidelines for research summaries and proposals—seem to require objectivity, they reenforce the underlying, subjective version of the organization's conduct. The pressure of this injunction to display an unquestioning, positive attitude while writing from the accepted point of view, stifles thinking and undermines the writing process and product.

Tact

Tact supports this commitment to an organizational stance: no matter what a writer has to say, it is almost always more important to say it in the *right* way than to say it at all. If there is room for negotiation or dialogue, it is initiated through an understanding of the vulnerabilities of those above and below the writer in the department or agency. Skill in working with others takes on a special quality in the day-to-day life of an organization because it signifies not only maturity but predictability. Professionally, for a writer, this predictability may become far more important than identifying problems, proposing alternatives, or resolving conflicts. But it is an expectation that wrecks havoc on his or her capacity to think clearly. Anyone with supervisory responsibility in small or large organizations understands the appeal of these attitudinal expectations. But when they pressure writers to contribute to an *illusion* of meaning, they create a communication problem that corrodes an organization from within.

Almost all of my writing clients from business and government have been convinced, at one time or another, that writing is a matter of affirming the values and goals of those above them in a hierarchy. The concept of "telling them what they want to hear" is reflected daily in a sluggish flow of documents, memos, and letters, few of which are worth reading, although we may be required to write or to respond to them. Such writing assignments contain no *argument;* no deliberation. We recognize useless writing, but we don't have a sense of how to make *good* writing matter in an organization. We feel helpless to change the environment.

People who complain about the writing in an organization generally mean one of two things: either the writing isn't clear, or the writing doesn't "look" professional. But "clear" writing about a point not worth making, or empty correspondence perfected by a word processor's spelling checker, deepens the communication problem. The self-defeating effects of inadequate or cynical writing expectations may be delayed, but they are inevitable.

As people who write, *we* all feel much more resilient and energetic than the uniform writing code permits. We have far more creativity than our organizations elicit from us. If an organization defines dependability for its writers as a positive, tactful, loyal approach and style, it cannot realistically expect writing to be clear, responsible, and powerful. Especially under pressure, the code conflicts with our struggle to write well because that struggle always requires the writer to take responsibility for his or her own words.

Brevity

In any organization, brevity is the familiar ideal of writing. People in general—especially people under pressure—admire writing that "doesn't waste words." These injunctions carry over from our school days: "Cut all excess words," or "Be concise." The theory makes sense: never use a word unless it contributes to meaning. But in practice, this expectation creates confusion and fear. People try to be brief instead of explaining what they mean, or struggling to put their ideas into words. In institutions, these fears are sometimes linked with professional survival. A cabinet secretary gave classic expression to the organizational ideal of brevity some years ago: "Anything that can't be put down on one page isn't worth reading.'' The damage within the department and outside resulting from people trying to live up to such a foolish ideal can be enormous.

Given the amount of writing we all read in organizations, this pressure for brevity is certainly understandable. Even before scientists, doctors, and scholars began to complain that they could no longer keep up with the literature in their fields, people in business and government were swamped by reports, memos, documents, and articles they were expected to read, analyze, and respond to every day. But if there is so much writing, and so little of it is worth reading, why hasn't the ideal of brevity been the obvious solution?

This paradox results from a failure to distinguish between causes and symptoms. Brevity is valued as an organizational skill to resolve the crush of time and information. But in a context in which a positive approach, tact, and loyalty to an organizational point of view are insistent pressures on the writer, brevity becomes a quality of *appearance,* not of meaning. Writers with a great deal of experience learn to squeeze as much meaning as they can into each word and sentence, and to take out every word that does not count. Their priority is not a word count, but fitting meaning to a structure that efficiently and adequately expresses it. But under the pressure to be brief, people without much writing experience fail to explain their ideas, conclusions, and recommendations. They leave out those connections that give their arguments coherence and unity. That is why we read so many memos, letters, and reports that make almost no sense at all. In some cases, they present the disconnected, fragmented surface of an unarticulated argument. In others, they offer little more than an assertion without evidence, compactly crafted to the assumed needs of a particular reader.

How could it be otherwise? Being brief is a blessing if writing has a point to make, and evidence to support it. But it may be impossible to make that point and support it in one page, or even in one hundred. I remember reading a one-page summary of a one-thousand-page report that had taken eight people six months to write. Even the summary, on the basis of which a cabinet secretary made his decision, consisted of half a page of titles and numbers identifying the manager of the project.

The report's recommendations, compressed into a few sentences, affected thousands of people, and involved a great deal of money and resources. But reduced to a few lines, these recommendations lost their rough edges, their indigenous warning signals, their internal conflicts. They were reduced to little more than a predictable slogan. The research may as well never have been done. You might argue that there is a time for thinking and a time for acting. But the appearance of efficiency was deceptive and dangerous. The cabinet secretary's demand for brevity neither ensured efficiency within his department nor, ultimately, hid the actual chaos from public view.

Instead of inspiring intense research, analysis, evaluation, and then compression, the writing code moves people in the opposite direction. If a writer knows that only his or her recommendation will be considered, and if he or she is expected to anticipate the recommendation the reader wants to hear, then the whole process of research and writing takes on the air of a performance, the appearance of thought, not the process and product of thinking. If the one-page summary of a report is written by someone who did not do the research, and who interposes his or her own interpretation of what the reader wants (or ought) to hear, then brevity reenforces confusion and even deception.

The pressure of time is not an excuse for incomplete, confusing, or misleading writing. The problem lies in a concept of efficiency that at best fails to take into account the nature of good writing, and at worst seeks to streamline policy around the values, goals, or work habits of a few people. It may be that describing gray areas, raising ethical questions, or analyzing the nature of a policy will slow readers down. But the virtue of good organizational writing is that it enables a reader who has delegated thinking and writing to others to examine material more thoroughly, and compare alternatives.

Brevity is not the cure for the communication problem. It may be one of the causes. As an abstract notion, brevity can be a comfort to people who don't want to read. But as a concrete reality of day-to-day life in organizations, where people inexperienced in the process and product of writing abide by a code of tact and team spirit, brevity becomes the resolution of an inability to think issues through.

"Tell Me What You Want Me to Say"

It's not surprising that writers in organizations struggle to discover what their readers want to hear. But in this kind of environment, communication becomes a self-sealed system with a restricted set of permissible statements within a rigid format. Thinking and meaning give way to affirmation and the appearance of meaning. Brevity, reenforcing appearance, narrows the channel of communication and clogs the flow of creative energy.

The effect of the code on the individual writer is equally dramatic. People who return to school for mid-career degrees or executive training programs find it hard to explain what they think, or to give evidence for why they think it. Habituated to the shorthand of their own organizations, and to meeting the narrow expectations of an audience which has power over them, they find it difficult to criticize underlying assumptions, consider other perspectives, or integrate their own experiences with new information and ideas. The uniform writing code has resulted in a model of communication that is the antithesis of discovering and presenting meaning. The writer's process of finding out what he or she thinks, of using writing as a way to discover meaning, has atrophied in the practice of telling readers what they want to hear. Once people realize they have more to offer, more to say, and better ways of saying it, they may try to change the environment to encourage better writing. But there is an even more fundamental problem, over which the individual writer has almost no control, contributing an enormous pressure against good writing: we all inherit the dialects of our organizations.

Dialects

Business, government, and the professions have idiosyncratic, exclusive languages of their own. These dialects are often the subject of satire. There are recurring lists of words that shouldn't be used, and classic examples of indecipherable or absurd memos. The older lists are fascinating as verbal nostalgia. Looking back over the last thirty years or so, we can review our history through the words, phrases, and metaphors that saturated the language: "separate but equal," "domino theory," "the light at the end of the tunnel," "trickle down." Contemporary lists of words that have lost their meaning may help make us self-conscious about the words we use. But while these lists give us insight into the relationship between language and culture, they remind us of how vulnerable we are in our search for meaning. Dead words replace living ones, and diminish our capacity to discover and present ideas. Dialects are not simply a shorthand to save time but a protective covering for the ingrained policies and practices of an organization.

To gain some perspective on the pressure these dialects exert on writers, imagine a software program that incorporated all current and past lists of dead words and rejected them in our letters, memos, and reports. This would certainly shorten our documents. What would remain?

Try this with one of the documents that comes across your desk. Cross out everything that doesn't convey fresh meaning, even the old standby, "You have asked me to . . ." (you knew that already, and so did the writer). Then read it again, to take out everything that the writer has put in simply to win you over to his or her message. Take out everything that merely demonstrates a positive attitude, tact, or loyalty to the organization.

There probably isn't much left over. But that's not as surprising as the *nature* of what remains. Instead of the kernel of an idea or the real substance of a proposal, as you might expect, there are only fragments of both. What's left is not the essence of the subject, but the introduction to an argument or an explanation. As the reader, you're probably left with a question like: "Yes, but *why?*" or "What do you mean?"

Perhaps you missed something. But if you go back over what you have cut, all you'll find are those dead words, the stroking words to win the reader over, the words and phrases that automatically accompany certain formats. For all its pragmatic appearance, the substance of much organizational writing is not explicit but implicit. What is explicit is not terribly complicated, and usually not very useful. What is implicit is fascinating, but often confused, fragmented, and unresolved. Part of the function of a dialect is to obscure this confusion. In any piece or organizational writing there is an inherent tension between surface appearance and inner turmoil.

It is extremely difficult to subdue language in this way and then try to use it to write with clarity and power. Writers in societies less free than our own have had to depend on allegory, symbolism, or fantasy to circumvent moral or political censorship. But the potential for discovering meaning is one of the main reasons why writing is useful, and why the possibility of sharing that discovery with a reader is so exhilarating. That is where jargon and dialect fit in, and why they survive. Dead words take up the space that thought might fill. In an environment in which fresh, passionate thinking is unsettling, adherence to a dialect constitutes a message in itself: that the usual rules apply, that the hierarchy is in order. This is how we know we can ignore almost everything that comes across our desk unless we find a key word or a phrase that calls for closer attention. We can skim with confidence because there will be no surprises, nothing fresh to respond to or consider.

The languages of organizations and professions are filled with indigenous jargon. Even highly trained writers who use a dialect self-consciously, with irony or chagrin, can be convinced of its necessity. People with little writing experience may not even question a dialect, struggling instead to become proficient at it. But organizational languages encode conflicting ideas, values, and goals, and cancel out meaning. Jargon survives not because we are lax in weeding out the latest vicious set of meaningless or overused words, but because the words we use often have more to do with survival in an organization than with fresh ideas, clarity of analysis, or meaning.

Our organizational dialects code the underlying assumptions of our work lives. They allow us the brevity, decisiveness, and loyalty we admire, under the pressures of time and the need for tact. But they make it difficult to *think* on paper; and in narrowing the basis for taking responsibility for meaning, they obscure information, confuse analysis, and undermine communication.

From "I" to "It"

Nothing points up the loss of the writer's responsibility for meaning more vividly than abandoning the first-person pronoun, "I," in favor of its organizational or professional surrogate, "It is to be believed. . . ."

Although it raises the word count, this shift to passivity, ostensibly to prevent subjectivity, saves the reader time. "I" can be troublesome, raising time-consuming questions: "Who is the writer? Has he or she any right to claim my attention? From what point of view is this person speaking?" By avoiding the first person pronoun, the writer participates in an institutional authority.

Science writing long ago established a peculiar authority for impersonality and the use of passive voice. This machine-style prose, apparently swept clean of personal interest, gathers into itself a kind of generalized responsibility: the *organization,* or the *profession* speaks, not the individual whose thoughts, whether superficial or thoroughly supported, are expressed in his or her own style and voice. The distinction between "It" and "I" is certainly not the difference between objectivity and subjectivity. In organizations, it is more often the difference between obedience and responsibility. With impersonal language that studiously avoids the first person pronoun, it is often impossible to trace the history of a decision in an organization. Instead of insuring objectivity, the shift from "I" to "It" protects subjectivity from analysis and judgment.

The impersonality of business, government, and professional writing not only inhibits thinking, but permits all kinds of mischief. In a discussion among collegues about the war in Vietnam an economist referred over and over again to "human resource units": moving two hundred thousand here, a hundred thousand there. I envisioned the docks in Oakland filled with neatly stacked cardboard boxes, but when I asked him about it, he seemed puzzled: "Human resource units? They're *people,* of course." I'm sure that is perfectly good professional shorthand, but it is an example of the mischief dialects create: language stripped of human reference, content devoid of humanity. The shift from "I" to "It" (reenforced by teachers at every level of education) pressures us to separate thoughts and feelings from responsibility.

The Ethics of the Uniform Writing Code

The uniform writing code at its best elicits compact, supportive writing. If it is done well, the writing will distinguish clearly among various perspectives, needs, or clients. But not many writers have the confidence and experience to overcome the destructive pressures of the code. In many organizational settings, conversations among people who work well together are more important than the written word in making policy or adopting new ideas. Writing, by people not involved in the original discussions, *follows* such conversations. The writer's work is seen as a service: making clear, lobbying for, or enforcing decisions made by others. Since there is almost always this distance between the decision-maker and the writer, it is easy to see why there is so much frustration and disappointment with organizational writing. If you are in a supervisory position, you want to hire people whose job is to write what you meant to say, or would say if you had the time or skill. But that service function as it exists throughout a hierarchy is neither one of the great virtues of good writing nor one of the inducements for writers to struggle to write well.

If writing in an organization was simply a service, all the books and courses on business or government writing would long ago have solved the communication problem. These textbooks are full of sample letters, and the courses are full of memo exercises. There is a general understanding that writing should be brief and clear. Yet much of the writing in organizations is unreadable because of the inherent conflict between obedience and responsibility.

The documentary history of the Vietnam war, the correspondence leading up to the Challenger tragedy, and the developing history of the Iran-Contra scheme provide countless examples of memos, position papers, reports, and letters that adhere to the uniform writing code. They are impersonal, decisive, and loyal, and they are expressed in the appropriate style and language. Yet they are terrible pieces of writing which have caused tremendous suffering. Reading them over, anyone would wonder how the gaps in thinking, the incompleteness of evidence, and the bizarre fantasies expressed in dehumanized dialects could have escaped notice. But the writing code forces words and responsibility apart. At any point in any of these tragedies, a writer could say, in his or her own defense, "I was only doing my job."

Obedience and good writing do not mix. Writing well, especially under pressure, requires hard thinking and taking responsibility for one's words. It's no wonder that so many people in organizations adopt the code without question. Responsibility for a policy, a product, or a program is a heavy burden, but the uniform code disperses responsibility: no one takes the blame. In such an environment bad writing drives out the good. If the environment hardens into regimented affirmation, implementing policies that cannot be continuously analyzed and evaluated, it becomes unstable and dangerous. When incomplete, biased, or deliberately misleading writing pierces the walls of an institution and enters the public domain, it causes considerable damage. Writing is no different from any other skill that combines carftsmanship and art. You can apply a brilliant coat of words to a deteriorating argument, or persuade people that a glittering fantasy is a sturdy possibility. But these uses of language undermine communication.

To say that we have a "communication problem" is to identify an issue much deeper than spelling, jargon, tone, or format. Writing is hard enough. To find ideas that matter and words to express them requires patience and confidence, in situations characterized by insecurity. Writing in organizations is embedded in still another layer of pressure: telling readers what they want to hear. How do people resolve these pressures and produce thoughtful, clear, powerful writing? Part of the answer is that writers in organizations have to do *more* of the kinds of things every writer must do: more preparing and planning, more sifting of thoughts, more balancing of writer's and reader's needs. Organizational writing has a responsibility to educate its readers. Writers have to press for authentic communication in everything they write, whether to a supervisor, a customer, or a colleague. To write well in an organization one must overcome a communal numbness to meaning, and begin to help people develop the skill of good reading. Writers and readers form the act of communication together, and they both must take responsibility for it. Our "communication problem" will not be solved by isolated writers toiling to find the right tone or the appropriate format, but by organizations transforming the pressures of the environment in such a way that effective communication is possible.

[ ]

Improving Organizational Writing

Improving the Process and Product

Enriching Writing

Keats' advice in a letter to Shelley to "load every rift in your subject with ore," may strike you as far removed from writing in organizations. But in their effort to pack meaning into the fewest words, and to be visually clear on the page, poets share a good deal with the writers of memos, letters and manuals. However different in motivation and in intended audience, good poetry and good organizational writing resemble each other in distilling meaning to evoke an immediate response. In poetry, the private effect on the reader can be enduring. In organizational writing, the effect is social and catalytic. But the compactness of poetry and of organizational writing creates an opportunity for immediate, powerful communication. Wallace Stevens, the American poet, worked as the vice president of an insurance company. He participated in both worlds: that of the patient writer creating art for its own sake, and that of the businessman admired for the clarity and power of his correspondence. The supreme poet as successful businessman points to an unexpected relationship between thought and action.

Imagination and Power

We rarely have an opportunity for self-expression beyond the circle of people close to us. Writing in an organization allows us to cross those boundaries, but this privilege exacts compromises. We are expected to adopt a non-critical approach, tactfully working within appropriate formats in a coded language. We rely on institutionalized authority. We substitute a superficial efficiency for ethical responsibility. In these ways, the potential power involved in writing for others is carefully circumscribed by the environment. Personal writing does not have the capability for immediate social impact, but it offers the writer the satisfaction of crafting a lasting product according to his or her deepest vision. The two kinds of writing exemplify complementary passions in a writer, one promoting action, and the other creating meaning.

But organizational writing is usually mission-oriented, emphasizing action and product over thought and process. That is why it is much harder to do well than most people realize, and is so difficult to improve. To write deeply and analytically about a subject without reference to a deadline would be futile in an organization: such writing would not find an eager audience, and would be rendered useless once the time for decision or action had passed. But to write shallowly and quickly, simply to fulfill the short-term needs of an intended reader, undermines the decision-making process. In both cases, the writer works from weakness, not strength. Yet the memos, letters, and proposals that people admire are almost always characterized by fresh ideas presented within a comprehensive perspective. In structure and content, they reflect Keats' ideal of loading every rift with ore; in organizations, where people chronically say too little, too much is never enough. To improve communication, thought and action, and imagination and power, must reenforce each other. In order to achieve this integration we need to redefine economy in writing.

Economy Follows Analysis

Economy is among the most sophisticated of writing skills. It does not result from cutting out all excess words, but from determining what you think, and then conveying it fully, forcefully, and efficiently to the reader. Poets, selecting as few words as possible, create symbols, and orchestrate the music of their words to make the transfer of meaning immediate. Metaphors, whether in poems or political speeches, connote much more to the reader or listener than their few words denote.

Economy *follows* analysis. You can only reduce something to its simplest terms if you understand it in its most complicated terms. The writer's private thought processes are made public through precise explanations that represent the twists, turns, and connections that led to meaning. Three different acts are involved: finding out what you think; clarifying the connections among your ideas; and integrating them for a reader. The first is an intense search, the second a disciplined analysis, and the third a focused presentation. Compression of meaning and adequate representation are not steps to be added on to the writing process as you might run a final draft through a word processor's spelling checker. For the poet, the risk of compression is that meaning becomes too private, inaccessible to readers. In organizations, the risk is that form substitutes for content. It *is* possible to write briefly and well, but it requires thorough analysis as the basis for clear, compact presentation.

Thinking in Three Dimensions

Authentic writing can only arise from fresh, comprehensive thinking, an activity that often has a low priority under the pressure of organizational life. Some people may not even realize what they are missing. By way of analogy, imagine that I have reached the end of a transatlantic flight to Boston. My fellow passenger slides his calculator into his attaché case and stands up, ready to leave.

"What's that book you've been reading?" he asks.

"*War and Peace*," I say.

"Oh, what's that about, anyway?"

The two of us move down the aisle with our bags. I only have a minute to characterize the novel, so there's not much hope. "It's about Russia," I say. "In the nineteenth century."

It sounds silly. I could have said something about the contrast between Pierre and Prince Andrei, or the relationship between the setting of the novel and its themes—something to give a sense of what the book says, or what it means to me.

"I don't read very much," the traveller says, putting on his jacket. "Skip the frills. Give me the main point."

I hold up the book, laughing: "This has lots of frills."

But the truth is that it *doesn't*. The characters, incidents, and themes are the main point, the "meaning" of *War and Peace*. They represent the resolution of years of struggle on the part of the writer, clearly revealed in wide manuscript sheets full of minute additions in both margins. That is why I feel silly about my two-dimensional response, locating the book in space and time as a dictionary might: "Russia, in the nineteenth century." How could I put my feelings about the one-thousand-page book into a few sentences? Could I have described the tremendous power that gathers as you read through the novel? If my fellow passenger had asked, "Why do people say it's such a great book?" or "Why do *you* like it?", it might have been easier. But the neutral question, "What's that about?", was an even greater burden because I felt a responsibility to be objective.

Now compare my ineffective answer to this imprecise question with the task of writing a one-page summary of a one-thousand-page report for a cabinet secretary. The writer of the summary may or may not have been one of the researchers, may or may not understand the conclusions, may or may not agree with them, and may simply decide to tell what he or she thinks the secretary wants to hear. Each of these margins for error regularly causes trouble in organizational writing, but they all matter less than the simple fact that you can only give a certain kind of answer in one line or one page, or in a particular format. Whether or not it is free of the familiar organizational tension between appearance and meaning, the answer will be linear and narrow. The secretary may want that, for the sake of efficiency, but will be getting the illusion of efficiency while sacrificing meaning. To improve communication in organizations we have to change the way we evaluate information, especially under pressure.

Analysis and the Axes of Evaluation

With questions that require quick thinking on issues affecting many people, what should you put in and what should you leave out? Suppose the question were: "What can we do to improve our students' writing?'', or ''What should we try to accomplish at the national conference?'', or ''How can we get the engineers to communicate with the sales force?'' Under the usual pressures, we would piece together an answer that: (1) would not cost the organization very much; and (2) would not upset anyone above us, or needlessly concern anyone below us whose support we need. But our answer might be entirely off the point of the question. Confined to the two dimensions of budget and teamwork, both filled with assumptions that may or may not be accurate, we get our part of the job done quickly and obediently, and pass on the responsibility for thinking hard about the real question to someone else.

''Who will this rub the wrong way?'' (or, at higher levels in the hierarchy, ''How will this make us look?''), and ''How much will it cost?'' are the organizational counterparts of time and space in my one-line description of *War and Peace*. They are the customary, easily identifiable axes for our two-dimensional, organizational answers. They don't arise from an analysis of a problem, but from an anticipation of how an answer will be evaluated, even when the basis for evaluation may be inadequate. These axes of evaluation confine answers to the needs of the immediate reader, a supervisor or section head, leaving little room for the analytical dimension and the long-range perspective that are the real strengths of good writing. I might have told my fellow passenger that *War and Peace* was about the way we shape our souls amid the pressures of the world. That's not so two-dimensional, and contains something of the struggle to bring my thoughts and feelings to bear on the question—the kernel of an idea. That answer might have prompted another question from the traveller, or at least given him a thought to ponder. It would not have given him the illusion of understanding something extremely complicated that is difficult to explain in a quick answer.

Any piece of writing that neglects an analysis of the question, and of the writer's assumptions about communicating meaning to the reader, will be narrow and incomplete. It will reveal a conflict between meaning and the appearance of meaning. Such writing is what people refer to when they say that there is a communication problem in their organization. They may not be experts in writing, but they know something is wrong. They may think it would be solved if people spelled and punctuated correctly, or if they cut out all excess words and got directly to the point. But the real problem is the writers' resignation, fear, or simple lack of experience in authentic writing for good readers. One page of timid writing, punctuated perfectly in a visibly structured layout, with a recommendation in boldface type, is still timid. Worse, it gives both writer and reader the illusion of communication. We know this as readers, but we don't know what to do about it. If the organization's lines of communication carry only narrow, predictable thoughts, writers become frustrated, and then cynical about their words.

The few good memos, letters, and reports we see are memorable. They are written on the assumption that there is much more to say than can ever be put down on paper. The writer's job is not to reduce all thought to dictionary-like definitions, but to represent as much meaning—and the context for that meaning—as possible. The organization's policy may or may not be based on the two-dimensional axes of "How does this make us look?" and "How much will it cost?" But the writer can address each question as a request, whether explicit or not, for fresh ideas and a comprehensive perspective.

On a grid composed of the two-dimensional axes of evaluation, ideas are guilty until proven useful by external pressures; even if a writer discovers something new there may be no audience for it. No one can write well, for long, under these conditions. If people resign themselves to this fate, their writing becomes detached, unimaginative, and predictable. They produce writing that does not need to be read. One of the functions of any piece of writing in an organization is to provide wisdom about the process of thinking questions through: not backing away from meaning, but arguing clearly and tactfully for it.

The Mask of Formality and the Conversational Ideal

When two people talk, they have the opportunity to build something together:

> "You know," one says, "we really should. . . ."
> "Yes," the other replies, "but to do that we'd have to. . . ."

"Well, couldn't we . . . ?"

"Sure, but we'd have to lay that out clearly."

"But could we afford . . . ?"

"It might cost less than what we're doing now. Let's not worry about that until we see if it works."

Some version of this conversation takes place in an office, corridor, or laboratory every day in every organization, accomplishing more in a few minutes than months of memos and reports. A flowing, energetic conversation is exciting because it involves both speakers in creating something useful. They can exchange perspectives, experiences, and ideas in building a concept that is stronger than what either of them could achieve alone. Such conversations depend on trust and directness, and freedom from the confined, coded language of formal written communication. For almost everyone in organizations, the ease and energy of such conversations make writing seem dull and tedious.

But if either participant in our imaginary dialogue wrote a memo about this conversation to a third person, it would contain little sense of excitement and enthusiasm. The memo would be formal, ostensibly objective, and written as if by a machine rather than a person with energy and ideas. The writer would struggle with appearance (headings, bullets, graphics), and format (should this be a "discussion" memo or an "action" memo?). He or she would be tactfully attentive to the particular needs of those above, below, and to the side of the organization. But the excitement of active participation in creating meaning would be missing.

Under the pressure of organizational writing, the formal appearance of our letters and memos gives them an illusory finality. That is why a proposal with completely unacceptable arguments, unsupported by any substantial evidence, in page after page of somber, formal language, will be received *as if* it were a serious document. Its immediate effect within the organization will be political. But the wasted effort in response to it diminishes trust and corrodes the lines of communication. In a conversation, such a tactic could be countered directly, but in dense, formal prose, it exerts a cumulative psychological impact far beyond what the individual writer imagines.

The quick, rough writing we do in organizations, lacking the clarity of a well-thought-out-argument, appears to be "final." Yet if formality masks lack of preparation or analysis under pressure, it drains the meaning and energy from carefully prepared arguments. The result is the flat, dull writing we all hate to waste our time reading. People in organizations take the life out of their writing because they think they are supposed to, in the same way that they think they are supposed to write oversimplified memos and reports. The remedy for this loss of vitality in communication is to restore the writer's responsibility.

Restoring the Writer's Responsibility

Good writing cannot exist in a vacuum, because it requires good readers. While readers are not likely to acknowledge this, writers know it and often feel hopeless about it. In many organizations, writers assume that the intended reader already has defined the issue. They assume their job is simply to meet the reader's needs in a dependable, predictable way. Books and courses about organizational writing reinforce this notion. They present models of tactful, formal letters, memos, press releases, and speeches in conventional formats, tailored to readers above, below, and to the side of the writer, and to people outside. But writing that makes a difference bursts the restraints of conventional formats with fresh analysis, genuine energy, and personal voice.

Good organizational writing takes little for granted. It poses useful questions and defines issues in incisive ways. Recommendations are presented in the context of enlightening comparisons, choices, and alternatives. The language is active, fresh, and alive, to engage a reader in completing the act of communication. To achieve these qualities, the writer must meet his or her reader's stated or assumed requirements, and go beyond, to broaden the channel for communication—whether or not the reader is aware of the need.

There are, of course, situations in which the writer is *not* expected to communicate as, for example, when issues are "staffed out" expressly to stall for time. Such bureaucratic strategies may be part of an administrative style, but it would be far better for the writer to turn in blank sheets, or printed flash cards reading "Not in today," than to short-circuit the communication process. (Consider what it would do to a tennis player to insist that she lose a match for the good of the team.) Even more oppressive is the common situation in which a writer has learned through repeated experience that no matter what he or she writes, the reader will ignore it, or use it within the organization for a different purpose from what the writer intended. It is useless to talk about improving writing if writers have so little control over the expectations for authenticity in an organization.

Good writers are flexible, and can adapt their skills to a variety of situations. The need for this versatility is nowhere so strong as in organizations. It is often necessary, under the pressure of time, to write from the materials at hand rather than from carefully sifted research. QWP can help the writer make the most of what he or she has to say, and to acknowledge what is missing. But skill and power are not at issue if someone is asked to do writing that is futile. Restoring a writer's sense of responsibility for his or her own words, in an authentic context, will help prevent the conflicts with the environment that inevitably erode skill and power.

Organizations conceive of writers as providing a neutral service in the same way that engineers contribute to projects whether or not they "approve" of the result. Writers are expected to do their jobs objectively, writing clearly and positively, on demand. The uniform writing code is the policy for implementing these expectations. Although engineers and scientists can refuse to work on certain projects, writers are expected to confirm and support. But if an organization defines the dependent writer's professional responsibility as a function of loyalty, then his or her writing cannot prevent mischief, and may instead contribute to it. It is neither "objective" nor "professional" to compose inaccurate or misleading writing. It may be "loyal," but it is unethical. One necessary step in improving writing in organizations is to open up communication so that people take responsibility for what they say, and what they refrain from saying. Then, there will be far less wasted time, and far more energy for genuine communication.

Power in writing can come from meaning; but it can also come from the appearance of meaning. In the long run, the appearance of meaning (a company "story" that is essentially a fantasy; an administrative line that has no connection with reality; a fraudulent research project) becomes an obsession, requiring disproportionate amounts of time and energy. Words can have power when they are separated from wisdom. Part of a writer's responsibility in an organization is to struggle constantly for meaning, and to try to make that struggle matter.

Improving the Environment

Providing a Framework for the Question

A quick, spontaneous page or two, written without reference to any sort of self-censorship, will help a writer sort through conflicting issues or constituencies. This initial "treatment" serves some of the same constructive purpose as talking through a subject with an objective listener. When you read it over, you find, especially in your own resistance to spontaneity, those unresolved issues that may create confusion. You will also discover ideas or examples that reveal your own interest in the subject.

This is particularly important when a question delivered to the writer under pressure is accepted as an ultimatum. Experienced writers shape questions so that the process of answering them will bring out their best energies. An initial treatment can give you a sense of the scope, scale, and style of the whole project, and a usable framework for generating the provisional thesis and *because*-clauses. Within the piece of writing, this framework, communicated explicitly to your reader, becomes a road map to your argument. Within the organization, this framework can become a new way of looking at the question.

Engaging Your Readers

If you talk through a writing assignment with your intended reader, you may discover that he or she has been unable to frame the question adequately. Especially under pressure, people often don't know what they really want to know, or they have an imprecise conception of what writing can and cannot do to clarify or solve a problem. Successful managers and supervisors in organizations have the tenacity and experience to stick with projects from inception to implementation. But while they may understand that policies or products arise from false starts, dead ends, redefinitions, and endless revisions, they may not be aware that writing is the same sort of process. The QWP writer who has prepared and planned a writing project is in an excellent position to help the intended reader redefine his or her question.

Such conversations enable writers and their readers to discover things they did not know about the issue, and about the way each sees the issue. They also discover how the project is important—its context. These discoveries sharpen the question and shape the design for the answer. As more and more people get a realistic sense of the value of authentic writing, and of how it is accomplished, they learn to frame questions more effectively, improving communication throughout the organization. QWP enables you to write well, on time; but under the pressures of organizational life, you have to engage others actively in the writing process, particularly those people who ask (or pay) you to write for them.

A Community of Readers

Where there is poor communication, there is not likely to be a habit of constructive feedback, of give and take about a piece of writing. Without the expectation of feedback, there is little impetus to improve writing. Writers in feedback-poor organizations withdraw from taking risks; eventually, *thinking* itself comes to be seen as a risk.

Constructive Feedback

Constructive feedback is different from "criticism." A person giving feedback is not expected to be an expert on style, or grammar, or even punctuation. He or she is simply expected to tell you how your writing "works." The importance of this distinction is worth exploring. A high school guidance counsellor was amazed when students brought in perfect drafts of their college application essays the day after they had discussed the rough drafts with him.

"I don't get that kind of dedication out of my creative writing class," he said.

I asked him to explain the difference between his teaching and his counselling.

"I told them how their essays *sounded,* what I thought they had left out. I didn't mention grammar or spelling until the end. The important thing was for them to say what they really wanted to say because that's what the colleges are looking for."

Student and counsellor could work together without the authoritarian pressure of the classroom on a genuine piece of writing that was worth the effort to improve. But in the process, the counsellor discovered the distinction between feedback and criticism. People criticize writing when they tell the writer what he or she has done wrong, what rules or conventions have been broken. A critic speaks as an authority, holding up an abstract although subjective (and perhaps inappropriate) ideal. But constructive feedback depends on *listening* for meaning, for interruptions in the flow of energy and in the clarity of the writer's voice. Criticism is an act performed by the critic without much regard for what comes after. Feedback, on the other hand, is a gift from the reader to the writer, with the intention of helping improve the writing process and product.

Such an exchange *is* possible in an organizational setting, but it is hard to establish an atmosphere that permits it. Few people see themselves as good writers. Even fewer see themselves as knowledgeable about writing. They are eager to disqualify themselves as constructive readers. But feedback depends only on a reader's willingness to help the writer see his or her work from a fresh perspective. In a community of people willing to help each other with writing, feedback helps define the question, resolve a writing block, improve a roughly final draft, or polish the final copy. In such an atmosphere, too, group writing projects derive the benefits of different perspectives and skills.

There is no surer way to improve writing in an organization than to create this atmosphere of constructive feedback. People come to expect it, and then to depend on it as a natural part of their writing process. In such an environment, energy and creativity flourish.

Authentic Writing in Organizations

I began Chapter 13 with the question: "What could be more exciting for a writer than discovering ideas and words that help people make decisions or take actions?" The few people who have this opportunity on a grand scale are usually well supported by staff and protected from pressure by hierarchy. But on a smaller scale, in our own daily opportunities to communicate, we can transform the task of writing what people may want to hear (which is usually far less than they need to know or we need to say), into a process of empowering ourselves and others to ask, to learn, and to say more.

Organizational writing is harder to do well than most other kinds of writing. In addition to the ordinary work of preparing, planning, generating, and producing, you have to educate your reader about writing. You have to hold the line against any sort of pressure from the environment that impedes thinking, and then try to reach a specific audience. You need to develop your own voice, your own writing process in the midst of arbitrary or self-defeating pressures for homogenization and routinization. And you have to do the best you can with your own writing to inspire and invigorate the reading and writing of others.

You may have to say "no" sometimes. You may have to acknowledge what you don't know, or haven't had time to find out. You may have to learn how some people mean what they say. You will have to help people understand what you have to say. Your work will then present an alternative to the common expectation of thin, dull, predictable writing, and your capacity to discuss ideas, talk through problems, and work with other people will reflect a more practical, more humane organizational style.

QWP and the Uniform Writing Code

Every piece of writing says something about its subject and about the possibility of authentic communication. Conscious writing that is carefully prepared, skillfully planned, freely generated, and energetically produced opens the channel for communication. If, in addition, you help clarify unrealistic expectations that either diminish or overwhelm the writing process and product, persuade others to discuss what good writing means, and apply steady pressure for constructive feedback, then you will contribute solidly to improving the environment.

The more people talk about the writing they do in organizations, the more likely they are to overcome the self-defeating limitations of the uniform writing code. No one should be let off the hook. Good writing is hard work, made all the harder in an environment in which people are allowed to plead ignorance, or say they haven't the time, or that it isn't their job. Good writing requires good readers. Moreover, involving readers in this search for meaning is humanizing work. Part of the excitement in writing for others is this opportunity for authentic contact.

Writing well under pressure in an organization means reuniting thoughts and words. It is a risky and difficult job. The goal may not always be the comfort of the reader. The capacity to write well is easily taken hostage under the pressure of a classroom, a profession, an organization, or a culture. People can be persuaded that they have no right to speak, that they should wait until they have the proper credentials, or that they ought to say only what others want to hear. They can be made to fear teachers, peers, supervisors, or colleagues. But what does "free speech" mean if we give only certain people the right to speak?

An Organizational Writing Agenda

   (A)   Preparing
          (1)   Resolve the writing issues raised by the uniform writing code: the abstract pressure for brevity; a professional dialect; the substitution of impersonal authority, tone, and style for individual responsibility and personal voice.
          (2)   Begin the writing process with a sense of responsibility for the meaning of your product.
   (B)   Planning
          (1)   Sketch out a realistic QWP timetable.
          (2)   Plan for too *much* material, in order to compress and represent analysis, evidence, and perspective adequately.
          (3)   Plan a structure that will portray the energy and dynam-

ics of a dialectical conversation instead of an abstract notion of formality.

(C) Generating

    (1) Engage the reader in the writing process; talk through the project, and establish a basis for feedback.

    (2) Create an adequate framework for longer projects by writing a quick "treatment" that sharpens the scope, scale, and style of the project, and will provide the reader with a road map of your argument.

    (3) Generate the provisional thesis and *because*-clauses freely, without self- or organizational censorship.

    (4) Analyze the clauses into the argument-outline of an authentic response to the question, acknowledging what you don't know or haven't had time to discover.

    (5) Include an authentic account of the "other side."

(D) Producing

    (1) Promote an atmosphere of give and take about writing in progress, especially after the first cuts on the raw draft.

    (2) Carry out the QWP agenda through the roughly final draft to the proofread final copy with as much feedback as possible.

    (3) Maintain a policy that authenticates your own work, and raises the level of literacy throughout the organization.

CHAPTER
15

Becoming Independent Writers

Integrating the Writing Process and Product

Writers constantly shuttle between extremes. In constructing a piece of writing the writer goes through two distinct forms of activity: the *process* of discovery, and the *presentation* of his or her discoveries. We know from experience that when our writing is going well these two extremes of spontaneity and conscious control reenforce each other. The goal of QWP is to help the writer achieve this integration.

The QWP system envisions writing as a continuous process, from defining the writer's relationship to the reader and the environment, to estimating the scope, scale, and style of each project. You plan the structure of your answer around meaning, transforming the topic or question into a provisional thesis supported by *because*-clauses. You gain perspective on gathering, selecting, and ordering material by talking through a subject or a writing block with an objective listener. For longer or more complicated projects, you start with a provisional "treatment" as a road map, or as a working framework for moving from argument-outline to raw draft. You record emerging themes in your research journal, and then gather them into theme-families in an argument-outline that will serve as an index to meaning in your long, raw draft. Cutting the draft for coherence and consistency produces a roughly final draft—an excellent opportunity for feedback before you polish the final draft into the unified product.

By contrast, the familiar compartmentalization of the writing process imposes upon it an unnatural sequence. Crucial writing issues arise at inconvenient times. If, out of anxiety over getting the job done, you wait until the final draft to find out what you are really trying to say, you can be sure your final product will lack balance and confuse the reader. If you try to perfect the elegance of a particular style from the moment you begin to work, you will waste a tremendous amount of time and energy on writing that is only useful for what it makes possible later.

Writing Polarities

A number of polarities accompany this familiar compartmentalization of the writing process. Instead of prompting useful writer's questions, they become writing problems, internal pressures that increase the writer's sense of helplessness. But by integrating process and product, QWP strengthens the writer's independence. For example, all good writing is characterized by the writer's discovery of how to present specific evidence with general application. Such writing not only informs readers, but invites them to integrate the material with their own ideas and experiences. This tension between the concrete and the universal is best resolved by making the level of generality a continuous concern throughout the QWP process. As you freely generate *because*-clauses, you work within a chosen range and order of material, from concrete examples to generalizations, that will help you build a bridge to your reader.

Some pieces of writing encompass both a public and a personal style. The material may suddenly shift from analytical to experiential with great effect. But analysis provides a basis for understanding new experiences, and the structure of your writing can elucidate that relationship. There are times in constructing a piece of writing when you feel the pull toward trusting your reader to understand more than you can put into words, and other times when the pull is in the opposite direction, toward meticulous explanation or documentation. Resolving these inherent tensions among possibilities is part of the underlying excitement of writing well. The more *consciously* you choose among the possibilities, the more independent you become as a writer. QWP encourages conscious choice.

Generations of teachers and books have presented writing as a limited series of immutable forms: "comparison and contrast," or "cause and effect." The familiar outlines to be filled in, the formats and conventions to imitate, inevitably erode the power of writing, substituting conventions for decisions. To become an independent writer, you need to think continuously about the process as you create it.

The Ultimate Pressure: Competing with Yourself

It is possible to immerse yourself in a project (even to make an outline first), finish thirty pages of an article or paper in a few days, and then discover on reading the draft that your tone is all wrong, or that the whole article is too abstract. When this happens, you may not have the time or the energy to fix it or begin again. It would be a great help if you could use what you have as a rough draft, written primarily to discover what you wanted to say, or how to say it; but a full, rough, exploratory draft is a luxury most writers under pressure cannot afford. Unpleasant rough-draft surprises almost always result from unresolved issues in your own writing process. For example, one of the fundamental tensions in writing is the expectation that the struggle to get it "right" will overwhelm the satisfaction of *doing* it. Few people are eager to write; almost everyone has some personal reason to dread it. Our teachers, organizations, and professions demonstrate that the act of writing is not very well understood or appreciated. Nor are the rules and conventions that have come to represent readers' expectations much help. We all know from the start that writing involves suffering. But if we begin every project on that assumption, everything we write will be out of balance, confused, or inappropriately addressed to the reader. You cannot write well with a divided spirit because the simple fact of not wanting to do it is immediately communicated to readers. Although they may not be able to articulate their response, readers stop after a first paragraph, or make a minor correction, or ask a rhetorical question containing their sense that your writing is working against itself.

Even when you free yourself of external pressures and self-censorship, there is still the internal pressure of competing with yourself. Shaking off the conventions and expectations that limit thinking—whether they are grades in school, or an organizational writing code, or

the criticism of someone close to you—brings you face to face with
your own confidence, experience, and values. This intensely personal
responsibility can become the most difficult pressure of all, involving
your deepest feelings. This is the tug of war at the heart of good writing,
the dynamic tension between security and risk, caution and confidence.

Struggling through these pressures one writing assignment at a time,
you create your own style. QWP assists you in that struggle by focusing
on meaning. You become independent as a writer when the personal
discovery of meaning coincides with your capacity to present it to your
readers.

QWP and the Common Effort for Change

Imagine, for a moment, a culture in which *everyone* wrote well and
easily, with the fun of spontaneity and the satisfaction of conscious
control. What would that mean in our work, our institutions, and our
society?

We would write the letters that are so pressing and yet so difficult.
We would send our proposals to organizations and administrators. We
would make the connections that are so difficult now because we do not
trust our capacity to express what we feel and think in a way that will be
understood as we mean it. Moreover, by taking responsibility for our
words, we would read with more energy. Instead of being a frustrating
burden, communication would become an exciting exchange of ideas
and feelings.

Instead, our writing on the job, in school, and in our professions is
characterized by the expectation of *not* being heard, of *not* being understood. As people who write, we learn early in school that our ideas,
experiences, and feelings will be squeezed through a template of "mechanical errors." We take those red marks in the margins to heart and
back down, protesting we have nothing to say or that someone else can
say it better. Some people keep diaries, but their discoveries are private,
and we all learn much too late in life that "other people felt the same
way, too."

Our fear of mechanical errors, reenforced by teachers in high school, college, and even graduate school, numbs the passion to write, and diminishes our capacity to understand others. That is a terrible personal loss, and a staggering deficiency for a society. Professions enforce these barriers, requiring that we write in a denatured, impersonal voice, as if our thoughts had been produced by a machine without passion or responsibility.

In most organizations, even a "low-profile accountability" is risky from the standpoint of survival. But disguising the decision-making process or obscuring the responsibility for final decisions exerts tremendous pressure against the authenticity of language. The paradoxes revealed in our daily papers continually erode our faith in communication. We reinterpret what we hear instantaneously, translating "We have never . . ." into "We already have . . .", or "We care . . ." into "We haven't the slightest interest. . . ." This resignation to the corruption of language, together with our experience that what *we* have to say does not matter, isolates us. We can slide by, privately making our peace with day-to-day life. But this will mean that we are oblivious to any common ground.

Our public communication, from whatever point of view, seems eerily consistent. Opponents in a public debate will share the same cynicism about the process of communication, and about the tolerance of their audience. Much public writing is barely readable in its dullness, not simply in style but in thought. As the audience, we have no way of demanding more, or of finding out what is true.

What if this were *not* the case? What if people expected that by writing clearly and forcefully, their readers would be eager to learn, or debate with them? We would know more about each other, and face each other with less illusions. We would certainly find that we are more resilient than we have been led to believe.

Time after time we find ourselves on the edge of this realization. A reporter asked a white teen-aged girl in Philadelphia why she participated in a riot to run a black couple out of her neighborhood.

"I wouldn't want my kids to get to know the blacks," she said.

"Why?" the reporter asked.

"Because they might get to like them!"

It isn't amazing that we accept our muteness and fragility. It's easy to see why we are willing to give up a voice of our own to save ourselves some of the pain of self-knowledge. But despite the conditioning of our schools, professions, and culture, there is a tug of war in each of us between ritualized thinking and genuine communication. We know that insisting on meaning would help us achieve power over our own lives; we know that that desire runs counter to what the culture expects from us. Sometimes this paradox comes crashing in on our numbness. But it is impossible to act on this knowledge in isolation.

Why would a society want the act of writing to remain difficult and mysterious? If everyone had a voice, if everyone believed in the value of authentic communication in personal, professional, and political life, we would continually transform our relationships, our institutions, and our vision of society. The more we tried to say, the less we would need to hide, and the less manipulable we would be by those to whom we habitually yield the power of communication.

Society tells its educators not to give everyone an equal share, but instead to "weed out the good from the bad." Experts in one field or another, whom we may mistrust but still obey, tell us: "This idea, this person is worthwhile, that one is not." Within such a culture, in which power is a reward for charm or obedience, communicating our own thoughts and feelings has a low priority. We dread the loss of freedom, but we have relinquished our right to communicate freely with each other.

The struggle to write well nourishes us, and gives us more power over our lives and our interactions with others. Through it we gather our best energies and make connections with the best energies of others. It is a private accomplishment for independent writers, and contributes to the common effort for change.